MULTIMEDIA SECURITY

Tools, Techniques, and Applications

AAP Research Notes on Optimization & Decision Making Theories

MULTIMEDIA SECURITY

Tools, Techniques, and Applications

Edited by

Bhaskar Mondal, PhD

Shyam Singh Rajput, PhD

AAP | APPLE ACADEMIC PRESS

First edition published 2025

Apple Academic Press Inc.
1265 Goldenrod Circle, NE,
Palm Bay, FL 32905 USA

760 Laurentian Drive, Unit 19,
Burlington, ON L7N 0A4, Canada

CRC Press
2385 NW Executive Center Drive,
Suite 320, Boca Raton FL 33431

4 Park Square, Milton Park,
Abingdon, Oxon, OX14 4RN UK

Apple Academic Press exclusively co-publishes with CRC Press, an imprint of Taylor & Francis Group, LLC

Library and Archives Canada Cataloguing in Publication

Title: Multimedia security : tools, techniques, and applications / edited by Bhaskar Mondal, PhD, Shyam Singh Rajput, PhD.
Names: Mondal, Bhaskar, editor. | Rajput, Shyam Singh, editor.
Series: AAP research notes on optimization and decision-making theories.
Description: FFirst edition. | Series statement: AAP research notes on optimization and decision-making theories | Includes bibliographical references and index.
Identifiers: Canadiana (print) 20240539206 | Canadiana (ebook) 20240539214 | ISBN 9781774915028 (hardcover) | ISBN 9781774915035 (softcover) | ISBN 9781003459040 (ebook)
Subjects: LCSH: Multimedia systems—Security measures.
Classification: LCC QA76.575 .M85 2025 | DDC 005.8—dc23

Library of Congress Cataloging-in-Publication Data

CIP data on file with US Library of Congress

ISBN: 978-1-77491-502-8 (hbk)
ISBN: 978-1-77491-503-5 (pbk)
ISBN: 978-1-00345-904-0 (ebk)

ABOUT THE BOOK SERIES: AAP RESEARCH NOTES ON OPTIMIZATION AND DECISION MAKING THEORIES

SERIES EDITORS

Dr. Prasenjit Chatterjee
Department of Mechanical Engineering, MCKV Institute of Engineering, Howrah, West Bengal, India
E-Mail: dr.prasenjitchatterjee6@gmail.com/prasenjit2007@gmail.com

Dr. Dragan Pamucar
University of Defence, Military academy, Department of Logistics, Belgrade, Serbia; E-Mail: dpamucar@gmail.com

Dr. Morteza Yazdani
Department of Business & Management, Universidad Loyola Andalucia, Seville, Spain; E-Mail: morteza_yazdani21@yahoo.com

Dr. Anjali Awasthi
Associate Professor and Graduate Program Director (M.Eng.), Concordia Institute for Information Systems Engineering, Concordia, Canada; E-Mail: anjali.awasthi@concordia.ca

Most real-world search and optimization problems naturally involve multiple criteria as objectives. Different solutions may produce trade-offs (conflicting scenarios) among different objectives. A solution that is better with respect to one objective may be a compromising one for other objectives. This compels one to choose a solution that is optimal with respect to only one objective. Due to such constraints, multi-objective optimization problems (MOPs) are difficult to solve since the objectives usually conflict with each other. It is usually hard to find an optimal solution that satisfies all objectives from the mathematical point of view. In addition, it is quite common that the criteria of real-world MOPs encompass uncertain information, which becomes quite a challenging task for a decision maker to select the criteria. Also, the complexities involved in designing mathematical models increase. Considering, planning, and appropriate decision-making require the use of analytical methods that examine trade-offs; consider multiple scientific, political, economic, ecological, and social dimensions; and reduce possible conflicts in an optimizing framework. Among all these, real-world multi-criteria decision-making (MCDM) problems related to engineering optimizations are categorically important and are quite often encountered with a wide range of applicability.

MCDM problems are basically a fundamental issue in various fields, including applied mathematics, computer science, engineering, management, and operations research. MCDM models provide a useful way for modeling various real-world problems and are extensively used in many different types of systems, including, but not limited to, communications, mechanics, electronics, manufacturing, business management, logistics, supply chain, energy, urban development, waste management, and so forth.

In the aforementioned cases, modeling of multiple criteria problems often becomes more complex if the associated parameters are uncertain and imprecise in nature. Impreciseness or uncertainty exists within the parameters due to imperfect knowledge of information, measurement uncertainty, sampling uncertainty, mathematical modeling uncertainty, etc. Theories like probability theory, fuzzy set theory, type-2 fuzzy set theory, rough set, grey theory, neutrosophic uncertainty theory available in the existing literature deal with such uncertainties. Nevertheless, the uncertain multi-criteria characteristics in such problems are not explored in depth, and a lot can be achieved in this direction. Hence, different mathematical models of real-life multi-criteria optimization problems can be developed on various uncertain frameworks with special emphasis on sustainability, manufacturing, communications, biomedical, electronics, materials, energy, agriculture, environmental engineering, strategic management, flood risk management, supply chain, waste management, transportations, economics, and industrial engineering problems, to name a few.

Coverage & Approach:

The primary endeavor of this series is to introduce and explore contemporary research developments in a variety of rapidly growing decision-making areas. The editors welcome proposals on the following topics:

- Crisp MCDM models
- Rough set theory in MCDM
- Fuzzy MCDM
- Neutrosophic MCDM models
- Grey set theory

- Mathematical programming in MCDM
- Big data in MCDM
- Soft computing techniques
- Modelling in engineering applications
- Modeling in economic issues
- Waste management
- Agricultural practice
- Material selection
- Renewable energy planning
- Industry 4.0
- Sustainability
- Supply chain management
- Environmental policies
- Manufacturing processes planning
- Transportation and logistics
- Strategic management
- Natural resource management
- Biomedical applications
- Future studies and technology foresight
- MCDM in governance and planning
- MCDM and social issues
- MCDM in flood risk management
- New trends in multi-criteria evaluation
- Multi-criteria analysis in circular economy
- Multi-criteria evaluation for urban and regional planning
- Integrated MCDM approaches for modeling relevant applications and real-life problems

Types of volumes:

This series reports on current trends and advances in optimization and decision-making theories in a wider range of domains for academic and research institutes along with industrial organizations. The series includes the following types of volumes:

- Authored volumes
- Edited volumes
- Conference proceedings
- Short research (thesis-based) books
- Monographs

Features of the volumes include recent trends, model extensions, developments, real-time examples, case studies, and applications. The volumes aim to serve as valuable resources for undergraduate, postgraduate, and doctoral students, as well as for researchers and professionals working in a wider range of areas.

CURRENT & FORTHCOMING BOOKS IN THE SERIES

- **Multi-Criteria Decision-Making Techniques in Waste Management: A Case Study of India**
  Editors: Suchismita Satapathy, Debesh Mishra, and Prasanjit Chatterje
- **Applications of Artificial Intelligence in Business and Finance: Modern Trends**
  Editors: Vikas Garg, Shalini Aggarwal, Pooja Tiwari, and Prasenjit Chatterjee
- **Advanced Computer Science Applications: Recent Trends in AI, Machine Learning, and Network Security**
  Editors: Karan Singh, Latha Banda, and Manisha Manjul
- **Decision-Making Models and Applications in Manufacturing Environments**
  Editors: Pushpdant Jain, Kumar Abhishek, and Prasenjit Chatterjee
- **Precision Agriculture for Sustainability: Use of Smart Sensors, Actuators, and Decision Support Systems**
  Editors: Narendra Khatri, Ajay Kumar Vyas, Celestine Iwendi, and Prasenjit Chatterjee
- **Attacks on Artificial Intelligence: The New Facets of Cyber Ecospace**
  Editors: Kukatlapalli Pradeep Kumar, Boppuru Rudra, Vinay Jha Pillai, and Prasenjit Chatterjee
- **Computational Optimization, Modeling, and Simulation for Engineering Applications**
  Editors: Anupam Shukla, Sourabh Rungta, Mohan Awasthy, and Rakesh L. Himte
- **Advances in Intelligent Systems: Paradigms and Applications**
  Editors: Amit Krishna Dwivedi, Manisha Guduri, Rajanikanth Aluvalu, and Uma Maheshwari V.

ABOUT THE EDITORS

Bhaskar Mondal, PhD
Assistant Professor, Department of Computer Science and Engineering, National Institute of Technology Patna, India

Bhaskar Mondal, PhD, is an Assistant Professor in the Department of Computer Science and Engineering at the National Institute of Technology Patna, India. He has more than 10 years of experience in academia during which he has worked at XIM University Bhubaneswar; Birsa Institute of Technology Sindri, Dhanbad; and National Institute of Technology Jamshedpur, all in India. He has published more than 40 research papers in journals and international conferences. He has delivered many expert talks on cybersecurity and machine learning. He is a senior member of IEEE and the Association for Computing Machinery, and a life member of the Computer Society of India and the Cryptology Research Society of India. He has served at several international conferences as session chair. He is the editor of CRC Press's Cyber Security book series and has acted as a guest editor (lead) for a special issue in the *International Journal of Computers and Electrical Engineering,* (Elsevier). His research interests include cybersecurity, machine learning, and quantum computing. He received his PhD from the National Institute of Technology Jamshedpur, India, and his MTech (CSE) from Kalyani Government Engineering College, India.

Shyam Singh Rajput, PhD
Assistant Professor, Department of Computer Science and Engineering, National Institute of Technology Patna, India

Shyam Singh Rajput, PhD, has more than six years of experience in teaching undergraduate and postgraduate classes. Presently, he is working as an Assistant Professor in the Department of Computer Science and Engineering at the National Institute of Technology Patna, India. His current research interests include image processing and computer vision. He has published more than 40 journal and conference papers and book chapters in the domain of image processing, biometrics, wireless ad hoc networks, and information security. He has published two Indian patents and edited a book with Elsevier. He is a member of IEEE and the Association for Computing Machinery. Dr. Rajput received a BE degree in Computer Science & Engineering (CSE) from R. G. P.V., Bhopal, Madhya Pradesh, India. He earned his MTech degree and PhD degree in CSE from ABV-Indian Institute of Information Technology & Management (ABV-IIITM), Gwalior, India.

CONTENTS

Contributors .... *xiii*

Abbreviations .... *xv*

Preface .... *xix*

1. Understanding Multimedia Security 1
Shubhamshree Avishek, Sikata Samantaray, and Sasmeeta Tripathy

2. Evaluation Metrics Used for Image Encryption to Analyze the Efficiency Toward Various Attacks 57
M. Devipriya and M. Brindha

3. Machine Learning Approaches to Malware Detection Systems 87
A. P. Sherly and P. Arun Raj Kumar

4. Securing Identities Through Biometric Template Security 115
Vivek Singh Baghel and Surya Prakash

5. Watermarking and Copyright Protection of Remotely Sensed Images 155
Rohit Thanki, Surekha Borra, and Purva Joshi

6. A Future Cyber Security for Secured Smart Home Applications Using the Internet of Things and Blockchain Technology 177
V. Arulkumar, Marlapalli Krishna, S. Jayachitra, and A. Prasanth

7. A Secure and Efficient Quantum Adopted Chaotic Cryptosystem for Video Surveillance Applications 203
T. Janani, Vinod Ramesh Falmari, and M. Brindha

8. Evaluation of Pansharpening Techniques for Multimedia Image Enhancement 229
Kuldeep Chaurasia and Shakti Sharma

9. Spam Remark Revelation Utilizing Supervised Machine Learning Approaches in Security Issues 255
Sohom Bhattacharya, Priyanka Dhara, Shubham Bhattacharjee, and Subir Gupta

10. Secure Chaos-Based Image Encryption Techniques Based on Feedback Convolution Method..289
R. Vidhya and M. Brindha

Index..329

CONTRIBUTORS

V. Arulkumar
School of Computer Science and Engineering, VIT University, Vellore, Tamilnadu, India

Shubhamshree Avishek
Department of Mechanical Engineering, Siksha 'O' Anusandhan Deemed to be University, Bhubaneswar, Odisha, India

Vivek Singh Baghel
Department of Computer Science and Engineering, Indian Institute of Technology Indore, Indore, Madhya Pradesh, India

Shubham Bhattacharjee
Department of Master of Computer Application, B. C. Roy Engineering College, Durgapur, West Bengal, India

Sohom Bhattacharya
Department of Master of Computer Application, B. C. Roy Engineering College, Durgapur, West Bengal, India

Surekha Borra
Department of Computer Science and Engineering (IoT, Cybersecurity and Blockchain Technology), K. S. Institute of Technology, Bengaluru, India

M. Brindha
Department of Computer Science and Engineering, National Institute of Technology, Tiruchirappalli, Tamil Nadu, India

Kuldeep Chaurasia
Department of Computer Science Engineering, Bennett University, Greater Noida, Uttar Pradesh, India

M. Devipriya
Department of Computer Science and Engineering, National Institute of Technology, Tiruchirappalli, Tamil Nadu, India

Priyanka Dhara
Department of Master of Computer Application, B. C. Roy Engineering College, Durgapur, West Bengal, India

Vinod Ramesh Falmari
Department of Computer Science and Engineering, National Institute of Technology, Tiruchirappalli, Tamil Nadu, India

Subir Gupta
Department of Computer Science and Engineering (Artificial Intelligence and Machine Learning), Haldia Institute of Technology, West Bengal, 721657, India

T. Janani
Department of Computer Science and Engineering, National Institute of Technology, Tiruchirappalli, Tamil Nadu, India

S. Jayachitra
Department of ECE, PSNA College of Engineering and Technology, Dindigul, Tamilnadu, India

Purva Joshi
Institute for Communications and Navigation, Deutsches Zentrum für Luft und Raumfahrt, Weßling, Germany

Marlapalli Krishna
Professor, Department of CSE, Sir C.R. Reddy College of Engineering, Eluru, Andhra Pradesh, India

P. Arun Raj Kumar
Senior Member, IEEE; Associate Professor, Department of Computer Science and Engineering, NIT Calicut, Kerala, India

Surya Prakash
Department of Computer Science and Engineering, Indian Institute of Technology Indore, Indore, Madhya Pradesh, India

A. Prasanth
Department of Computer Science and Engineering, Vel Tech Rangarajan Dr. Sagunthala R & D Institute of Science and Technology, Chennai, Tamil Nadu, India

Sikata Samantaray
Department of Mechanical Engineering, Siksha 'O' Anusandhan Deemed to be University, Bhubaneswar, Odisha, India

Shakti Sharma
Department of Computer Science Engineering, Bennett University, Greater Noida, Uttar Pradesh, India

A. P. Sherly
Graduate Student Member, IEEE; Research Scholar, NIT Calicut, Kerala, India

Rohit Thanki
CTO & Senior IEEE Member, DetMedX, Germany

Sasmeeta Tripathy
Department of Mechanical Engineering, Siksha 'O' Anusandhan Deemed to be University, Bhubaneswar, Odisha, India

R. Vidhya
School of Computer Science and Engineering (SCOPE), Department of IoT, Vellore Institute of Technology, Katpadi, Vellore, Tamil Nadu, India

ABBREVIATIONS

AES	advanced encryption standard
AFIS	automatic fingerprint identification systems
AHP	analytical hierarchy process
AI	artificial intelligence
ANN	artificial neural network
AUC	area under the curve
BC	blockchain
BER	bit error rate
BTP	biometric template protection
C&C	command and control
CFG	control flow graph
ClaMP	classification of malware with PE headers
CNN	convolutional neural network
CNN-LSTM	convolutional neural network-long short-term memory
CRC	cyclic redundancy check
DBN	deep belief network
DCT	discrete cosine transform
DDoS	distributed denial of service
DES	data encryption standard
DFT	discrete Fourier transform
DITOM	densely infinite-to-one mapping
DN	digital numbers
DNN	deep neural network
DoS	denial of service
DRM	digital rights management
DT	decision tree
DWT	discrete wavelet transform
ECG	electrocardiogram
EER	equal error rate
EHNFC	evolving hybrid neuro-fuzzy classifier
ELF	executable and link format
EO	earth observation
FAR	false acceptance rate
FCG	function call graph

FCMR	false cross-match rate
FDA	Fisher discriminant analysis
FHT	fast Hadamard transform
FIFO	first-in-first out order
FL	fuzzy logic
FRQI	flexible representation of quantum image
FRR	false rejection rate
GAR	genuine acceptance rate
Gdeg	gray difference degree
HCS	hyperspherical color space
HPF	high pass filter
HRMX	high-resolution multispectral
HSRP	high spatial resolution panchromatic
IFOV	instantaneous field of view
IoT	internet of things
IP	intellectual property
IPsec	internet protocol security
KLT	Karhunen Loeve transform
KNN	K-nearest neighbors
k-NNS	*k*-nearest neighborhood structure
LBP	local binary pattern
LDA	linear discriminant analysis
LDP	local directional pattern
LFSR	linear feedback shift register
LRMX	low-resolution multispectral
LSB	is the least significant bit
LSTM	long short-term memory
M2M	machine-to-machine
MAE	mean absolute error
MCC	minutia cylinder-code
MD5	message-digest algorithm 5
MDBUTMF	modified decision-based unsymmetrical trimmed median filter
MIHS	modified intensity hue saturation
MitM	man-in-the-middle
ML	machine learning
MLC	multi-line code
MLP	multi-layer perceptron
MS	multispectral
MSBs	most significant bits

MSE	mean square error
MVD	minutia vicinity decomposition
NB	*Naive Bayes*
NDSI	normalized difference snow index
NDVI	normalized difference vegetation index
NDWI	normalized difference water index
NEQR	novel enhanced quantum representation
NIR	near infrared
NLP	natural language processing
NPCR	number of pixel change rate
n-PIR	non-plain image related
NSO	Niv, Shalev, and Omri
OpCode	operation code
OSINTs	open-source intelligence tools
OTK	one-time-key
PC	principal component
PCA	principal component analysis
PE	portable executable
PIR	plain image-related
POW	proof of work
PSCS	parametric switching chaotic system
PSNR	peak signal-to-noise ratio
PVD	pixel value differences
RCK	ransomware construction kit
RDQT	randomized dynamic quantization transformation
RDWT	redundant discrete wavelet transform
RF	random forest
RMQ	random multispace quantization
RNG	random number generation
RNN	recurrent neural network
RSA	Rivest Shamir Adleman
SEF	system calls execution fingerprints
SES	simple encryption scheme
SGD	stochastic gradient descent
SIFT	scale-invariant feature transform
SLA	service level agreement
SMOT	synthetic minority oversampling technique
SNR	signal-to-noise ratio
SNS	social network service

SPGK	shortest path graph kernel
SSIM	structural similarity index measure
SVD	singular value decomposition
SVM	support vector machine
TDK	trojan development kit
TM	thematic mapper
TNR	true negative relationship
UACI	unified average change intensity
UAV	unmanned aerial vehicles
UIQ	universal image quality
UPS	uninterruptible power supply
USB	universal serial bus
VNS	Voronoi neighborhood structure
WCT	WIPO copyright treaty

PREFACE

Multimedia security has become one of the most important topics for organizations and individuals. A huge amount of multimedia data like images, audio, and videos are generated every day, and they are transmitted over the Internet. Any user with ill-intent may intercept these data and misuse, tamper, or modify them for personal gain and cause severe damage to the victim. This book on security is thus extremely relevant to the present times. This book presents state-of-the-art multimedia security techniques and algorithms along with the recent advancements in the field. This book consists of 10 chapters, which are organized as follows.

Chapter 1 provides a detailed explanation of the overall aspect of multimedia security which includes the definition of multimedia security, the types of multimedia security, data hiding, data encryption, and analysis and evaluation of the data. Moreover, certain mathematical models used in data hiding and data evaluation techniques have been stated and discussed thoroughly to develop a good understanding of the principles of multimedia security.

Chapter 2 presents a detailed study of evaluation metrics used for image encryption to analyze the efficiency of various attacks. The pixel values of the encrypted image are altered through the use of an encryption algorithm, resulting in a randomized outcome that conceals any patterns or features. A good encryption algorithm ensures that the correlation coefficient between adjacent pixel values in the encrypted image is close to zero. Visual examination is the primary method of assessing the encrypted image, with the objective of ensuring that the features of the plain image are not visible. However, visual examination alone is not sufficient for evaluating security efficiency, and therefore additional metrics are employed to measure the randomness and effectiveness of the encryption.

Chapter 3 demonstrates an AI-based malware analysis using freely available open-source intelligence tools (OSINTs). Analysis shows that malware detection in stego images is indeed challenging due to its hidden behavior and malicious scripts.

Chapter 4 provides a detailed discussion of all aspects of biometric template security. Biometric cryptosystems are designed to enhance the security of original biometric templates by combining them with cryptographic keys or generating keys directly from the templates. This ensures that the original biometric data remains secure. There are some essential properties, such as revocability, diversity, security, and performance, which need to be followed to determine the overall effectiveness of a biometric template protection (BTP) technique.

Chapter 5 describes an overview of different watermarking schemes for copyright protection of remotely sensed images. Two types of watermarking techniques, in the spatial domain and transform domain for the remotely sensed image, are discussed in detail in this chapter.

Chapter 6 presents a blockchain (BC) based security model for smart homes that combines blockchain technology with cloud computing in an effective manner. The model uses a multivariate correlation analysis and identification of correlations between traffic functions to guarantee the safety of the smart home network.

Chapter 7 presents a Quantum-based video cryptosystem to encrypt the digital video frames to ensure the privacy of surveillance video. Video surveillance systems are commonly adopted for real-time event detection and video analytics. The proposed system ensures secure video surveillance against cryptanalytic attacks.

Chapter 8 focuses on the evaluation of the six pansharpening approaches for satellite image enhancement and attempts to determine the best method for land use classification using the Indian Remote Sensing dataset. In this chapter, the high-resolution color imagery is obtained by using Resourcesat-1 satellite images acquired by the LISS-IV sensor and IRS-P5 (Cartosat-1) images acquired by the PAN sensor. The study also identifies the land use characteristics that are improved through pansharpening techniques. These findings are valuable in assessing the suitability of the pansharpening method for different geospatial applications.

Chapter 9 expands on different well-known directed machine learning approaches in detecting the spam and ham comments.

Finally, an efficient encryption scheme using convolution operation is proposed in Chapter 10, which is applied to the spatial information of the plain image for generating the keys of the encryption process. The one-time key is also utilized for achieving Shannon's perfect secrecy. The proposed chaotic feedback convolution is proven against various test cases using simulations.

CHAPTER 1

UNDERSTANDING MULTIMEDIA SECURITY

SHUBHAMSHREE AVISHEK, SIKATA SAMANTARAY, and SASMEETA TRIPATHY

Department of Mechanical Engineering, Siksha 'O' Anusandhan Deemed to be University, Bhubaneswar, Odisha, India

ABSTRACT

For decades, we have extensively researched and marked certain innovations in the field of Computer Security in regards to System and Network Security. With the advancement in technology, we have witnessed the importance of Cyber Security protecting our systems and network from getting exposed globally without our prior notice. And as all these evolve with time, the need of securing information from a single media as an image to multimedia has grown immensely. With the transfer of tons of terabytes of data each day via different forms like web, wireless, physical transfer, etc. It has now become of utmost importance to secure it from unwanted intruders. The term Multimedia covers graphics, animation, audio, video, and other digital content. As we are moving towards a digital world rapidly, the prime significance of multimedia security is parallelly increasing. Hence, the technique of Data Encryption and Data Hiding has become challenging due to the broad segregation of data and level of vulnerability possibilities. In this regard, to understand the fundamentals of Media Security the current chapter has been made taking a linear learning curve into account that would help to create a base for initial understanding and would further assist in delving deeper into the topic. A detailed explanation of the overall aspect of Multimedia Security has been discussed which includes the definition of

Multimedia Security: Tools, Techniques, and Applications. Bhaskar Mondal and Shyam Singh Rajput (Eds.)

Multimedia Security, the types of Multimedia Security, Data Hiding, Data Encryption, and Analyzing and Evaluation of the Data. Moreover, certain mathematical models used in data hiding and data evaluation technique has been stated and discussed thoroughly to develop a good understanding of the principles of Media Security. The chapter also gives a brief explanation of the types of Media Security like the fundamentals of Steganography, Digital Watermarking, Digital Scrambling, and Digital Surveillance and Authentication. Additionally, a brief idea of digital rights management (DRM) and the application has been discussed with certain significance to the world of digital media security. This would lead to a path of opportunity for further innovation in the field of Multimedia Security with advanced techniques in the long run. Lastly, this chapter would further help the reader to understand the various aspects of the current context by laying off a good base to start from, and it would facilitate a detailed insight into the topic, which would be beneficial to understand the dedicated advanced topic in the upcoming sessions.

1.1 INTRODUCTION

With the advancement in technology, we have shifted ourselves into a digital space and are constantly evolving the world around us with technological brilliance. Over the decade, we have pushed our knowledge and exploration ahead of the bounds and marked significant excellence in various fields, including the creation of a digital hub around us. We are all a crucial part of this digital ecosystem where tons of terabytes of data worth billions of dollars revolve around us and play a vital role in running the world. This information in various forms, which we potentially term as Multimedia, is one of the key factors for which the entire world depends on today's date. From data of an individual to the most confidential data of a country, all are available in this digital space, and as it evolves ahead becoming more accessible, the need for Information Security rises exponentially. According to the study, each year, billions of data get breached and exposed publicly by hackers and potential intruders in the ecosystem [1].

Due to such a rise in hampering of private space and confidential information, the significance of Multimedia Security is booming each day. These intrusions/leakages not only invade personal privacy but are an overall serious threat to society as they account for countless losses for a nation by publicly exposing highly confidential details while breaking national security. In this regard, this present chapter has been curated to lay off a basic understanding of the meaning of Multimedia, the significance of Multimedia, and lastly discuss the types of Multimedia Security [2, 3]. This chapter would help to understand the multiple Multimedia Security methods and the general concept behind the techniques which would potentially ease out to grasp basic knowledge required for the advanced upcoming sessions and practical applications. Lastly, this chapter would also cover the role of digital rights management (DRM) and the Multimedia Security laws implemented in various countries.

1.2 WHAT IS MULTIMEDIA AND WHY MULTIMEDIA SECURITY?

Multimedia is a form of information as digital content that covers graphics, visuals, animations, music, and many other digital aspects of data [4, 5]. These are the same data that we share on social media with our friends, and these are the same data used by big companies and government officials to work in the practical field. According to the survey, there are tons of terabytes of data present on the internet and counting as we become more accessible to technology. But the question arises here: are all of our data safe? Are these data safe from unwanted intruders? The general and most obvious answer would be No unless all these data are under security. This is where the role of Cybersecurity, especially Multimedia Security, comes into play, where we tend to protect these highly confidential and private data from unwanted invaders.

Over time it has been analyzed that the traditional techniques used with the help of computer and network security are not enough to deal with content encryption and can be altered using modern tools, hence the need for content security is of utmost requirement to secure this data. Multimedia Security further helps in adding a highly protective shield to the high-level content and introducing encryption with signal processing for higher security prospects. However, this gets complicated over time as the ease of piracy and intrusion takes over different forms. Detecting threats like replication without any information loss and manipulation of data without any possibility of detection are a few top-grossing points in the field of multimedia [6–8]. As these techniques grow rapidly and take different forms, it becomes even harder to secure such data and prevent it from data piracy methods. In the field of software, the developers encrypt the original software with the Serial Number which plays the role of Passkey that allows a user to install it on his device, this is one of the most traditional ways of preventing software piracy and keeping the ecosystem clean. But as the internet becomes more accessible, these Serial Numbers have been seen being shared on the internet and hackers started to decode the software to make it available for free while sharing it over in cyberspace. For digital content like images, Digital Watermarking plays a crucial role in preventing unwanted copies and has been the primary way of securing images for big commercial companies for decades [9]. Most often, in the corporate field, a company logo is used to protect the digital image from any copyright and is a way of proving ownership, likewise, government e-document is also protected using the Digital Watermarking systems [10]. This same technique was also adopted in the field of videos, which was later found to be not enough, and a more secure level of encryption was needed. As technology advances with modern tools and techniques, the possibility of synthesizing multimedia becomes easier and highly accessible. This becomes a significant challenge in the field of Multimedia Security and remains the most widely researched field.

1.3 MEDIA EVALUATION CRITERIA

Detection and Evaluation of the media is a crucial field in Multimedia Security where we tend to evaluate the media and search for any tampering. Depending upon the level of data manipulated, the criteria of the media evaluation are decided [11]. Also, the evaluation of the media depends on the media file we have chosen to detect and vary accordingly. If the manipulated or the tampered data has a greater variance as compared to the original data, which could be visually noticed easily, then the Subjective criteria fits the best. The subjective criteria can be determined by the help of the degree of resemblance or the clear understandable difference. These methods are primarily helpful when a significant change can be noticed visually by a human.

The degree of resemblance can be further accounted for by the amount of information that has been manipulated from the original source or the distinct difference between the original and manipulated media that can be obtained by human perceptual verification. Alternatively, if the altered media is hard to notice any significant change, then we opt for the Objective Criteria where a thorough analysis of the altered media is done and it is then compared with the original image. In this method, several techniques are used to get the correspondence for the level of alteration, like for example, the use of Peak Signal Ratio. The altered media can also be known with the help of Color Space criteria, where we tend to check the altered image's color space with the original image. Here, in this method, we find out the deviance of color space between the two images using the Distance of two-color spaces. However, this method could also be a part of the Subjective criteria of Media

Evaluation if the changes in the altered media file are significantly distinct, but if we fail to determine the difference of the color spaces of the two images perceptually then we could go for the distance measurement of the two images color spaces.

1.3.1 PERCEPTUAL/SUBJECTIVE CRITERIA

Subjective criteria of Media Evaluation is the basic technique of evaluating a media and search for any alteration using the human perceptual system. This includes checking the level of alteration of the media by comparing it to the original media and checking for various properties like quality, technical perspective, and other significant changes. But this way of evaluating the media is not the most promising as several factors depend while evaluation. For example, the evaluating person might have some problems with the vision and the result may vary from person to person. The age of a person and Understanding of a particular also might play a crucial role here.

However, one of the ways of getting the accurate result using the Subjective criteria of Media Evaluation would be averaging out the results got from a few observers and using the average result as the final outcome. This reduces the chances of error and ensures the result is much more accurate than a single observer. The mathematical expression is given by Praum, Hoppe, & Finkelstein [11]:

$$C = \frac{\sum_{i=1}^{k} n_i c_i}{\sum_{i=1}^{k} n_i} \tag{1}$$

where; C from the Eqn. (1) denotes the level of media alteration; c_i is the score of Class I; and n_i is the number of people evaluating which belonging to Class I.

But as we have stated above, this method of evaluating the media has many limitations, and the results vary a lot depending upon the situation. Therefore, a new way of evaluating media, the Objective criteria of Media Evaluation has been introduced.

1.3.2 CALCULATIVE/OBJECTIVE CRITERIA

As the Subjective criteria failed to give a precise result and vary frequently depending upon the observer, the Objective Method of evaluating media has been introduced where the precision of the result is high independent of the perception of the observer. This way of evaluation is more reliable and promising as it proves the difference between the original media and the altered media.

The objective criteria of media evaluation can be done by comparing the data of the two media with the help of the Definition of Reality ($\in$) and the peak signal-to-noise ratio (PSNR) [11–13]. The Definition of Reality ($\in$) will give us the amount of data that has been altered among the two images, hence the greater values of $\in$ correspond to higher originality and less tampered data. Similarly, the PSNR is also used to determine the level of alteration. A high value of PSNR corresponds to more similarity with the original media.

To demonstrate the two methods of objective evaluation, let's consider the intensity of an original image's pixel at F(x,y) and altered image's pixel at F'(x,y), where 0<F(x,y)<255. For the grayscale image, the range of the intensity of a pixel varies from 0 to 255, where 0 gives us a black value and 255 gives us the white value as shown in Figure 1.1.

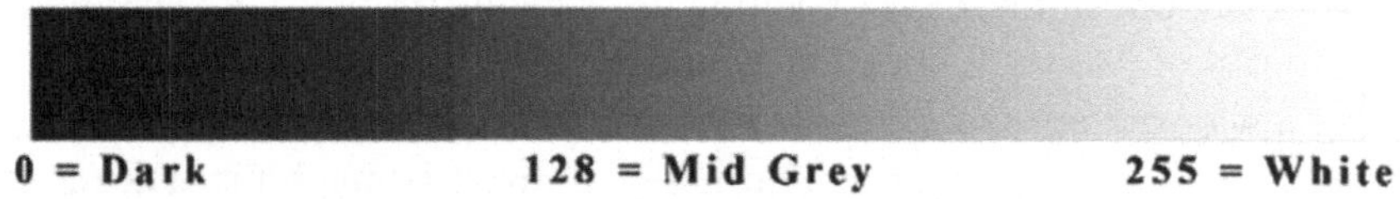

FIGURE 1.1 Grayscale image.

Now, the difference between the original image and the altered image can be given by the following two methods:

1. Reality Definition

$$\in = \frac{\sum_{i=1}^{M}\sum_{j=1}^{N}[\sigma\{f(i,j)-\sigma\{f(j,k)\}\}]^2}{\sum_{i=1}^{M}\sum_{j=1}^{N}[\sigma\{f(i,j)\}]^2} \tag{2}$$

where; σ{.} is the probability Expectation in Eqn. (2).

2. Peak Signal to Noise Ratio (PSNR)

$$PSNR = 10\log\left(\frac{m.n.P^2}{RMS}\right) \tag{3}$$

where m and n are the number of rows and columns of pixels of that input image, respectively.

$$MSE = \frac{1}{M * N} \sum_{i=0}^{M-1} \sum_{j=0}^{N-1} [\{I(i,j) - \{K(j,k)\}\}]^2 \tag{4}$$

$$RMS = \sqrt{MSE} \tag{5}$$

Among the above-mentioned methods, the PSNR given in Eqn. (3) is one of the most widely used techniques. This allows us to understand the capacity of the information that can be possibly hidden in an image. The greater the PSNR values, the higher is the similarity with the original media.

1.3.3 COMPARING THE VARIANCE IN COLOR SPACES

One of the ways of detecting the difference between the two images is by checking the difference between the two pixels from the images to be compared. This allows us to ensure the exact data on pixel change that regulated the image properties [86].

This method is generally carried out by checking the two images with various color spaces, the distance of the measure of color spaces would give us the result on any physical change. In order to account for this method, first, we check the distance in the YIQ color space, where Y represents the Intensity (Brightness), I and Q represents the color information. It is important to check for the YIQ color space as it signifies the luminance of an image, and on that note, our human eye is more sensitive to the luminance change than hue and saturation.

The RGB to YIQ conversion is defined as:

$$\begin{bmatrix} Y \\ I \\ Q \end{bmatrix} = \begin{bmatrix} 0.299 & 0.587 & 0.114 \\ 0.596 & -0.275 & -0.321 \\ 0.212 & -0.523 & 0.311 \end{bmatrix} \begin{bmatrix} R \\ G \\ B \end{bmatrix} \tag{6}$$

Now, the difference between the two given images can be measured by:

$$RMS = \sqrt{\frac{1}{M * N} \sum_{i=0}^{M-1} \sum_{j=0}^{N-1} [\{I(i,j) - \{K(j,k)\}\}]^2} \tag{7}$$

where; I and K from the Eqn. (7) belong to two different images; and m and n is the resolution in the vertical and horizontal direction, respectively.

For more accurate results, we can also consider the LAB color space, which is regarded as the most accurate color space. As this works like a human eye, checking the two images with the help of LAB color space would be highly accurate. The LAB color space is a 3-axis color system where the L component is the Lightness, and A and B stand for the dimension of color. As it includes all the color in the spectrum as well as color outside of the human visional perception and is independent of the device, this is considered the most highly accurate color model. Although it is the most accurate model and highly efficient, it is still not used commonly as the monitors and printers we have can use either three or 4 colors to represent any image, i.e., RGB or CMYK.

To use the LAB color space we can convert the RGB color space to XYZ color space and then convert again the XYZ color space to LAB.

To begin with, convert the RGB space to CIE XYZ space as shown in Eqn. (8):

$$\begin{pmatrix} X \\ Y \\ Z \end{pmatrix} = \begin{pmatrix} 0.43 & 0.34 & 0.17 \\ 0.22 & 0.70 & 0.07 \\ 0.02 & 0.13 & 0.93 \end{pmatrix} \begin{pmatrix} R \\ G \\ B \end{pmatrix} \tag{8}$$

Now, convert from XYZ space to L.A.B. Space as given in Eqn. (9):

$$L = \begin{cases} 116(\frac{Y}{Y'})^{\frac{1}{3}} - 16 & if \frac{Y}{Y'} > 0.0088 \\ 903.3\frac{Y}{Y'} & if \frac{Y}{Y'} < 0.0088 \end{cases} \tag{9}$$

$$RMS = \sqrt{\frac{1}{M * N} \sum_{i=0}^{M-1} \sum_{j=0}^{N-1} (L_{ij} + L'_{ij})^2 + (a_{ij} + a'_{ij})^2 + (b_{ij} + b'_{ij})^2} \tag{10}$$

where; (L_{ij}, a_{ij}, b_{ij}) and (L'_{ij}, a'_{ij}, b'_{ij}) from Eqn. (10) belongs to two separate images having the same resolution. M and N correspond to the vertical and horizontal resolution of the image.

So, here are all the ways to evaluate the media and detect the manipulation of the original media. However, the techniques discussed in the Objective criteria are weak analysis of the media alteration as it only accounts for the physical change of properties depending upon the content of the image. Thus, it nowhere tells us about the image dimensional change, such as the image rotation and translation. Due to these, further techniques have been introduced ahead in upcoming sections that could account for such alteration in the original image.

1.4 MATHEMATICAL MODELS OF DATA HIDING

Data Hiding in multimedia is a technique of embedding data into multimedia within the capabilities of the targeted media to hold the data without significant noticeable change. From a technical perspective, Data Hiding is a technique of hiding the signal or the signature by embedding it into the targeted media to get a composite signal [14]. Several researchers have tried various implications to make the Data Hiding more effective, for example, Data Hiding technique using the pixel value ordering with higher capacity, using the channel models for higher capacity data hiding, Data Hiding in the Gray Scale Image using the PVD, using Multi-Directional PVD on color images and other such techniques [15–18]. Figure 1.2 shows a general embedding and extraction process that happens in the Data Hiding process. While hiding the data 3 major parameters rules out the success of Data Hiding, namely, perceptual transparency, robustness, and the overall storage capacity [19].

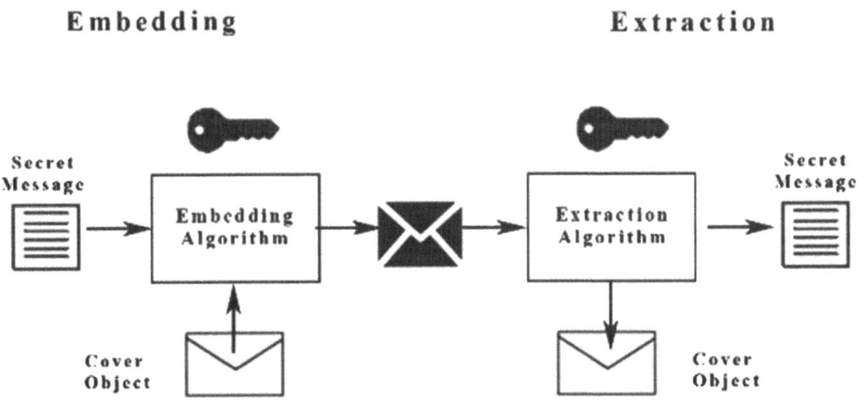

FIGURE 1.2 Embedding and extraction process in data hiding.

This means that composite media after data hiding should be quite transparent and should not get into the perceptual vision, the embedded data should be robust enough to sustain various attacks or try to capture the signal, and lastly, the data embedded in a media depends upon the capacity of it to store the information. But what's the maximum capability of a media to store the data? Though the goal is to store the maximum bits possible, the limitation of data holding capacity before it could entirely distort the media or could easily be retractable is the main hurdle. Therefore, to understand the maximum capacity a media can potentially store the data can be visualized through the mathematical models [88].

1.4.1 MATHEMATICAL MODEL BASED ON COLOR SPACES

The data can be hidden in a pixel of the image, but with due condition, it doesn't come under the human visual system. If it successfully passes the human visual detection, then the Subjective criteria of the media evaluation don't get applied. Human eyes are less sensitive to high grayscale or bright pixels but have a fairly good sensitivity to changes made within a medium level of luminance. So, any difference in the color tone of 16 continuous tones, specifically in the higher bright pixels, cannot be detected by the HVS system. Thus, any information can be hidden in these pixels successfully. Thus mathematically:

$$[f(x,y)-f'(z,y)| \leq 16 \, and \left|g(x,y)-g'(x,y)\right| \leq 16] \tag{11}$$

So, any information within this range can be stored and can surpass human detection.

1.4.2 MATHEMATICAL MODEL BASED ON PSNR

Essential information can be hidden in the color images by utilizing the bit substitution pattern [20]. In order to detect the hidden information or the alteration of the image, PSNR is one of the most common approaches in the field of Multimedia Security and especially in the Watermarking and the Stenography field. This plays a crucial role in determining the level of alteration made to an image with reference to the original image. This also quantifies the reconstruction quality for the media.

Mathematically, it can be represented as:

$$PSNR = 10\log\left(\frac{m.n.P^2}{RMS}\right) \tag{12}$$

$$MSE = \frac{1}{M*N}\sum_{i=0}^{M-1}\sum_{j=0}^{N-1}[\{I(i,j)-\{K(j,k)\}\}]^2 \tag{13}$$

$$RMS = \sqrt{MSE} \tag{14}$$

where I(i, j) from Eqn. (13) is the original grayscale image; K(i, j) is the value from the reference image containing the hidden information; and (i, j) represents the coordinates of the pixel on each image.

A higher value of PSNR depicts a better quality and less alteration of the media as compared to the original one. As this accounts for the minimal difference between the two images, a higher PSNR or lower MSE is always preferred. A lossless compression tends to have a high PSNR value as the mean square error (MSE) in a lossless compression is quite less. For a lossy compressed image, the PSNR values were found out to be within 30–60 for an 8-bit grayscale image.

From Table 1.1, it can be seen that the 5 lowest bits can be accounted for the Data Hiding as the PSNR values tend to be higher than 28 or 30 in an 8-bit grayscale image.

TABLE 1.1 Individual Bit Substitution and its Corresponding PSNR and MSE Value

Method	PSNR	MSE
Least bit substitution, 1st bit	55.87	0.16
2nd bit substitution	49.79	0.68
3rd bit substitution	43.93	2.62
4th bit substitution	37.85	10.65
5th bit substitution	31.97	41.29
6th bit substitution	26.04	161.55
7th bit substitution	19.81	679.05
8th bit substitution	14.34	2.39e + 006

1.5 DESIGN CONSIDERATION FOR DATA HIDING

While embedding data into a media, a few factors are required to be considered for an effective composite signal as an output, whereas failing to do so would result in easy detection of the media manipulation [21]. So here, a brief description of the design consideration of the Data Hiding concept has been summarized.

1.5.1 TRANSPARENCY

Perceptual transparency is the basic ground for all the data hiding techniques. While embedding data into a particular media, the basic goal always stays to make the output signal perceptually transparent. This makes the altered media pass through human verification and can only be detected upon calculative methods.

1.5.2 ROBUSTNESS

The embedded data must be robust enough to sustain the attacks made to capture or remove the embedded information from the media. It is impossible to make a security system sustain every form of attack; the main goal of a robust algorithm would be sustaining many common forms of attack done both intentionally and non-intentionally.

1.5.3 CAPACITY

Though embedding as many bits as possible is the prime goal of the data-hiding technique, this is fairly done while keeping other constraints, such as Perceptual Transparency and Robustness, in account. However, while embedding information into a media, the maximum bits that can be stored is evaluated first to get rid of any distinct distortion of the output media.

1.5.4 STATISTICAL CORRECTNESS

Statistical Correctness is a must requirement while considering a system for secret message communication. This is done by altering the host media to the minimum possible change during the embedding process.

1.5.5 COMPUTATIONAL COMPLEXITY

While designing the data hiding system, the computational complexity also plays a significant role. It is highly efficient to have faster encoding and decoding.

However, all these design considerations vary with the applications in the practical world. For nearly all the applications, the Perceptual Transparency stays intact, but other parameters change according to the needs of the situation. For example, a particular system might not require too much data to be hidden in a media but requires to be robust to survive against different forms of attacks, well, in such case the Robustness of the Data Hiding technique needs to be well-taken care of rather than considering higher bits to be stored in the media.

1.6 TYPES OF MULTIMEDIA SECURITY

So far, we discussed the importance of Multimedia, the Data Hiding procedures, the way we can check the alteration of the media, and the design aspect of the Data Hiding technique. It is quite evident that the importance of securing the data is high enough, and encrypting it with desired techniques while keeping the media safe is of utmost requirement. In this regard, here in this section, a brief of different Multimedia Security techniques has been discussed that are being used in today's world to secure media content.

The way of securing the data depends upon the media file that is to be encrypted. The basic goal of multimedia security is to protect the media in such a way that the attacker would not be able to intercept the data, and this is all done by keeping the original data safe. The most prominent solution of multimedia security that we have been using for decades [22, 23] are discussed in subsections.

1.6.1 CRYPTOGRAPHY

Cryptography is the way of securing the data or the communication through the use of secret codes in such a form that only the intended person can understand it. The word itself means "hidden writing" which is derived from the Greek word "Kryptos" meaning hidden. Generally, in today's world, Cryptography is accounted with mathematical concepts and certain algorithms to secure the information and make it harder to decipher [24–26]. It has a broad application in the practical field such as securing credit card transactions, emails, and many web-related fields.

1.6.2 STENOGRAPHY

Stenography is a method of securing the data or the communication by hiding the information either in a fake form or hiding the data within another message, ordinary file, or other media. It is a sub-discipline of Cryptography but potentially has a greater hand over it due to the ability to get unnoticed or not get anyone's attention and successfully pass the information without being exposed [27–30]. The word Stenography is derived from the Greek word, "Steganos" meaning hidden or covered and the word "graphy" means writing.

1.6.3 DIGITAL WATERMARKING

Digital Watermarking is the process of embedding data into the media content in order to secure the media in the form of ownership claims, copyright content, or hidden communication. It is also termed as hiding the data in the form of a digital signal within the signal itself. Though Digital Watermarking is based on the concept of Stenography itself, the goal of both techniques is what makes them separate. The principle of Stenography is based on maintaining Perceptual Transparency whereas, Digital Watermarking solely sticks to high robustness [31–33].

1.6.4 DIGITAL SCRAMBLING

Digital Scrambling is the process of manipulating the signal before transmission in such a way that the output signal loses its semantic data and is hard to retrace and cannot be decoded easily. This is often used in the television industry to hide information. It ensures the data to be scrambled in order to protect the media and get it restored back to the original through the process of descrambling without any quality loss [34, 35].

1.6.5 DIGITAL AUTHENTICATION FORENSICS

Digital Authentication is a major field in Multimedia Security as it ensures the authentication of the media content and proves whether the media has tampered with or has been ever altered by any means. This technique of authentication of the media plays a vital role in the practical field especially for the Security Agencies and legal prospects [36].

These multimedia security techniques discussed above are a few promising ways that have been used for years to protect the media content from any alteration and are widely used for information hiding. In the upcoming sections, a detailed overview of all the techniques with their respective technical aspects has been discussed, which would help to get a quick insight into these topics.

1.7 CRYPTOGRAPHY

Cryptography is the art of hiding the information into a ciphertext (encrypted text) before transmission which can only be decrypted by the intended person. The word Cryptography is derived from the Greek word "Kryptos" meaning hidden and "graphy" meaning writing. The first-ever known evidence of Cryptography was traced from Egypt, where Egyptians used to communicate secretly with the help of Hieroglyph. Later it has been also used as a form of hidden communication in the time of World War 2 and has later taken into a modern form of data hiding technique. Ahead of time, various ways have been researched to enhance the security of the multimedia, which includes the combination of Cryptography and Digital Watermarking, strengthening the security of the key, and the advancement in the asymmetric key cryptography by using the composite chaotic systems [24, 37].

In today's world, the Cryptographic technique is ascertained with mathematical models and certain algorithms to hide the information successfully. The efficiency of the encryption or the Cryptographic technique can be found out by checking if the attacker is able to decrypt the ciphertext. The stronger the algorithm is, the larger is the difficulty for the attacker to decipher the text.

A few terminologies used in cryptography:

1. **Plaintext:** The original message.
2. **Ciphertext:** The encrypted or transformed message.
3. **Cipher:** Algorithm used to encrypt the original message in such a form that it cannot be understood by an unintended person.
4. **Key:** Data used during the encryption and decryption of the message.
5. **Encipher:** The technique of encoding the plaintext to ciphertext using a cipher and a key.
6. **Decipher:** The technique of decoding the ciphertext back into plaintext using a cipher and a key.
7. **Cryptanalysis:** Also known as Code Breaking is the study of principles and techniques of conversion of an unintelligible message back into an intelligible message.

Mathematically, this can be written as:

$$C = E_k(P)$$
$$P = D_k(C)$$

where P is the plaintext, C is the ciphertext, E is the encryption method, D is the decryption method, and k is the key.

There are five key objectives of the Cryptography technique that ensure a high level of media security, as given below:

1. **Confidentiality:** The information is highly confidential and cannot be understood by any other person except for whom this has been created.
2. **Integrity:** The Information stays original throughout the transmission and the alteration of the signal is restricted.

3. **Non-Repudiation:** A process of proof of the sender that he has acknowledged that he has sent the signal/message.
4. **Authentication:** The method of proving the identity of the sender and the receiver.
5. **Key Exchange:** The process in which the Crypto Key gets shared between the sender and the receiver.

Cryptosystem adhering to some or all the above-mentioned objectives helps to secure the media through predefined protocols and human behaviors. Cryptosystem uses the Cryptography algorithm with the human behavior protocols to secure the computer system and the communication to encrypt and decrypt the media content. Generally, the process carries an algorithm for encryption, digital signing and authentication, and key exchange. Taking all these processes and the objectives into consideration, the enforcement of securing the data is injected and transmitted to the target.

1.7.1 CRYPTOGRAPHY TECHNIQUES

Cryptography techniques are algorithm on which encrypting and decrypting is based, generally, there are three types of cryptography techniques, secret key, asymmetric key, and hash functions [25, 26].

1.7.1.1 SECRET KEY CRYPTOGRAPHY/SYMMETRY KEY CRYPTOGRAPHY

Secret key cryptography, also known as symmetry key cryptography, is the process where both the transmitter or the sender and the receiver share a single key to encrypt and decrypt the ciphertext [38, 39]. The key used by the sender to encrypt the data is also used by the receiver to decrypt it for further processing. In this method, a fixed length of bits, which is known as block cipher, is created with the secret key by the sender and is transmitted to the receiver, where the receiver tends to decrypt the information with the same key that has been used for encryption. This is a faster and simpler method of data encryption but then the problem that arises here is to send securely the key to the receiver. Due to the problem of the same key distribution securely to the receiver, this Symmetry Key Cryptography was found to be inefficient and therefore later Asymmetric Key Cryptography was designed. It is mostly used to comply with the objective of confidentiality. The data encryption standard (DES) is a widely used Symmetrical Key Cryptographic system but later got replaced by the advanced encryption standard (AES), which is more efficient and can handle longer key lengths like 128-bit, 192-bit, and 256 bits for preventing several brute force and attacks. Other types of symmetric key cryptography are [40]:

i. Block cipher;
ii. Blowfish;
iii. Stream cipher;
iv. RC2;
v. IDEA.

1. **Block Cipher:** It is the process of taking a block of original plain text bits and creating a block of ciphertext bits of the same size. This means a plain text of 64-bit would have an output of 64-bit ciphertext. It uses the Symmetric Key algorithm to secure the media content.

$$E_k(P) := E(K,P) : \{0,1\}^k \times \{0,1\}^n \rightarrow \{0,1\}^n \tag{15}$$

   It has the advantage of having high resistance to tampering without getting noticed but has relatively slow encryption since the entire block is required to be accounted for in order to process it ahead for encryption or decryption. A few prominent examples of Block Cipher are DES, Triple DES, AES, IDEA [41].
2. **Stream Cipher:** In the Stream Cypher, instead of accounting for the whole block for the encryption process, each bit of the plain text is encrypted one by one with the corresponding cipher digit stream also known as a keystream. Steam Cipher is also known as the State Cipher, as each digit that is to be encrypted depends on the present situation or state of the cipher. As a key it uses a Pseudorandom Bits generator to generate a key, a longer key would not only make cryptanalysis more difficult but also secure the data from brute force attacks and other attacks. The most common Stream Ciphers present are RC4, Salsa20, ChaCha, HC-256, etc.
3. **DES:** The data encryption standard (DES) algorithm is a part of the Symmetric key block cipher created by IBM in the early 1970s, which was later adopted by NIST, formerly known as NSB [42]. The DES developed by IBM was the modification of Lucifer Design which was the first proposal of the company and later got into several modifications which created DES. This is the practical implementation of

the proposed Feistel Cipher named after the researcher Horst Feistel, consisting of 16 round Feistel Structure where each round has a different unique key. The DES algorithm takes a 64-bit plain text and converts it into ciphertext using a 56-bit key. Generally, it has 64 bits but every 8th bit of the key is discarded for the parity check and is the XOR of the previous 7 bits, hence, this produces a key length of 56-bits.

As it's a Symmetric-key algorithm, it uses the same key for the encryption of the message and decryption of the message. The two fundamentals of DES are substitution and transposition; based on these basics, the algorithm follows these steps:

i. Plain text handed to initial permutation;
ii. Initial permutation on the plain text is performed;
iii. The IP creates two halves, left plain text and right plain text;
iv. Each of the halves created goes through 16 rounds of encryption process known as DES Functions;
v. Final permutation is formed by rejoining the left and right plain text;
vi. The result gives us the same 64-bit cipher text.

4. **Initial and Final Permutation:** The initial permutation is the first thing that happens before the rounds which play the role of transposition. Here, the first bit of the original plain text gets replaced by the 58th bit of the original text, the second bit gets replaced by the 50th bit, and so on, and at last, the last remaining bit of the original plain text tends to be the 7th bit of the plain text as shown in Table 1.2.

TABLE 1.2 Initial Permutation Table

58	50	42	34	26	18	10	2	60	52	44	36	28	20	12	4
62	54	46	38	30	22	14	6	64	56	48	40	32	24	16	8
57	49	51	33	25	17	9	1	59	51	43	35	27	19	11	3
61	33	45	37	29	21	13	5	63	55	75	39	31	23	15	7

The Final Permutation is just the inverse of the Initial Permutation; however, these permutations have no Cryptography significance in the DES as they don't have any particular reason for their existence in the DES model. Later once the Initial Permutation has been completed the result of the 64-bit text block is then divided into two halves having 32-bit each and goes for a 16 round of DES Functions.

5. **Rounds:** The two halves created after the Initial Permutation go through 16 rounds of encryption process which is also known as DES function, the heart of the Cipher. The process of expansion and swapping goes on in this DES function, and the encrypted 64-bit Ciphertext is the output. Each round contains two invertible cipher elements, i.e., mixer, and swapper. The DES function uses a 48-bit key to the rightmost 32-bit data to give an output of 32-bit data as a key transformation process. After the expansion, the data is undergone with swapping and the substitution process, and once it completes expansion and swapping, the result undergoes Final Permutation.

   Expansion permutation D Box – as the rightmost part has an input of 32-bit data and we have a key of 48-bit we need to process the input to 48-bit through the expansion using the permutation logic (Table 1.3).

TABLE 1.3 Expansion Table

32	01	02	03	04	05
04	05	06	07	08	09
08	09	10	11	12	13
12	13	14	15	16	17
16	17	18	19	20	21
20	21	22	23	24	25
24	25	26	27	28	29
28	29	31	31	32	01

6. **XOR Whitener and Swap:** Now the XOR operation and the round key are used on the expanded permuted input. As both the input and the key are now 48-bit, we can go with this process.

   S Boxes Substitution box – the substitution box or the S – box carries the mixing operation where a total of 8 S boxes are used and for each S box, 6-bit Input and 4-bit output is processed. At last, before moving to the Final Permutation the result of all the 8 S-boxes is combined into a 32-bit channel for the Final Permutation.

7. **Final Permutation or Straight Permutation:** Finally, the 32-bit input is permuted to get a 32-bit output following the rule same rule

of the 7th bit of input gets into the 2nd bit of output and hence Table 1.4 given is as follows.

TABLE 1.4 Final Permutation Table

40	08	48	16	56	24	64	32	39	07	47	15	55	23	63	31
38	06	46	14	54	22	62	30	37	05	45	13	53	21	61	29
36	04	44	12	52	20	60	28	35	03	43	11	51	19	59	27
34	02	40	10	50	18	58	26	33	01	41	09	49	17	57	25

8. **Blowfish:** It was originated in 1993 by the researcher Bruce Schneier as an alternative to the DES and Triple DES algorithm [43]. It has prime significance over the DES algorithm as this is fast and has an efficient encryption rate, freely available for all to use. This uses the concept of the Symmetric Key algorithm and was a successful encryption technique than the DES algorithms. Later, the successor Two Fish and finally, AES evolved, which played a major role in the security of network systems. It uses a block size of 64-bit, but unlike the DES algorithm, the size of the key varies from anywhere between 32 bits and 448 bits. Just like the DES, it is also based on two fundamentals, expansion, and data encryption while using the Feistel Cipher with 16 rounds.

   The steps involved in the Blowfish algorithm can be well related to the DES [83]:

   i. The subkeys are generated, 18 subkeys are generated that would help while encryption and decryption technique. These subkeys are then stored in a P-Array having a 32-bit entry.
   ii. Now 4,256-bit entry S-boxes are used in both the encryption and decryption process with 32-bit entry points each. This S-box would accept an 8-bit input while giving a 32-bit output.
   iii. The encryption is done in 16 rounds. Each round, one P array is used, and the last one is used for the XOR operation.
   iv. After this, the left and right blocks are swapped.
   v. Lastly, they are combined to get the Cipher text.

   As it could be seen that the working principle of Blowfish is based on DES and Feistel encryption, it is still having a greater advantage of being faster than DES and Triple DES as it uses a variable key anywhere between 32-bit and 448 bits. However, one major drawback of Blowfish was the small block size that can account for only

64-bit, which later became the primary reason to create a stronger algorithm and the rise of AES.

9. **AES:** As a replacement to the previous DES and Triple DES algorithm, the most secured and efficient algorithm was designed especially for government officials to secure their confidential data, known as AES, which was 6 times faster than previous algorithms [44]. AES is a Symmetric block cipher that uses a key of 128/192/256 bit to encrypt the plaintext to ciphertext. As it supports larger key sizes, unlike the DES system, which supports 64 bit as the maximum key size, the AES algorithm is quite stronger and provides higher encryption than conventional ways used before [81]. It came around the early 2000s as an alternative to the old DES and Triple-DES algorithms and was found out to be more secure as it uses 128 bits blocks and a key of 1,28,192,256 bits. It is widely used in network security, especially wireless security, file encryption, SSL/TLS, and securing the website connection over the network.

Unlike the DES algorithm that uses the Feistel Structure and bit-level access, AES relies on the substitution-permutation network on the byte level to encrypt and decrypt the data. It can use 128-bit, 192-bit, 256-bit keys to encrypt the plain text to ciphertext and decrypt vice versa. While the swapping and mixing process, instead of swapping the data of each block, here the entire row and column are shifted. This ensures a higher degree of encryption while giving more advanced security. From Table 1.5, it can be observed that for 128-bit Key 10 rounds has to be processed, for 192-bit 12 rounds and 256-bit 14 rounds have to be processed, once the rows and columns are shifted the addition of the round key is done at the last.

Each round except for the last round involves:

i. Substitution Byte Transformation: S box is used to substitute byte by byte.
ii. Shift rows Transformations: Permutation is done by shifting cyclically the last three rows.
iii. ShiftColums Transformations: Using the Galois arithmetic the substitution is done.
iv. Add Round Key Transformation: XOR operations are used with the key.

However, unlike DES, AES doesn't follow the same steps for decryption, instead, the decryption involves in the inverse steps.

TABLE 1.5 Key Length and Number of Rounds for Different AES Systems

AES	Key Length	Block Size	Number of Rounds
128	4	4	10
192	6	4	12
256	8	4	14

Several modifications to the existing techniques have been done to increase the Robustness, for example, using a stronger algorithm for shifting row and S box transformation, and using the sub-bytes with the shift rows operations [45, 46]. This ensures very well-organized encryption for the overall data security.

1.7.1.2 PUBLIC KEY CRYPTOGRAPHY

Public key cryptography is a form of asymmetric key sharing technique where two independent keys are used, namely, a public key and the private key to encrypt and decrypt the message. The public key can be distributed freely and is required for the encryption process, whereas the private key is kept secure and is required for decrypting the signal. The encrypted data with the public key can only be decrypted with the associated private key, this means to send a message to person A requires the message to be encrypted with the A's public key and can be decrypted by only A as he has the private key. This technique is one of the most robust and is primarily used for a high level of authentication, non-repudiation, and key exchange. The most commonly used public key cryptographic system is the RSA, used for the Internet, and DSA for the digital signatures.

1.7.1.2.1 RSA

The Rivest-Shamir-Adleman RSA is an asymmetric algorithm widely used for the secure transmission of data [47]. As it is based on Asymmetric key cryptography, this model has two keys for the encryption and decryption process: Public Key and Private Key. The message is generally encrypted with the Public Key and can only be decrypted using the Private Key having some identical properties. This ensures, Confidentiality, Authenticity, and Integrity of the message. In the RSA algorithm, a Public Key consisting of two large prime numbers with an auxiliary value is created for the Public

Key. The Private Key holds the prime number, hence, anyone knowing the prime numbers can decrypt the message. It is comparatively very slow and is not mostly used to encrypt user data. Primarily, it is used to secure the transmission of the shared key in the Symmetric Key Cryptography technique.

The RSA involves four steps:

1. **Key Generation:** Selection of P and Q prime numbers where P is not equal to Q. Calculation n = P * Q, Calculate Ø(n) = (p – 1) × (q – 1). Select integer e such that gcd(Ø(n),e) = 1 and 1<e< Ø(n). Then Calculate d, d ≡ e – 1 (mod Ø(n)) or d.e ≡ 1 (mod Ø(n)) Public Key = {e, n} Private Key = {d}
2. **Key Distribution:** Public Key = {e, n} (distributed openly). Private Key = {d} (kept secret). The public key allows the sender to perform encryption, while only the private key holder can decrypt the message, ensuring secure communication.
3. **Encryption:** Plaintext: M < n Ciphertext: C = Me mod n.
4. **Decryption:** Ciphertext: C Plaintext: M = C d mod n.

1.7.1.3 HASH FUNCTIONS

Also known as One Way Cryptography, Hash Functions on the other hand is quite different than the Symmetry and Asymmetry Key Cryptography as no key in this algorithm is used [48, 49]. Instead of using a key, a fixed length of the hash value is computed so that the plain text completely lost its original form and can't be recovered. This is mainly used in password encryption, where a HashMap is used to encrypt and store the password without exposing it publicly. This technique maintains the data integrity of the system. The mathematical models used to encrypt the data are signed by the digital signature and are ensured that it is irreversible. In this technique, the media can be authenticated easily, as tampering with the digital signature would make the alteration attempt highlighted. The general Hash Functions that are used in the practical field are SHA-1, SHA-2, SHA-3, and SHA-256.

1.7.1.3.1 SHA 256

SHA 256 belongs to the SHA-2 family and the successor of the previous SHA-1 family is one of the most prominent and strongest Hash Functions available to date. Supporting the AES, the SHA-256 is a good partner solution to maintain a high level of security [50, 51]. As the Hashing itself is an irreversible and one-way process, it is quite impossible to reconstruct the initial data from the Hash Value. Also, a 256-bit output is as hard as attempting about 2^{256} times to generate the initial data, which itself is out of our practical capacity. Also, there are no chances of any data collision for the same Hash values, which makes it completely unique and super secure to use. Cryptocurrency like Bitcoin uses this SHA-256 for securing their transactions, and most of the password protection solutions with the digital signing of the content uses this algorithm.

The steps involved in SHA 256 hashing are:

- Preprocessing;
- Initialize hash value/buffers;
- Initialize round constant K;
- Block/chunk loop;
- Message schedule W;
- Compression;
- Final hash.

A typical "Hello World" message after Hashing would look like:

a 5 9 1 a 6 d 4 0 b f 4 2 0 4 0 4 a 0 1 1 7 3 3 c f b 7 b 1 9 0 d 6 2 c 6 5 b f0bcda32b57b277d9ad9f146e.

Among all these cryptographic techniques, symmetric key cryptography is the simplest and the fastest method for carrying a piece of large information, whereas Asymmetric Key Cryptography can only handle small pieces of data, less than the key size. Therefore, often the Asymmetric Cryptography is used to secure the Symmetric Key that can encrypt larger data, the same concept is also applied to encrypt the digital signature for the Hash Functions.

1.8 STENOGRAPHY

Stenography is the process of hiding the data within a signal in such a way that it remains undetected and can only be extracted by the intended person. The word stenography is derived from the Greek word "Stenos" meaning hidden and "graphy" meaning writing. Stenography, when combined with other forms of encryption process, can result in stronger security of the media content. The main advantage of Stenography over Cryptography is that the signal does not attract any attention and goes undetected throughout the transmission [27, 28].

Just like Cryptography, the use of Stenography has a historical existence; it was reported to be used in Greek times when messengers used to hide their message on their shaved heads and let the hair grow to keep it a secret. The use of Stenography was also found out to be highly entertained during World War 2, where German spies were sending a message containing a hidden secret. These translations are very hard to understand and detect. For example, during that time, the use of invisible ink was used to hide secret information which seems like blank paper to normal human but the intended person could use it to view the hidden message.

Previously for the Text Stenography, the techniques were common in nature, mostly there were a few generalized techniques used to hide the secret message [30]. This includes the First Letter Algorithm, n-Letter algorithm, and altering the amount of the white space. With these variations, the entire text message stays away from any human perception and it gets successfully transmitted to the receiver, therefore the Text Stenography was largely used in the time of World War 2 for secret communication. Recently, the Stenographic techniques have seen a surge in development as they attracted the researchers towards it proving a better modality to the Cryptographic techniques. Various research work has been carried out to improve the Stenographic technique which includes, high capacity information hiding scheme using the optimized LSB substitution, using new methods such as hybrid transforms and vector machines, and using the minimum modified bits with LSB substitution [51–53].

Due to the ability to stay undetected, this was largely used in public messages without getting caught by the attackers perceptual. The media could entitle any form like image, video, audio, or any other form of digital content. It has been used both in the form of physical and digital to hide secret information with or without encryption, depending upon the situation and degree of security requirement. Physical Stenography is used to hide a message within the wax tables, hide a message using the secret inks, and a message written in Mohr's code, whereas Digital Stenography accounts for the information hiding in the image or audio having the lowest bits, modifying the echo of the sound, embedding data into the least viewable section. Figure 1.3 shows the embedding and extraction process in Stenography, generally, a pseudo media is formed when a protected media contains secret information, this can be simply represented by:

Protected data + Embedded data + Pseudo Key = Pseudo Media

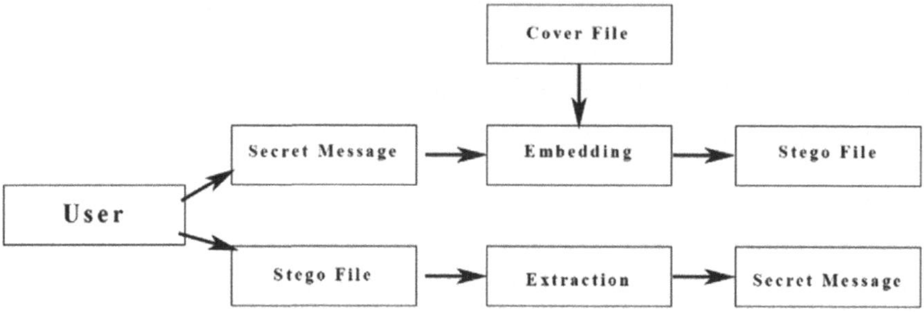

FIGURE 1.3 Embedding and extraction process in stenography.

This way, the protected data is sent over to the receiver, successfully adhering to the perceptual transparency. Even If the interceptor knows that the data has been embedded into the media, the algorithm should be robust enough to sustain many forms of common attacks. Depending upon the situation and the use case the stenography techniques are employed either on the Spatial domain or the Frequency domain.

Due to the ease of invading through the attacker's vision, Stenography has been adopted in various fields to keep the secrecy in the communication and data hiding methods. This allows the sender to entertain the opportunity of not censoring the information, instead using Stenographic techniques to hide the secret message that is not detectable. On pairing with the existing communication models and other encryption techniques, this could be a super secretive communication within the government officials, which are both perceptually transparent and highly encrypted.

1.8.1 TYPES OF STENOGRAPHY

Stenography helps in protecting various digital content which includes, text, image, video, and protocol [54]. The process of hiding the information quite depends on the type of cover media and varies accordingly.

1.8.1.1 TEXT STENOGRAPHY

Text-based stenography is the process of hiding secret information within a text file. This is done in various ways, like introducing random character sequences, altering the words within a sentence, using the nth character algorithm, etc. This can be done by Random statistical generation, format-based methods, or linguistic methods. The use of spaces within a sentence has also been seen in the literature to hide messages in the binary form where single space represents 0 and double spaces represent 1 [55]. The main goal is to store the information in the text in such a way that the text file itself doesn't look like it's storing any hidden information. However, storing hidden information in the text file is not as efficient as other media files, and this is due to the limitation of the storage capacity.

1.8.1.2 IMAGE STENOGRAPHY

Embedding secret information inside an image is one of the most widely studied fields. In this method, the hidden information is stored using an algorithm and a key. Though the use of a key is not mandatory, however, using a secret key would ensure high encryption of the hidden data. The encrypted image is said to be a stego image, which is the output that is ready to be transmitted. Once it is received by the intended receiver, it undergoes with certain extraction algorithm with the associated key to deriving the hidden message. The prime goal here is to make the stego image perceptually transparent so that the attacker would be unable to detect any stored hidden message. Even though the attacker detects the existence of the hidden message, the hidden data cannot be extracted due to the encryption using the keys. The use of images for hiding secret information is considered the most in the area of Digital Stenography is due to the presence of a huge number of bits, which are the containers of the data that is to be embedded.

1.8.1.3 VIDEO STENOGRAPHY

In this technique, secret data of any kind is embedded into the video file. Just like the Image Stenography, the video also has a large capacity to store hidden data. One more advantage of using Video Stenography is the possibility to store more than one form of digital media, this could be also termed as a combined form of both Image Stenography and Audio Stenography. This can be done either by embedding the hidden information in the uncompressed raw video and then compressing it later or by embedding data directly into the direct stream. The most common formats used are the H264, MP4, and AVI.

1.8.1.4 AUDIO STENOGRAPHY

In the Audio Stenography, the hidden message is embedded into the audio signal altering the binary sequence of the signal and storing the information without being detected [56]. Comparatively, storing any hidden information in the audio is quite more complex than any other digital media. The methods used are Parity coding (Low Bit Encoding), Spread Spectrum, and Phase Coding.

1.8.1.5 NETWORK STENOGRAPHY

In the Network Stenography, the cover media is taken to be the network protocol to hide the secret information. In this process, the secret data are kept hidden in the least accessible part of the network protocol.

While designing the Stenography Model, it is highly essential to consider the factors affecting the efficiency of the stego media. Like Cryptography, Stenography also has a few objectives that should be considered to make the Stenography technique highly effective, such as:

1. **Perceptual Transparency:** As the principle of the Stenography algorithm is to be perceptually transparent, therefore, the algorithm used should adhere to this objective primarily. The data embedded into a media should be detectable, and the stego media should not be suspicious of hidden information.
2. **Robustness:** This of the technique is highly required as it ensures the stego media can sustain common attacks made intentionally and accidentally, this includes transformations, tampering, etc.
3. **PSNR:** Peak signal-to-noise ratio helps to detect the quality of the two media content. Higher PSNR values refer to higher quality and higher similarity of the two images. So, the stego media should have a higher PSNR in order to be visually undetectable and securely store and transmit the hidden data.

Based on these objectives, a stenography algorithm can be checked for its effectiveness, and further implications could be decided. Considering these fundamentals of Stenography, the upcoming section briefly shows how the information can be hidden in various media.

1.8.2 STENOGRAPHY APPROACHES

The methodology accounted for the data hiding solely depends on the form of the media and the requirements. Thus, in this section, individual data hiding techniques of various media content have been discussed briefly.

1.8.2.1 IMAGE STENOGRAPHY

Over the decade, various research work has been done in the field of Image Stenography to make data hiding more efficient. With a goal to make it impossible to intercept, the techniques from the Spatial domain to the Frequency domain have been constantly enhanced. Generally, to hide a secret message inside an image, there are a couple of methods that can be adopted: spatial domain algorithm, frequency domain algorithm, and masking and filtering [57].

In the Spatial Domain Stenography, the data is stored by modifying the pixel values of the cover image. The selected pixels can store the data until they are not visually interceptable. The most common approach in Spatial Domain Stenography is the LSB. In the Transform Domain Stenography, the hidden data is stored in the transform coefficient of the cover image. This technique hides the message in significant areas of the cover image or can be applied to the entire image. This is comparatively more robust than the Spatial Domain Stenography. Apart from this the Masking and Filtering helps to hide the data by marking an image. This is also a basic concept of Watermarking, which helps to hide the data in the form of a marker on a significant area within a 24-bit grayscale cover image.

1.8.2.1.1 Spatial Domain Stenography

1. **LSB:** The least significant bit (LSB) is the simplest and most common stenographic approach to embed the data in form of bits into the LSB of the cover image in a deterministic sequence [57, 58]. The major advantage of using this method is, it does not come

under the human visual system as the change is very small to notice. The higher bit-sized image can store larger information as the bits in one pixel are higher. So, the capacity of the data-hiding technique depends on the bit size of the entire image; for example, a 24-bit image has the potential to store higher information than an 8-bit image. The first proposal to the LSB substitution showed how the data can be embedded using the LSB method using the Eqn. (16):

$$I_{Si} = I_{Ci} - I_{Ci} \, mod \, 2^k + d_i \tag{16}$$

where; I_{Si} is the ith transformation of the pixel value of the stego image; d_i is ith k-bit secret data; and k represents a total number of bits embedded in each pixel.

The extraction process of the stego image is basically a mod of stego image pixels and 2n. However, the quality of the stego image directly depends on the bits plane used for the extraction process. The higher the bits plane used, the lesser would be the quality of the stego image. This could be accounted for by the result of PSNR. While using a 1-bit plane, the PSNR turns out to be 51dB, whereas using 4LSB planes gives a value of 31 dB. The less PSNR depicts the quality degradation of the stego Image. One recent methodology has proposed implementing the modified sign digit to maintain a PSNR of 51dB and have a very low computation time [59, 60].

2. **PVD:** Pixel value differences (PVD) is the technique used to store the secret data to the adjacent neighbor pixel that has an edgy region in the cover image [61, 62]. In this process, the cover image is first divided into blocks of two pixels, where the blocks are categorized into ranges R_i according to their Delta value.

   The capacity of each block is given by Eqn. (17):

$$c = log_2^{u_j - I_j + 1} \tag{17}$$

   The extraction of the secret bits can be done by:

$$b = \begin{cases} Diff^* - L_j & if \; Diff^* \geq 0 \\ -Diff^* - L_j & if \; Diff^* < 0 \end{cases} \tag{18}$$

The PVD scheme is a promising technique to improve the data holding capacity and a good quality stego image, but they ultimately lack security as these are quite vulnerable to steganalysis easily. Therefore, the improvements in the security of these techniques have been in constant development, and recent schemes tend to solve the issue of security. For example, introducing the pseudo-random methods to select the blocks, the use of patched reference tables, and the use of Hilbert curve-based embedding have proven higher security and higher data hiding capacity.

1.8.2.1.2 *Transform Domain Stenography*

1. **DCT:** Discrete cosine transforms (DCT), also known as JPEG Data Hiding, is one of the most extensively studied and has wide applications in image processing. First, the image is transformed from the Spatial domain to the Frequency domain, and it is segregated by the spectral bands. Mathematically, it can be represented as:

$$C(u,v) = \alpha(u)\alpha(v)\sum_{x=0}^{N-1}\sum_{y=0}^{N-1} f(x,y)\cos[\frac{(2x+1)u\pi}{2N}]\times cos[\frac{(2x+1)v\pi}{2N}] \quad (19)$$

   The basic Flow diagram of the DCT-based Stenography is as follows.

   Initially, the cover image is transformed into an 8 * 8 block of non-overlapping pixels. Then, it is further converted into the DCT coefficients, which are generally floating numbers, and lastly, the quantization is done with the help of a quantization table. The values derived are used ahead for the data embedding process. However, the selection of the coefficients for the data embedding process is crucial, as any unintentional alteration of the data would result in the corruption of the stego image. Finally, to get the stego image the inverse of the DCT is carried out.

   To increase the embedding capacity and the speed, a new method is proposed by using a 16 * 16 block instead of 8 * 8. Another method that uses a linear chaotic map-based random LSB embedding has been seen to have a PSNR value of 59 which sustains well against many steganalysis approaches.

2. **DWT:** Discrete wavelet transforms (DWT) help to provide both the Spatial and the Frequency data of the image [63]. This technique has higher security and is used often to separate the low and high frequency of the cover image. Mathematically, 1D DWT can be represented as:

$$a_{j+1}[p] = \sum_{n=-\infty}^{\infty} l[n-2p]x_j[n] \quad (20)$$

$$d_{j+1}[p] = \sum_{n=-\infty}^{\infty} h[n-2p]x_j[n] \quad (21)$$

The general process of DWT is as follows.

In this technique, first, the decomposition of the cover image using the DWT is done. This can be generally done using the 1D DWT transformation with the help of rows and columns. Where in the row processing, each of the rows is applied with Low Pass Filter (l[n]) and high pass filters (h[n]) followed by the Down Sampler. As the following operation divides into low-frequency and high-frequency parts, it stores relevant data for further processing. The low-frequency stores the approximation of the image, whereas the high-frequency stores the details of the image. Similarly, the column processing also divides the column into low frequency and high frequency but with two more sub-bands. The low-frequency band is the first sub-divided into a low-low band and low-high sub-bands; the high-frequency band is similarly divided into its respective sub-bands. Now this process, 2D DWT divides the image into four equal-sized sub-bands. Further, for the decomposition, the input is signal is elected as the LL sub-band. This LL sub-band is responsible for holding the approximation details of the image, which is very crucial for the reconstruction of the image. Any minor change in the approximation detail would result in a permanent alteration of the image.

The major problem with the DWT approach is the high fractional values coefficient. For efficient Data Hiding, the truncation of these data needs to be done, which would result in information loss for the reversible stenography system. As most of the information embedding algorithm is worn on integers, for DWT, truncation of the fractional values is the only way solution. However, this was effectively tacked with the introduction of IWT that solves the issue of reversible stenography system.

3. **IWT Transforms:** To tackle the problem of DWT fractional coefficient, the IWT transform has been introduced that uses lifting schemes like floor and ceil to achieve the integer values without the requirement of quantization. The 1D wavelet transform can be represented by:

$$d_k = x_{2k+1} - x_{2k} \tag{22}$$

$$s_k = x_{2k} - \frac{d_k}{2} \tag{23}$$

Several types of research have been carried out to improve the security of the IWT by using randomly selected high-frequency coefficients using graph theory, using the arithmetic coding for high embedding capacity, and by changing the transformed image histograms. Also, recently, a new approach has been proposed to hide data and its application in the field of medical images.

1.8.2.2 AUDIO STENOGRAPHY

Audio Stenography is the process of hiding the data of any digital form inside an audio signal that cannot be detected [64]. It is comparatively harder to store data inside audio as there are several factors that play a crucial role, for example, the signal to noise ratio (SNR) should be greater than 20 dB, the quality of the signal should not deduce and the power range of the sound should be greater than 109:1. Again to get unnoticeable to the human ears the noise embedded should be low than 70 dB without hampering the original signal. The Audio Stenography can be divided into the Spatial domain (LSB) and the Transform Domain (Parity, Phase Coding, and Spectral System).

1.8.2.2.1 Low Bits Encoding

Just like the LSB techniques used in the Image Stenography, it could be applied the same way to hide information in the LSB of the carrier. Generally, we can store an image in the form of bits to the LSBs. Due to the less complexity and less consumption time, this is generally a simpler data hiding technique. To increase the security of the data, the XOR operations are used to encrypt the data.

1.8.2.2.2 Parity Coding

In the parity coding, the signal is broken into different samples, and each bit of the secret message from the parity bit is embedded. The main disadvantage of Parity Coding is that, if the data is resampled then some of the information might get lost.

1.8.2.2.3 Phase Coding

In the Phase coding technique first, we convert the message into blocks and then embed it into a certain phase, this is done by replacing the initial phase with another phase having secret data. Considering the SNR value, this is considered to be efficient but tends to be affected by the phase dispersion when a large set of variations undergoes between the phase and frequency component. The major disadvantage of this process is the ineffective payload.

1.8.2.2.4 Spread Spectrum

Spread Spectrum is a good way of providing higher security to the transmission. In this process, the hidden data gets transferred over the wide frequency bandwidth. It is a robust method used in highly confidential communication where secrecy is the main objective. The main idea of Spread Spectrum is to send the hidden data with the frequency having very high noise. As the hidden data is spread over each band, the data can only be recovered after extracting the whole band.

1.8.2.2.5 Echo Hiding

As the name suggests, in the Echo hiding an echo is implemented in the audio signal by the means of amplitude modulation and implementing a delay in the signal. Each echo in the cover signal stores the hidden information, to restrict the overlap the delay has been used. In the Echo hiding process, first, the audio is segmented into pieces, and within each piece, the data is being hidden. After the completion of the embedding of the data, all the pieces are joined together and transmitted. This technique is mainly used due to its robustness and larger data transmission rate.

1.9 DIGITAL WATERMARKING

Digital Watermarking is a process of embedding data in the form of bits or markers into media such as image, audio, and video to protect the ownership of the digital media content. It has been widely researched in the field of media security and plays a crucial role in protecting intellectual property (IP). It is primarily used to authenticate the media content and maintain the integrity of the signal. Apart from all these, Digital Watermarking can be used to track the source, video, and image authentication, screen casting, ID security, tamper detection, content management, broadcast monitoring, and a lot more [31, 65]. Though it might seem like it's equivalent to Cryptography but no, the principles of Watermarking are way different than Cryptography and Stenography.

Once the signal has been decrypted in the Cryptographic algorithm, the data can be modified, and the tracking of the legitimate use of the data cannot be done, which is the main aim of Watermarking. So, Digital Watermarking can help to protect the content even after it has been decrypted. The concept of Stenography is quite similar to that of Watermarking with a slight difference; in Stenography, the major goal is to hide a message in such a way that the attacker is unable to detect any information hiding inside, whereas in Digital Watermarking, the data is also hidden in such a way that it's not only untraceable but also restrict the removal and replacement.

In 1999, Sony and other 4 leading companies, including IBM, decided to unify their approach by adapting Invisible Watermarking for their digital products like movies and videos. This is the point where Digital Watermarking was officially adopted by the big firms, and the Copyright Acts were practiced. At the same period, Adobe came up with its software Photoshop 4.0 with a digital watermarking feature to protect the original work of the creators. This was the period when constant advancement in the area of Digital Watermarking and the enforcement of Copyright Laws were implemented and adapted in various countries.

1.9.1 WATERMARKING CLASSIFICATION

Digital Watermarking can be classified into various forms which are dependent on the way of the watermarking algorithm [66]:

i. Visible and invisible watermarking;
ii. Robust and fragile;
iii. Spatial and frequency domain.

1.9.1.1 VISIBLE AND INVISIBLE WATERMARKING

According to human perceptuality, Digital Watermarking has been categorized into two basic divisions, visible and invisible watermarks. As the name suggests, the Visible Watermark is a marker embedded into digital media such as image and video which is easily visible and is in the range of human perceptuality. This embedded digital signal is mostly a logo that has been inserted to claim the ownership of the digital content.

Whereas Invisible Watermarking is a process of embedding information into a digital signal in such a way that it is not imperceptible and passes away unnoticed. As Invisible Watermarks are not visible, it is widely used to store hidden information and data like identifiers, ownership information, metadata, and other forms of data that are hard to detect and remove. It is also used to communicate secretly between the mediums.

Both Visible and Invisible Watermarking are used to authenticate the media content; however, even after the signal has been decrypted, the signal can still be monitored and could restrict any form of alteration as the presence of a digital watermark is still intact. The applications of Visible Watermarking can be found in the television industry, image galleries, stock images, official videos, and brand media content. These are primarily used to restrict unauthorized access, copying, and modifying the original content as a form of protection of IP. A concept of Dual Watermark where both the Visible and Invisible Watermark is also used to increase the security of the media file as even if the attacker tends to remove the Visible Watermark, the digital signal still stays the safe and easy validation is possible through the Invisible Watermark contained.

1.9.1.2 ROBUST AND FRAGILE WATERMARKING

Robust Watermarking is the process of embedding watermarks to digital media in such a way that even if the media has been modified and tampered the watermark can still be recovered from the altered media. Robustness is one of the most required features in the entire Data Hiding field, although, there's no universal security that could handle all the forms of attack, robustness potentially helps to secure from many attacks up to a certain degree. Fragile Watermarking, on the other hand, helps in detecting the integrity of the media content. Fragile Watermarking is widely used to authenticate and authorize the data as it significantly detects any minor changes in the digital media. Comparably, it is easier to devise than Robust Watermarking as it requires less computational complexity.

1.9.1.3 SPATIAL AND FREQUENCY DOMAIN

The classification of the Watermarking can be in the working domain basis, dividing it into two categories: the Spatial domain and the Frequency domain. Embedding Digital Watermarks to the Spatial domain refers to embedding watermarks into the pixels of the visual media in various color spaces. The data is hidden within the pixels in various color spaces that make it difficult to view and remove. The Spatial domain watermarking is used as it has low computational complexity and is less time-consuming. There are various techniques used in Spatial domain watermarking such as LSB, SSM Modulation, Patchwork, etc. The most important and widely studied method among all these techniques is the LSB.

In the Frequency domain watermarking, the watermark is applied to the spectral coefficient of the digital image. As the Human Visual System is more prone to spectral coefficient regions, the Frequency domain watermarking is used. The common transforms used in Frequency domains are discrete cosine transformation (DCT) and discrete wavelet transformation (DWT).

1.9.2 WATERMARKING REQUIREMENTS

There are certain design considerations for Digital Watermarking which is required while using the watermarking algorithm to make it more secure and efficient. On a general basis, the Digital Watermark is expected to sustain normal distortions and a low level of attacks. However, an ideal Watermarking is made in a way that adheres to the following requirements.

1. **Imperceptibility/Fidelity:** Imperceptibility refers to the perceptual transparency of the watermark which goes undetected. This process ensures that the original image and the watermarked image remain similar and the quality of the original image is preserved.
2. **Robustness:** It is the prime feature of secure Watermarking. It ensures that the watermark sustains all forms of common attacks and does not get affected by any means while saving itself from degrading the quality of the watermark.
3. **Blind Detection:** The system is a process where the original data is not required for the watermarking, this is also known as a Public Watermarking system. In the non-blind/Informed detection, the original data is required for the analysis of the watermarking. This is also known as the Private Watermarking System.
4. **Capacity:** It defines the quantity that can be embedded in a media for successful detection at the time of the watermark extraction process.
5. **Security:** It allows the watermark to be only accessible by the authorized intended persons. Various algorithms, including Cryptographic techniques, could be used to secure the watermark.

6. **Computational Cost:** It is dependent directly on the way or watermarking technique that is to be adopted. A simpler Visible Watermark could cost comparably less than the Invisible and Robust Watermarking system. The more complex the algorithm is, the more resources are required for watermarking.

1.9.3 WATERMARKING ATTACKS

After the watermark has been embedded into the signal, any form of tampering or manipulation of the watermark can be termed as a Watermark Attack. The sole intention of the attack is to destroy the watermark, which could be both intentionally made or could be due to an accident. With technological advancement, several forms of attacks have been recorded in order to tamper the watermark embedded, a few of them are [67]:

1. **Removal Attack:** These are the attempts made to remove the watermark from the digital signal or attempts to remove the hidden data from the watermark. In this process, the key is not necessarily needed, and the watermark can be removed without tackling the security system. The main operations involved in the Removal attack are the quantization, denoising, re-modulation, and collusion attack.
2. **Oracle Attack:** It is a Cryptographic attack that aims at the security flaws of the Watermark System. In this attack, the attacker is able to manipulate the watermark with the help of the Watermark Detector. First, a Sensitivity Analysis attack is carried out, which results in the presence of the watermark in the binary form; once the confirmation of the watermark has been stated, the gradient attack helps in changing the values of the detected statistics.
3. **Ambiguity Attack:** These are the protocol attacks that tend to confuse by adding a fake watermark signal with the help of the Invertible Watermark. Using this concept, the attacker tries to replicate the output signal with a fake watermark, this makes the ownership of the digital signal uncertain.
4. **Geometrical Attack:** These are made by changing the transforms of the watermark and the digital signal by various means, like cropping the media. This does not entirely rely upon removing the watermark, rather distorting the watermark.
5. **Interference Attack:** In this attack a noise is intentionally embedded into the watermark, for example, compression.

6. **Attacks in the Wavelet Domain:** This is one of the advanced attacks on watermarking as it tends to reject the robustness of various watermarking methods. In this attack, the attacker finds the flat regions of the watermark and manipulates the coefficient of each region to destroy the watermark with the least detection to the HVS.
7. **Self-Similarity Attack:** In this attack, a few regions of the image get substituted with other parts of the image which look similar within a good PSNR range. This makes the removal of the watermark successful without being perceptually visible.

1.9.4 WATERMARKING EVALUATION AND COMMON PROCESS

The evaluation of the watermark conveys the quality and the tampering evidence of the digital media. These quality factors help us to determine the level of alteration a digital image has been undergone. Basic quality factors like MSE, PSNR, SNR, BER help us to determine the sensitivity, quality loss, amount of alteration, and other such properties [68–70]. These can be mathematically written as:

1. **MSE:** The mean square error (MSE) as shown in Eqn. (24) can be used to check the distortion between the original cover image and the watermarked image. This allows us to detect any change in the watermarked image.

$$MSE = \frac{1}{M*N}\sum_{i=0}^{M-1}\sum_{j=0}^{N-1}[\{I(i,j)-\{K(j,k)\}\}]^2 \quad (24)$$

$$RMS = \sqrt{MSE} \quad (25)$$

2. **SNR:** The signal-to-noise ratio (SNR) can be used to check the sensitivity of the images, which is given in Eqn. (26).

$$SNR = 10\log\left(\frac{P_{signal}}{P_{noise}}\right) \quad (26)$$

3. **PSNR:** Peak signal-to-noise ratio (PSNR) is widely used to determine the similarity between the original and the watermarked image. The larger PSNR values define the higher similarity of the two images.

$$PSNR = 10\log\left(\frac{m.n.P^2}{RMS}\right) \quad (27)$$

4. **Euclidean Distance:** It is a common distance between the two points in the Euclidean space. It can be used to embed a sequence of data

in the form of bits with the help of the vertex position using the Euclidean Distance [84, 85].

$$ED(X,X') = \sum_{i=1}^{M}\sum_{j=1}^{N}[\{X_{(i,j)} - X_{(i,j)}'\}]^2 \tag{28}$$

5. **BER:** The bit error rate (BER) is the ratio of the number of error bits received to the total number of bits received.

$$BER = P(\frac{H}{W}) \tag{29}$$

6. **SSIM:** The structural similarity index metric (SSIM) helps to determine the similarity of the two images based on the perceived changes in the structure of the image. This is however better alternative to the PSNR technique where the change is accounted for the specific pixels instead of taking the neighboring pixels into consideration where vital data about the structure of the image might be found.

$$SSIM(X,Y) = \frac{(2\mu_x\mu_y + c_1)(2\mu_{xy} + c_2)}{(u_x^2 + \mu_y^2 + c_1)(\sigma_x^2 + \sigma_y^2 + c_2)} \tag{30}$$

With the help of these metrics, the quality of the watermarked images can be determined. A general watermarking process has 3 steps, Embedding, Extraction, and Detection. First, the perceptual mask is calculated where the amount of certain pixels that can be altered is found out; once this is done, the information is spread using the key, and the watermark is embedded in the signal.

1.10 DIGITAL SCRAMBLING

Digital Scrambling is the process of altering the original data with the help of noise and disorder so that the original data lose its form and becomes difficult to view. This makes the original data secure from the attackers as the scrambled data has no resemblance to the original message. It is often found from the literature that the word scrambling is also used as the synonym for the word encryption, however, the goal of both the process is the same, that is to protect the original data but the ways of resulting is different. Digital Scrambling helps us to scramble images, audio, and videos while securing them from the wrong hands. The fundamentals on which digital scrambling is based are given as follows:

i. The lowest level of distinction and understanding;
ii. High level of security;
iii. No change in the original signal and hence keeping the original signal completely safe.

Hence, the goal of digital scrambling is to primarily generate a high-quality output by scrambling the original message using the Scrambling algorithm such that it doesn't alter the original data, maintains high secrecy of the original data, and hence makes it quite impossible for the attacker to understand the output data after digitally scrambled.

1.10.1 IMAGE SCRAMBLING

The first time the concept of Scrambling was used in the field of television where the signal was scrambled so that the output is not properly visible to the viewers and only the operator knows the original signal. In this regard, it became an extensive field of research where we can secure our cable television signal transmission and can make it reach the other end who is intended to get the message after descrambling without any quality loss. The process of descrambling would make the altered media transformed into its original form and hence it would become accessible. The major advantage of this process is that, even if the output is shared with the attacker, it's quite difficult and impossible to descramble the image into its original form, and hence the data security is well protected here.

In the Image Scrambling process, the scrambled image can only be recovered using the key and the proper algorithm technique; failing to do so would result in failure of getting the original data. The size of both the original image and the altered image should be the same in order to descramble the media to get the information. The image could be a good way of storing large information than the text as it is made up of large sizes. However, decoding an image to gather information requires prior knowledge and a strong algorithm, however, the process itself is very hard for the attacker if he wishes to derive every bit of information from each pixel. From the previous section, we could analyze that the information could be easily hidden in the pixel and hence the maximum number of data that could be hidden could be found out with the help of PSNR and the Comparison of Distance among color spaces. In this regard, various Image Scrambling algorithms have been proposed by the researchers to ensure a high level of media protection.

A few common techniques used in the Data Scrambling process are Arnold Transforms, DES-based scrambling, and use of chaotic map [71–73].

1.10.1.1 ARNOLD TRANSFORMS

Arnold Transform uses a periodicity concept and is based on a matrix and a correlation equation where the matrix defines the position of the pixel from the original image and the altered image which tend to satisfy an equation of reversibility [74]. It can be applied to both pixels and the blocks of the image for higher security and robustness; however, in the Arnold Scrambling, the number of iterations is used as a key and is used extensively for tamper detection and other cryptographic techniques. A two-level Arnold Transforms are also used for enhanced robustness [82].

A 2D Arnold Transform is given by:

$$\begin{bmatrix} x' \\ y' \end{bmatrix} = \begin{bmatrix} a00 & a01 \\ a10 & a11 \end{bmatrix} \begin{bmatrix} x \\ y \end{bmatrix} ModN \tag{31}$$

1.10.1.2 XOR OPERATION BASED SCRAMBLING

The XOR operation plays a major role in the area of encryption and multimedia security [90]. It has very significant evidence in the field of security and is efficient enough to encrypt any form of given media. Just like the XOR function was used in the DES cryptography algorithm in the round functions, it is also used in the image scrambling process. The XOR operations are as follows:

1 XOR 1 = 0

1 XOR 0 = 1

0 XOR 1 = 1

0 XOR 0 = 0

The primary advantage of using the XOR operations in encrypting the data is the possibility of reversibility with robustness. This means a message can be encrypted using the XOR operation, and upon using the XOR operation again, it can convert it to the original form. Though it depends on the algorithm followed the basis of the security process remains the same. This same methodology is applied in the case of Image Scrambling, where an original Image and a randomly generated image as a key are taken, and an

XOR operation is used to scramble them; now to get the original image back, the scrambled image is XORED with the randomly generated image to get the original image. However, a double-direction XOR operation is used for higher effectiveness as the quality of the scrambled image decreases while using a single-direction-based XOR operation [77].

1.10.1.3 USE OF CHAOTIC MAP

Chaos is a non-periodic system of generating a pseudo-random definite sequence using a nonlinear dynamic system [89]. The Logistic map is considered as the general Chaotic Map, which has been extensively researched due to the high complexity behavior.

1.10.2 AUDIO SCRAMBLING

Audio scrambling is a process of altering the original audio in a form of altered audio where the original data is overridden with a noise that hides up the main content to ensure data confidentiality. For the Image Scrambling the pixels were taken into account for the alteration, here in the Audio Scrambling, technical properties like frequency, time, and amplitude are required for the further scrambling process. This can be done using the Frequency Domain Scrambling process.

1.10.2.1 FREQUENCY DOMAIN SCRAMBLING

In the Frequency Domain Scrambling, the low frequency of the signal is taken for the scrambling process, and the alteration in the frequency is processed. Asymmetric Matrix is taken into consideration using the Swapping Matrix and Unit Matrix.

$$M = \begin{bmatrix} X & X & 0X \\ X & X & 0X \\ 0 & 0 & I\,0 \\ X & C & 0\ \ X \end{bmatrix} \tag{32}$$

where; I from Eqn. (30) is the unified matrix; X is the standard swapping matrix; and 0 is the zero matrix.

1.11 MULTIMEDIA AUTHENTICATION

With the exponential increase in technology, the surge for Multimedia Authentication in the field of Forensics and Security has been increased rapidly. The authentication process helps to determine any tampering, trustworthiness, and integrity of the media when received by a user. As media plays a crucial role as a shred of evidence, it is important to authenticate the media before we could process it as legal proof. The process of authentication depends on the type of multimedia and has prime significance in the field of media security [91]. In this section, we would account for different media authentication like Video Authentication, and Printed Media Authentication.

1.11.1 PRINTED MEDIA AUTHENTICATION

The field of research on Printed Media Authentication lacks thorough research as compared to Digital Media protection. Various Media Security techniques, such as Cryptography, have solved the issue of Digital Media protection, but there is a lack of advanced solutions for printed media. Though digital content has taken over the world as a significant medium pri,nted media will never leave its existence in the corporate, government, and financial world. Hence, protecting the printed media with promising techniques is of utmost requirement in today's date. One of the ways adopted to secure the originality of the printed media is using special papers and an original sign of the intended person, this plays a major role in the authenticity of the printed media. Other ways of securing the printed media are [75]:

1. **Use of Signature:** Signature can be used to protect the originality of the document and restrict any further duplication.
2. **Use of Fingerprint:** This was one of the old conventional methods of protecting the document's originality by adding physical fingerprints which would make the document unique in itself. Wagner [81] introduced the idea, and later, various applications of using Fingerprints for digital media have surged.
3. **Use of Original Embedded Watermarks:** Just like the digital watermarking for the Digital Media, the watermarking techniques can also be integrated here to protect the document from any copying and alteration of ownership. These are widely used by the brands

to preserve the ownership of the media and copyright it from any infringement.

4. **Using Special Papers:** These made by high-quality printers which are kept inaccessible from the local markets and public and are only available for the high-security mediums. This is the process used by the government to print currency, cheques, government papers, etc.
5. **Barcodes:** These can be entitled into printed media to implement a digital signature with a unique identifier that protects the media from any alteration. A Barcode reader is hence required to decode the information hidden and could be used for the authentication.

The authenticity of any of the printed media can be ascertained by comparing the level of information and the properties of any document. The Digital Signature extracted from the media could give the Digital Signature and the unique identifier that could help to determine any possible modification chances. The extracted information is when compared with the decrypted data, the results of similarity would give the final call if the media has been modified.

1.11.2 VIDEO AUTHENTICATION

As digital surveillance has improved and videography has been primarily adopted in various fields, including the security departments, as a form of proof, authenticating video has taken a great rise over the decades. A general Video Encryption and extraction process is shown in Figure 1.4. In the process of Video Authentication, the main goal is to detect any modification or tampering that has been conducted on the original image in order to certify the integrity of the video [76]. Not just in the field of security but also it defining the reality of a happening; for example, a modified video shown by a news channel could create a bad perspective on a particular topic even if the original video has different content. This is a form of cheating and misleading content, which are also taken as a criminal offense by several countries.

The Video Authentication process, just like any other digital Cryptography technique, obeys the same principle to encrypt and verify the integrity of the original media. For the encryption process, the feature extraction and the algorithm for the authentication are undergone for a video and are finally encrypted with a key to get the signature. To validate the integrity of the video, the original video goes again with the feature extraction, and the signature is used to check if the video has been ever modified [87].

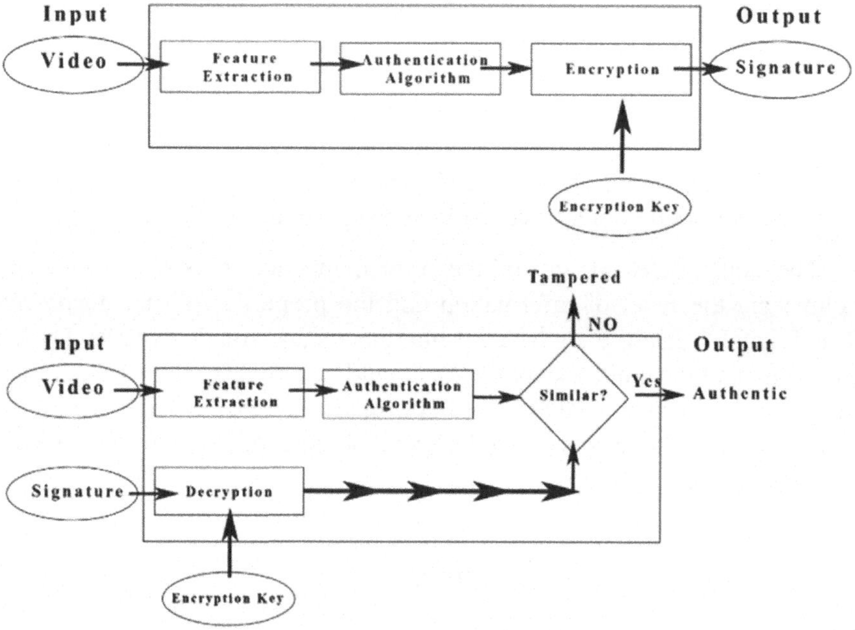

FIGURE 1.4 Video encryption and extraction process.

1.12 DIGITAL RIGHTS MANAGEMENT (DRM)

Digital rights management (DRM) is the way of protecting digital content and restricting any form of copying and data infringement. It is the way of enforcing the copyright owner of the digital content so that piracy could be limited to the lowest [76]. DRM is a crucial part of Multimedia Security which ensures media security through different algorithms and techniques as discussed so far [76, 78–80]. It is widely used today to protect the user rights over the digital media so that the original stays original forever and there's no point where piracy could happen.

It helps to protect:

- From copy;
- Editing and saving;
- Sharing and illegal distribution;
- Print proof;

- Prevents from screen capture;
- Revoking user access;
- Watermarking.

The most use cases of DRM could be seen majorly in the field of:

- Securing the Android and iOS applications while uploading to a marketplace.
- Artists restrict unauthorized access to their work.
- Streamlines the distribution of the product.
- Helps big firms to transmit data securely while ensuring confidentiality and detecting any data leaks.
- It protects the intellectual rights of any entity for whom security is a major concern.
- Securing the files shared socially.

DRM uses restrictive license agreements, laws, and certain multimedia security measures in order to protect digital content. Various ways DRM can be enforced are:

- Limitation on user accessibility;
- Product keys use case;
- Authentication through online;
- Encryption through any of the multimedia security technique;
- Copy, edit, save, share restriction;
- Watermarking on the digital media;
- Anti-tampering features;
- Use of metadata.

Certain laws have also been brought up to protect user information and privacy on the International Level. Though each country has its own data security law, there are certain international laws too that protect our content on the internet.

1.12.1 DRM LAWS IN DIFFERENT COUNTRIES

To protect the world from piracy and to secure the media content over the internet, several laws have been enforced by different countries. The first-ever requirement of such laws was seen to surge during 1996 when WIPO Copyright Treaty (WCT) asked all the nations to come with the strongest laws to entertain DRM. This is the period when several countries like the United States, European Union, and other countries started to enforce dedicated laws to protect user rights and media content over the network. Starting with DMCA by United Nations, other countries took this path and curated certain regulations for the right use of digital content. A few countries and the laws they enforced are discussed in subsections.

1.12.1.1 INDIA

India has had certain predefined laws since 1957 that have protected user rights over digital content. The Copyright Act 1957, stated in Sections 65A and 65B, defines protection of technological measures and protection of rights management, respectively. It states most of the part for the ownership of copyright and hence aims at protecting the digital content from any infringement.

India has seen to be strict for the Copyright laws and digital media protection for years as over time several laws to protect user privacy have come up and have been entertained. Recently in 2012, the latest amendment to the Copyright Act saw the implementation of DRM as the fundamental of Digital Security. India entered into the treaty for ensuring digital privacy and securing multimedia in 2013 and is now a crucial part of the most important international conventions running the copyright laws area which includes, Universal Copyright Convention, TRIPS, WCT, Berne Convention, and WIPO.

1.12.1.2 UNITED STATES

Under the WIPO Treaties Implementation Act, the United States in 1998 came up with the DMCA as the regulatory amendment to criminalize any unauthorized alteration and digital content piracy in the form of illegal creation and dissemination of the digital content. This act has been widely implemented and is protecting digital content over the internet since then. There are several exceptions to the Act, which allows the user to reverse engineer under specific conditions. Also, research work has been given relaxation from this activity with a certain condition, however, cryptanalytic research is still subjected to the infringement of the act.

1.12.1.3 EUROPEAN NATIONS

As an implementation of the WIPO Treaty, the European Union passed an amendment in 2001, the Information Society Directive, to address several issues on copyright infringement, digital media security, and unauthorized access to digital content. The implementation of the Information Society Directive was quite similar and identical to the existing DMCA; however, after a while, many significant changes to the amendment were made and were widely implemented starting from 2007.

1.13 CONCLUSION

As we are moving towards the digital space and creating a virtual hub for all our needs, the importance of protecting our privacy increases with it. Tons of thousands of data are stored and are getting transacted each day in the form of texts and media files like images, videos, documents, and many more. From personal data to government highly confidential data, they are all over the cloud and in the system, which could be exposed if not taken good care of. All these are termed multimedia, and the importance of protecting them is increasing exponentially. Several measures, from individual prospects to government agencies, have been regularly entertained each day to keep our data secured. In this context, the present chapter has been curated to lay off a base to understand the meaning of Multimedia, the importance of Multimedia Security, and the ways of protecting Multimedia from a broader perspective.

Initially, the Media Evaluation criteria help us to understand the detection of any tampering and alteration of data in the multimedia, which further helps in checking the integrity of the data. The Media Evaluation Criteria are subjected to the Data Hiding Techniques, and to get a deep insight into any minor changes in the response, certain mathematical models used for data hiding are analyzed. These mathematical models help us to hide a certain level of data in the multimedia while taking care of the design considerations required to maintain the efficiency of the technique. The most basic design consideration for the Data Hiding Technique is Robustness, Transparency, Capacity, Statistical Correctness, and Computational Complexity. Taking all these points into consideration, the Data Hiding process is done to hide sensitive information in one media file.

There have been several ways developed so far to protect Multimedia from any intentional intrusion, and these techniques are constantly in research and development to increase its efficiency even more. Several Multimedia Security techniques like Cryptography, Steganography, Digital Watermarking, Digital Scrambling, and Digital Authentic Forensics are used to keep the multimedia safe from any data alteration and keep away the secured data from the unintended person looking to gain access to the secret message transferred. All these security techniques have been in a constant development cycle for decades and differ in their prime use cases. Depending on the media file and the level of protection required, the ways of securing the data change. For example, if the Copyright is to be maintained and the ownership of the media is to be entertained, then Digital Watermarking would be a great choice for any image and video. Further, in this chapter, brief information on the Data Hiding Techniques and various Multimedia Security techniques have been discussed to gain a quick insight into the topic. The chapter also explains well the previous literature work, techniques, types, approaches, and tradeoffs of all the covered Information Hiding techniques. Additionally, the combination of security techniques has also been discussed for improved protection, where it has been seen that the Multimedia Scrambling process can play a key role. Lastly, a brief discussion on DRM with its importance and role in different countries and platforms has been made. The National Laws for Multimedia Security have seen an exponential surge in the past decades and are all set to tighten up ahead in a motive to protect cyberspace from unwanted intruders. Therefore, the significance of Media Security is booming with each increasing year, and in-depth research on security techniques is of utmost requirement to tackle any unintended intrusion.

The chapter takes a linear learning curve into account, which would further help the readers to dive deeper into the topic slowly while understanding the overall concept of the Multimedia field and preparing for further deeper sections ahead in this book.

KEYWORDS

- **cryptography**
- **data encryption**
- **data hiding**
- **digital authentication**
- **digital rights management**
- **digital scrambling**
- **multimedia security**
- **stenography**
- **watermarking techniques**

REFERENCES

1. Jang-Jaccard, J., & Surya, N., (2014). A survey of emerging threats in cybersecurity. *Journal of Computer and System Sciences, 80*(5), 973–993. ISSN 0022-0000, https://doi.org/10.1016/j.jcss.2014.02.005.
2. Furht, B., (2008). *Encyclopedia of Multimedia*. doi: 10.1007/978-0-387-78414-4.
3. Kulkarni, N. S., Balasubramanian, R., & Indra, G., (2009). Multimedia encryption: A brief overview. *Studies in Computational Intelligence*, 417–449. doi: 10.1007/978-3-642-02900-4_16.
4. Reena, J. S., & Bhavna, K. P., (2014). Multimedia security techniques. *International Journal of Innovative Research in Electrical, Electronics, Instrumentation, and Control Engineering, 2*(5).
5. Giri, K. J., Shabir, A. P., Rumaan, B., & Khan, M., (2021). *Multimedia Security, 1,* 3–24. doi:10.1007/978-981-15-8711-5_1.
6. Jean-Paul, A. Y., et al., Cyber-physical systems security: Limitations, issues, and future trends. *Microprocessors and Microsystems, 77*, 103201. doi: 10.1016/j.micpro.2020. 103201.
7. Lu, Y., (2017). Industry 4.0: A survey on technologies, applications, and open research issues. *Journal of Industrial Information Integration, 6*, 1–10.
8. Ashibani, Y., & Qusay, H. M., (2017). Cyber physical systems security: Analysis, challenges, and solutions. *Computers & Security, 68*, 81–97.
9. Salman, Y., & Al-Haj, A., (2021). Copyright protection of 3D images using digital watermarking. In: *2021 7th International Conference on Information Management (ICIM)* (pp. 82–86). IEEE.
10. Al-Haj, A., & Hussam, B., (2017). Copyright protection of e-government document images using digital watermarking. In: *2017 3rd International Conference on Information Management (ICIM)* (pp. 441–446). IEEE.
11. Praum, E., Hoppe, H., & Finkelstien, A., (1999). Robust mesh watermarking. In: *SIGGRAPH '99: Proceedings of the 26th Annual Conference on Computer Graphics and Interactive Techniques* (pp. 49–56). New York, NY, USA Press/Addison-Wesley Publishing Co.
12. Singh, H. V., Ashutosh, K. S., Balasubramanian, S. K., & Anand, M., (2006). Minimizing security threats in multimedia systems. In: *The 2nd International Conference on Distributed Frameworks for Multimedia Applications* (pp. 1–5). IEEE.
13. Castlemen, K., (1979). *Digital Image Processing*. Prentice Hall.
14. Kaushal, S., (2005). *Multimedia Data Hiding: From Fundamental Issues to Practical Techniques.* PhD Thesis, University of California, Santa Barbara.
15. Kaur, G., Singh, S., & Rani, R., (2020). *A High-Capacity Reversible Data Hiding Technique Based on Pixel Value Ordering Using Interlock Partitioning*, 727–732. 10.1109/SPIN48934.2020.9071330.
16. Briffa, J. A., & Manohar, D., (2003). Channel models for high-capacity information hiding in images. In: Mark, S. S., (ed.), *Mathematics of Data/Image Coding, Compression, and Encryption V, with Applications.* doi: 10.1117/12.451259.
17. Shah, K., Lakshmi, G. V. R., & Phirke, A., (2011). *ICTSM 2011, CCIS* (Vol. 145, pp. 27–33). 2011. © Springer-Verlag Berlin Heidelberg.
18. Pyung-Han, K., Eun-Jun, Y., Kwan-Woo, R., & Ki-Hyun, J., (2019). Data-hiding scheme using multidirectional pixel-value differencing on color images. *Security and Communication Networks, 2019*, 1–11. doi: 10.1155/2019/9038650.

19. Wu, M., & Liu, B., (2003). *Multimedia Data Hiding*. 10.1007/978-0-387-21754-3.
20. Kaw, J. A., Shabir, A. P., & Bhat, G. M., (2021). Information hiding in color images utilizing multi-bit substitution patterns. In: *Multimedia Security* (pp. 131–155). Springer, Singapore.
21. Ali, M., Ahn, C. W., & Pant, M., (2016). *Data Hiding Schemes: A Survey*. 10.15579/gcsr.vol5.ch1.
22. Shih, F. Y., (2017). *Multimedia Security: Watermarking, Steganography, and Forensics*. CRC Press.
23. Nithya, M., (2005). *Multimedia Security*. IIT Bombay.
24. Al-Haj, A., Hussein, N., & Abandah, G., (2016). Combining cryptography and digital watermarking for secured transmission of medical images. In: *2016 2nd International Conference on Information Management (ICIM)* (pp. 40–46) doi: 10.1109/INFOMAN.2016.7477531.
25. Rountree, D., (2011). Cryptography. *Security for Microsoft Windows System Administrators,* 29–69. doi: 10.1016/b978-1-59749-594-3.00002-8.
26. Dhanalaxmi, B., & Tadisetty, S., (2017). Multimedia cryptography — A review. In: *2017 IEEE International Conference on Power, Control, Signals, and Instrumentation Engineering (ICPCSI)* (pp. 764–766). doi: 10.1109/ICPCSI.2017.8391817.
27. Gurunathan, K., & Rajagopalan, S. P., (2019). A stegano – visual cryptography technique for multimedia security. *Multimedia Tools and Applications, 79*(5, 6), 3893–3911. doi: 10.1007/s11042-019-7471-1.
28. Kessler, G. C., & Chet, H., (2011). An overview of steganography. *Advances in Computers*, 51–107. doi: 10.1016/b978-0-12-385510-7.00002-3.
29. Dey, N., & Santhi, V. (2017). *Intelligent Techniques in Signal Processing for Multimedia Security: Studies in Computational Intelligence*, 660. doi: 10.1007/978-3-319-44790-2_4.
30. Kesa, N. R. K., (2018). *Steganography A Data Hiding Technique*. Culminating projects in information assurance, 75. https://repository.stcloudstate.edu/msia_etds/75 (accessed on 28 September 2023).
31. Anoop, K. C., & Piyush, K. S., (2018). Data hiding and watermarking techniques: A survey. *International Journal of Computer Applications, 179*(32), 28–36.
32. Rashid, A., (2016). Digital watermarking applications and techniques: A brief review. *International Journal of Computer Applications Technology and Research, 5*, 147–150. 10.7753/IJCATR0503.1006.
33. Mohanarathinam, A., Kamalraj, S., Prasanna, V. G. K. D., Renjith, V. R., & Manikandababu, C. S., (2019). Digital watermarking techniques for image security: A review. *Journal of Ambient Intelligence and Humanized Computing 11*(8), 3221–3229. doi: 10.1007/s12652-019-01500-1.
34. Agarwal, S., (2018). A review of image scrambling technique using chaotic maps. *International Journal of Engineering and Technology Innovation, 8*, 77–98.
35. Li-Hua, G., Jin, D., Jing, W., & Nan-Run, Z., (2021). Image encryption scheme based on block scrambling, closed-loop diffusion, and DNA molecular mutation. In: Angel M. D. R., (ed.), *Security and Communication Networks* (Vol. 2021, pp. 1–16). doi: 10.1155/2021/6627005.
36. Vivek, S., & Neelesh, K. J., (2015). Digital image forensics in multimedia security: A review. *International Journal of Engineering Research & Technology (IJERT), 4*(5). http://dx.doi.org/10.17577/IJERTV4IS051057.

37. Dhanalaxmi, B., & Srinivasulu, T., (2017). A novel image self-adaptive encryption algorithm based on composite chaotic systems. *International Journal of Electronics Engineering (IJEE) – CS Journals, 9*(1), 160–164.
38. Yassein, M. B., Aljawarneh, S., Qawasmeh, E., Mardini, W., & Khamayseh, Y., (2017). Comprehensive study of symmetric key and asymmetric key encryption algorithms. In: *2017 International Conference on Engineering and Technology (ICET)* (pp. 1–7). doi: 10.1109/ICEngTechnol.2017.8308215.
39. Delfs, H., & Helmut, K., (2007). Symmetric-key encryption. In: *Introduction to Cryptography* (pp. 11–31). Springer, Berlin, Heidelberg.
40. Thakur, J., & Nagesh, K., (2011). DES, AES, and Blowfish: Symmetric key cryptography algorithms simulation-based performance analysis. *International Journal of Emerging Technology and Advanced Engineering, 1*(2), 6–12.
41. Daemen, J., & Vincent, R., (1998). The block cipher Rijndael. In: *International Conference on Smart Card Research and Advanced Applications* (pp. 277–284). Springer, Berlin, Heidelberg.
42. Standard, Data Encryption, (1999). *Data Encryption Standard, 46*(3). Federal Information Processing Standards Publication.
43. Alabaichi, A., Faudziah, A., & Ramlan, M., (2013). Security analysis of blowfish algorithm. In: *2013 Second International Conference on Informatics & Applications (ICIA)* (pp. 12–18). IEEE.
44. Pitchaiah, M., Philemon, D., & Praveen, (2012). "Implementation of Advanced Encryption Standard Algorithm," *International Journal of Scientific & Engineering Research, 3*(3), 1–6.
45. Riyaldhi, R., & Aditya, K., (2017). Improvement of advanced encryption standard algorithm with shift row and *S. box* modification mapping in mix column. *Procedia Computer Science, 116*, 401–407.
46. Abikoye, O. C., Ahmad, D. H., Abdullahi, A., Noah, O. A., & Emmanuel, O. A., (2019). Modified advanced encryption standard algorithm for information security. *Symmetry 11*(12), 1484.
47. Milanov, E., (2009). *The RSA Algorithm* (pp. 1–11). RSA Laboratories.
48. Preneel, B., (1994). Cryptographic hash functions. *European Transactions on Telecommunications, 5*(4), 431–448.
49. Sobti, R., & Ganesan, G., (2012). Cryptographic hash functions: A review. *International Journal of Computer Science Issues (IJCSI), 9*(2), 461.
50. Rachmawati, D., Tarigan, J. T., & Ginting, A. B. C., (2018). A comparative study of message digest 5(MD5) and SHA256 algorithm. *Journal of Physics: Conference Series, 978*, 012116. doi: 10.1088/1742-6596/978/1/012116.
51. Yang, H., Xingming, S., & Sun, G., (2009). A high-capacity image data hiding scheme using adaptive LSB substitution. Radio engineering. 18.
52. Sukumar, A., Subramaniyaswamy, V., Vijayakumar, V., & Logesh, R., (2020). A secure multimedia steganography scheme using hybrid transform and support vector machine for cloud-based storage. *Multimedia Tools and Applications, 79*(15, 16), 10825–10849. doi: 10.1007/s11042-019-08476-2.
53. Zakaria, A., Mehdi, H., Ainuddin, W., Mohd, I., Norli, A., & Ki-Hyun, J., (2018). High-capacity image steganography with minimum modified bits based on data mapping and LSB substitution. *Applied Sciences, 8*(11), 2199. doi: 10.3390/app8112199.

54. Surana, J., Aniruddh, S., Bhavesh, J., Deepesh, S., & Nilesh, C., (2017). Steganography techniques. *IJEDR, 5*, 989–992.
55. Banerjee, & Indradip, (2011). A survey of steganography and steganalysis technique in image, text, audio, and video as cover carrier. *Journal of Global Research in Computer Science, 2*(4), 116.
56. Aswatha, K. M., et al., (2012). *Proceedings of ICAdC, AISC* (Vol. 174, pp. 699–706).
57. Ahmad, S., Thanikalselvan, V., & Rengarajan, (2017). Data security through data hiding in Images: A review. *J. Artlf. Intel, 10*, 1–21.
58. Saraswati, K., & Sumit, S., (2020). *A Literature Survey on Stenography Approach Based on Different LSB Technique, 10*(2).
59. Gupta, S., Ankur, G., & Bharat, B., (2012). Information hiding using least significant bit steganography and cryptography. *International Journal of Modern Education and Computer Science, 4*(6), 27.
60. Kavitha, K. K., Ashwini, K., & Priya, D., (2012). Steganography using least significant bit algorithm. *International Journal of Engineering Research and Applications (IJERA).* ISSN: 2248–9622.
61. Mandal, J. K., & Khamrui, A., (2011). A data-hiding scheme for digital image using pixel value differencing (DHPVD). In: *2011 International Symposium on Electronic System Design*. doi: 10.1109/ised.2011.37.
62. Hashad, A. I., Ahmed, S. M., & Wahdan, A. E. M. A., (2005). A robust steganography technique using discrete cosine transform insertion. In: *2005 International Conference on Information and Communication Technology* (pp. 255–264). IEEE.
63. Reddy, H. S. M., & Raja, K. B., (2009). High capacity and security steganography using discrete wavelet transform. *International Journal of Computer Science and Security (IJCSS), 3*(6), 462–472.
64. Rekik, S., Driss, G., Habib, H., & Sid-Ahmed, S., (2012). Audio steganography coding using the discrete wavelet transforms. *International Journal of Computer Science and Security (IJCSS), 6*(1), 79–93.
65. Gurpreet, K., & Kamaljeet, K., (2013). Digital watermarking and other data hiding techniques. *International Journal of Innovative Technology and Exploring Engineering.*
66. Tiwari, A., Arya, A., Shubham, S., Maharishi, V. V., & Nagar, Y., (2015). *Digital Watermarking Encryption and Decryption Using DWT, 2*(2).
67. Tanha, M., Seyed, D. S. T., Mohd, T. A., & Fazirulhisyam, H., (2012). An overview of attacks against digital watermarking and their respective countermeasures. In: *Proceedings Title: 2012 International Conference on Cyber Security, Cyber Warfare and Digital Forensic (CyberSec)* (pp. 265–270). IEEE.
68. Setiadi De, R. I. M., (2020). PSNR Vs SSIM: Imperceptibility quality assessment for image steganography. *Multimedia Tools and Applications, 80*(6), 8423–8444. doi: 10.1007/s11042-020-10035-z.
69. Naveed, A., Yasir, S., Nisar, A., & Aasia, R., (2015). Performance evaluation and watermark security assessment of digital watermarking techniques. *Science International, 27*(2).
70. Yang, C. Y., (2015). Robust high-capacity watermarking scheme based on Euclidean norms and quick coefficient alignment. *Multimedia Tools and Applications, 76*(1), 1455–1477. doi: 10.1007/s11042-015-3065-8.
71. Yu, X. Y., et al., (2006). *J. Phys.: Conf. Ser.,* 48065.

72. Adrian-Viorel, D., Alexandru, C., & Marius-Aurel, C., (2014). Color image scrambling technique based on transposition of pixels between RGB channels using knight's moving rules and digital chaotic map. *Mathematical Problems in Engineering, 2014*, 1–15. doi: 10.1155/2014/932875.
73. Zhang, X., Lingfei, W., Guangzhao, C., & Ying, N., (2019). Entropy-based block scrambling image encryption using des structure and chaotic systems. *International Journal of Optics, 2019*, 1–13. doi: 10.1155/2019/3594534.
74. Min, L., Ting, L., & Yu-Jie, H., (2013). *Arnold Transform Based Image Scrambling Method.* 10.2991/icmt-13.2013.160.
75. Huang, S., & Jian, K. W., (2007). Optical watermarking for printed document authentication. *IEEE Transactions on Information Forensics and Security, 2*(2), 164–173.
76. Upadhyay, S., & Sanjay, K. S., (2012). Video authentication: Issues and challenges. *International Journal of Computer Science Issues (IJCSI), 9*(1), 409.
77. Singh, P. K., Singh, R. S., & Rai, K. N., (2015). An image encryption algorithm based on XOR operation with approximation component in wavelet transform. In: *2015 Fifth National Conference on Computer Vision, Pattern Recognition, Image Processing and Graphics (NCVPRIPG)* (pp. 1–4) doi: 10.1109/NCVPRIPG.2015.7490013.
78. Zeng, W., Heather, Y., & Ching-Yung, L., (2011). *Multimedia Security Technologies for Digital Rights Management.* Elsevier.
79. Moskowitz, S., (2006). Introduction—digital rights management. In: *Multimedia Security Technologies for Digital Rights Management* (pp. 3–22). Academic Press.
80. Ma, Z., Ming, J., Hongmin, G., & Zhen, W., (2018). Blockchain for digital rights management. *Future Generation Computer Systems, 89*, 746–764.
81. Abdullah, A., (2017). *Advanced Encryption Standard (AES) Algorithm to Encrypt and Decrypt Data, 16,* 1–11.
82. Wu, L., Jianwei, Z., Weitao, D., & Dongyan, H., (2009). Arnold transformation algorithm and anti-Arnold transformation algorithm. In: *2009 First International Conference on Information Science and Engineering* (pp. 1164–1167). IEEE.
83. Manku, S., & Vasanth, K., (2015). *Blowfish Encryption Algorithm for Information Security, 10*(10).
84. Ben, A. Y., Trabelsi, I., Dey, N., & Bouhlel, M., (2016). Euclidean distance distortion based robust and blind mesh watermarking. *International Journal of Interactive Multimedia and Artificial Intelligence, 4.* doi: 10.9781/ijimai.2016.428.
85. Asamoah, D., Oppong, E., Oppong, S., & Danso, J., (2018). Measuring the performance of image contrast enhancement technique. *International Journal of Computer Applications, 181*, 6–13. 1 doi: 0.5120/ijca2018917899.
86. Al-Saleem, R. M., Al-Hilali, B. M., & Izz, K. A., (2020). Mathematical representation of color spaces and its role in communication systems. *Journal of Applied Mathematics, 2020*, 1–7. doi: 10.1155/2020/4640175.
87. Fadl, S. M., Qi, H., & Qiong, L., (2017). Authentication of surveillance videos: Detecting frame duplication based on residual frame. *Journal of Forensic Sciences, 63*(4), 1099–1109. doi: 10.1111/1556-4029.13658.
88. Zaidan, B. B., Zaidan, A. A., Abdul, K. H., & Ahmad, N. N., (2017). A new approach based on multi-dimensional evaluation and benchmarking for data hiding techniques. *International Journal of Information Technology & Decision Making*, 1–42. doi: 10.1142/s0219622017500183.

89. Pan, H., Yongmei, L., & Chen, J., (2018). Research on digital image encryption algorithm based on double logistic chaotic map. *EURASIP Journal on Image and Video Processing, 2018*(1). doi: 10.1186/s13640-018-0386-3.
90. Sammons, J., (2014). Digital forensics. *Introduction to Information Security* (pp. 275–302). doi: 10.1016/b978-1-59749-969-9.00013-4.
91. Sharma, P., Shrivastava, R., Sarthi, V. K., & Bhatpahri, P., (2018). Security analysis of XOR based ciphered image. *Asian Journal of Computer Science and Technology, 7*(1), 55–60. https://doi.org/10.51983/ajcst-2018.7.1.1829.

CHAPTER 2

EVALUATION METRICS USED FOR IMAGE ENCRYPTION TO ANALYZE THE EFFICIENCY TOWARD VARIOUS ATTACKS

M. DEVIPRIYA and M. BRINDHA

Department of Computer Science and Engineering, National Institute of Technology, Tiruchirappalli, Tamil Nadu, India

ABSTRACT

Image encryption requires evaluation metrics to analyze whether there is any data leakage or possibility of discovering the key in the encrypted image. Among the various metrics used to analyze image encryption, some are essential for security assessment, others for performance evaluation, quality preservation, and resistance to attacks. This chapter concentrates mainly on attacks related to security assessment, quality preservation, and resistance to attacks. Permutation techniques are used to change the position of pixels, which is one of the processes in image encryption. Chaotic maps are used to perform both permutation and diffusion. Various chaotic maps and permutation techniques are discussed to provide insight into chaotic encryption.

2.1 INTRODUCTION

The pixel values of the encrypted image are changed compared to the plain image with the application of an encryption algorithm. The encrypted image should form a randomized result and should not reveal any pattern or features.

Multimedia Security: Tools, Techniques, and Applications. Bhaskar Mondal and Shyam Singh Rajput (Eds.)

The correlation coefficient between the adjacent pixel values of an encrypted image should be nearer to zero for a good encryption algorithm. The first and foremost metric to examine the encrypted image is a visual examination. The features of the plain image should not be seen in the encrypted image. But visual examination alone does not provide efficiency in security, and hence, other metrics are used to measure the randomness and effectiveness of the encryption. There are two main stages in the encryption process, and they are permutation and diffusion. The permutation is the process of changing the position of pixels to make the image random and to reduce the correlation. Diffusion is the process of substituting the pixel values for another value using a function to improve the efficiency of the encryption. The diffusion should be in such a way that a single bit change in the original image should change at least half of the pixels in the encrypted image.

2.2 ENCRYPTION EVALUATION METRICS

The statistical analysis or the pattern analysis is done by the correlation coefficient, histogram analysis, chi-square analysis, and so on. The differential analysis is done by NPCR and UACI values, respectively. The evaluation metrics used to analyze/measure the randomness and efficacy of the encryption are explained in detail.

2.2.1 KEY SPACE

The key space [1] is the set of possible values that can be used to encode the data. Usually, it should be larger than 100 bits to overcome the exhaustive attack. In image encryption, it is measured using the significant decimal digits for real values. The standard IEEE precision is 10^{-15}, and it is used if the key value is a floating point number. If the key value is an integer, then the number of bits in the power of 2 is determined as key space.

2.2.2 CORRELATION COEFFICIENT

The degree of likeness between the pixel values is calculated by the correlation coefficient. The correlation coefficient [2] value ranges from –1 to 1. The values that are nearer to zero tell the correlation between the pixels is less, and the value nearer to one tells the correlation between the pixels is high. In general, images have a strong correlation between the pixel values. In order to secure the image, this correlation should be decreased as much as possible in the encrypted image to resist the attacks. The correlation is measured between the two sets of values using Eqn. (1):

$$CC_{a,b} = \frac{cov(a,b)}{\sqrt{F(a)}\sqrt{F(b)}} \tag{1}$$

$$cov(a,b) = \frac{1}{X}\sum_{l=1}^{X}(a_l - Me(a))(b_l - Me(b)) \tag{2}$$

$$F(a) = \frac{1}{X}\sum_{l=1}^{X}(a_l - Me(a))^2 \tag{3}$$

$$Me(a) = \frac{1}{X}\sum_{l=1}^{X}a_l \tag{4}$$

where; a_l and b_l are the value of the pixels; X is the number of pixels totally in a set and $Me(a)$ is the mean of a_l. cov(a,b) is the covariance of a and b set of pixel values.

2.2.3 CORRELATION COEFFICIENT GRAPH

The relationship between two variables is represented as the correlation. Usually, scatter plot is used to determine the correlation. If the points in the scatter plots form a straight line and are closer to each other then it is termed as high correlation. If the points are scattered along with the space, then it is termed as less correlation. Figure 2.1 shows the correlation graph of random plain image horizontal values. Figure 2.2 shows the correlation graph of cipher image horizontal values. Algorithm 2.1 shows the MATLAB code for correlation graph scatter plot of horizontal, diagonal, and vertical values of the image.

2.2.4 HISTOGRAM GRAPH

The histogram plot determines the frequency of occurrence of each pixel value in the image. Histogram of a plain image mainly has large frequency for some values and small frequency for some values. But the cipher image has equal frequency of occurrence for each pixel. The possibility of attack is high if the distribution of pixels in the cipher is skewed, and the possibility of attack is negligible if the pixel distribution is uniform. Histogram graph

ALGORITHM 2.1 Horizontal, Diagonal, and Vertical Correlation Graph

Input: [M,N]=size(image); I =image;

Output: correlation graph

```
Horizontal correlation graph
For a=1: M
    For b=1:2: N
        If k<10000
            xhor(k)= I(a,b);
            yhor(k)= I(a,b + 1);
            k=k+1;
        End
    End
End
scatter(xhor, yhor);
Diagonal correlation graph
For a=1: M
    For b=2:2: N
        If k<10000
            xdia(k)= I(a,b – 1);
            ydia(k)= I(a + 1,b);
            k=k+1;
        End
    End
End
scatter(xdia, ydia);
Vertical correlation graph
For a=1: M
    For b=1:2: N
        If k<10000
            xver(k)= I(b,a);
            yver(k)= I(b + 1,a);
            k=k + 1;
        End
    End
End
scatter(xver, yver);
```

provides the visual inspection-based analysis to ensure the uniformity in the distribution of pixel values. Figures 2.3 and 2.4 show the histogram plot of a random plain image and its cipher image. imhist(image) is the function in MATLAB that gives the histogram graph of the image.

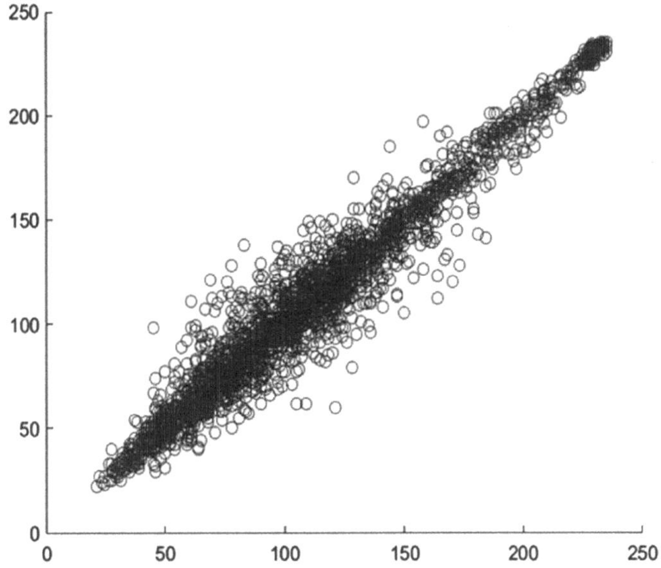

FIGURE 2.1 Correlation graph of plain image.

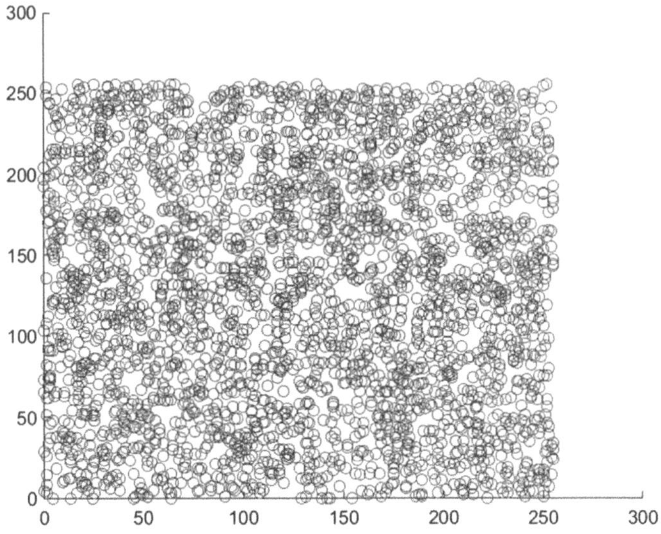

FIGURE 2.2 Correlation graph of cipher image.

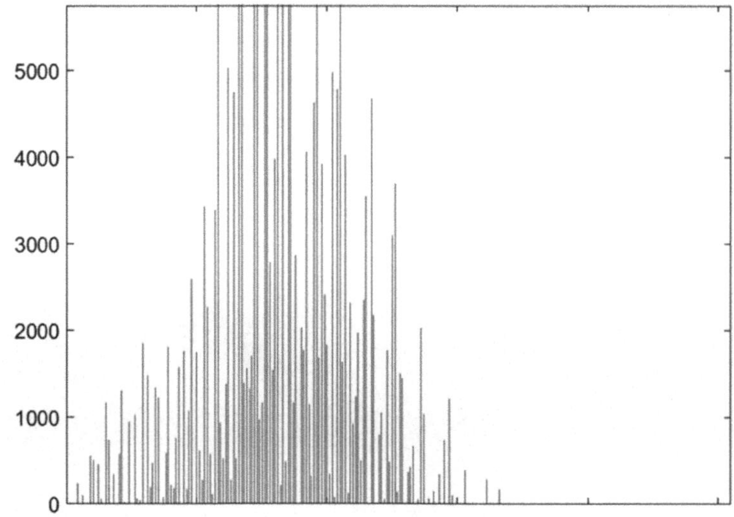

FIGURE 2.3 Histogram graph of plain image.

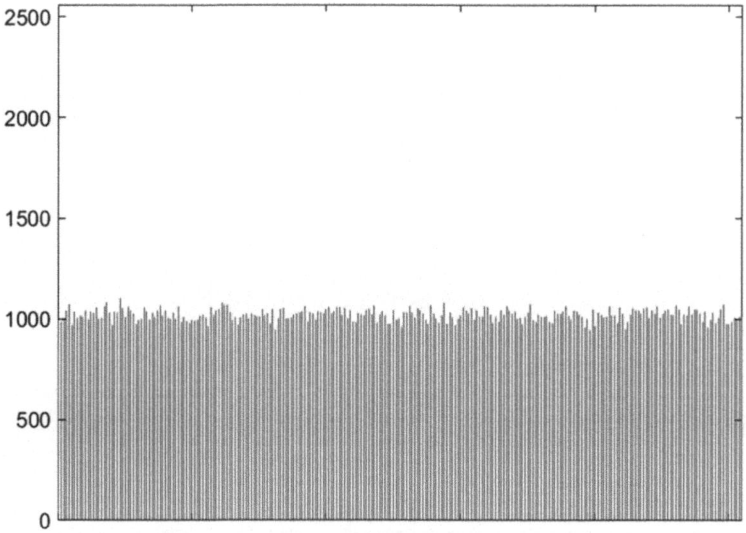

FIGURE 2.4 Histogram graph of cipher image.

2.2.5 KEY SENSITIVITY

A small change in the encryption key should give a large change in the resultant cipher [3]. The cipher image C obtained from key k_1 and C_1 is the cipher obtained using key k_2 with a negligible difference between k_1 and k_2. The difference between the pixels of C and C_1 should give a high value if key sensitivity is high. Similarly, a small change in the decryption key should not decrypt the cipher image. The decrypted image using negligible change in the key should give a new scrambled image.

2.2.6 HISTOGRAM VARIANCE

The histogram variance [4] is the quantitative measure of the distribution of pixel values. Smaller the histogram value means the uniformity in pixel distribution. The lesser the value of histogram variance means more uniformity in the pixel distribution. It is expressed mathematically by Eqn. (5):

$$var = \frac{1}{256}\sum_{x=1}^{256} kk_x - \overline{kk} \tag{5}$$

where; kk_x is the number of incident of the gray value between 0 and 255. $\overline{kk}$ is the mean value of the histogram.

2.2.7 CHI SQUARE VALUE

The Chi-Square [5] is another statistical measure used to calculate the uniformity in the distribution of pixels. The chi-square value should be less than the significant value 293, the p value of the chi-square should be greater than 0.01. The p value ranges between [0,1] and the ideal value is 1. If the chi-square value is nearer to 1, then the uniformity in the pixel distribution is high, and if the value is nearer to zero, then the distribution of pixel values is non-uniform. The chi-square is calculated using Eqn. (6):

$$chi = \sum_{s=0}^{255} \frac{(p_s - c_s)^2}{c_s} \tag{6}$$

where s represents the gray levels; p_s and c_s are the observed and expected values of each gray level (0 to 255).

2.2.8 PEAK SIGNAL TO NOISE RATIO (PSNR) AND MEAN SQUARE ERROR (MSE) VALUE

The distortion level of the encrypted image is compared with the original image to measure the quality of encryption [6]. The lesser PSNR value represents the high quality of encryption. MSE value should be greater to get the noise-like image as an encrypted image. PSNR and MSE are calculated using Eqns. (7) and (8):

$$P1 = 10 \times \log_{10}\left(\frac{255 \times 255}{M1}\right) \quad (7)$$

$$M1 = \frac{1}{K \times L}\sum_{x=1}^{K}\sum_{y=1}^{L}\left|i(x,y) - j(x,y)\right|^2 \quad (8)$$

where; *i* is the plain image; and *j* is the cipher image and (x, y) is the index value for each pixel value. *K* and *L* are the image dimension. A good encryption should have less PSNR value and high MSE value.

2.2.9 MEAN ABSOLUTE ERROR (MAE)

It is a measure of the difference between two images [7]. Larger the value between plain and cipher image, the quality of encryption is good. It is calculated using Eqn. (9):

$$Me = \frac{1}{K \times L}\sum_{x=1}^{K}\sum_{y=1}^{L}\left|i(x,y) - j(x,y)\right| \quad (9)$$

where; *i* be the plain image; and *j* be the cipher image. *K* and *L* are the image dimension. (x, y) is the index for each pixel value.

2.2.10 GLOBAL INFORMATION ENTROPY

In 1948, Shannon introduced the metric called the global information entropy [8]. The degree of randomness in the distribution of pixels is measured using global information entropy. The quality of the confusion is measured using the global information entropy value. The greater the global entropy value, the better the confusion is better. It is calculated using Eqn. (10):

$$HH(R) = -\sum_{j=1}^{n} p(R_j)\log_2\frac{1}{p(R_j)} \quad (10)$$

where n is 255 and is the highest value of pixel R_j and $p(R_j)$ is the probability of occurrence of each pixel value. The pixel value distributed in the image is in the range [0,255] and the ideal value of the information entropy is 8.

If the information entropy value of the cipher is close to 8, then it has good randomness in the encryption.

2.2.11 LOCAL SHANNON ENTROPY

It measures the entropy using the local blocks of an image to test the randomness [9]. With non-overlapping blocks T_1, T_2, T_3, ... T_K and L is number of pixel values in a block, Local Shannon entropy is calculated using Eqn. (11).

$$LE_{K,L}(T) = -\sum_{j=1}^{K} \frac{LE(T_i)}{K} \tag{11}$$

The local Shannon entropy value should be between 7.901515698 and 7.903422936 with a significant level α =0.001.

2.2.12 GRAY DIFFERENCE DEGREE (GDEG)

To measure the quality of scrambling, gray difference degree (Gdeg) is used [10]. The gray difference of each pixel with the neighborhood pixel values is defined in Eqn. (12):

$$Gr(a,b) = \frac{1}{4}\sum_{a'b'}\left[p(a,b) - p(a'b')\right]^2 \tag{12}$$

where; $(a'b') = \{(a-1,b), (a+1,b), (a,b+1), (a,b-1)\}$, $p(x,y)$ denotes the pixel value at position (a,b).

The average neighborhood gray difference is calculated using gray difference of every pixel values except at edge pixels by Eqn. (13):

$$E(Gr(a,b)) = \frac{\sum_{x=1}^{K-1}\sum_{y=1}^{L-1} Gr(a,b)}{(K-2)\times(L-2)} \tag{13}$$

The Gdeg value is calculated using Eqn. (14):

$$Gdeg = \frac{E'(Gr(a,b)) - E(Gr(a,b))}{E'(Gr(a,b)) + E(Gr(a,b))} \tag{14}$$

where; $E(Gr(a,b))$ and $E'(Gr(a,b))$ denote the average neighborhood gray differences of the input and scrambled images. Gray difference degree

(*Gdeg*) value is within [–1, 1] and the value is nearer to 1 if the quality of scrambling is good.

2.2.13 MAXIMUM DEVIATION

Maximum deviation [11] is another measure for the quality of the encryption. The maximum deviation is measured as follows:

- The histogram of the plain image and the cipher image is calculated.
- The absolute difference between the histogram of plain and cipher image is denoted as *m*.
- The amplitude of the histogram at the index *l* is m_l, and the sum of deviation is calculated using Eqn. (15):

$$M = \frac{m_0 + m_{255}}{2} + \sum_{j=1}^{254} m_l \tag{15}$$

where m_0 and m_{255} are histogram difference values at gray index 0 and 255, respectively. If the value of M is higher then the encrypted image is more deviated from the original image.

2.2.14 IRREGULAR DEVIATION

The pixel values are randomized in distribution after encryption. The change in the pixel values between the plain and the encrypted values should not be larger for one pixel and smaller for another pixel. The encryption should treat all the pixels similarly and distribute them uniformly. The irregular deviation [12] is a measure to analyze the uniformity in the distribution of pixel values. The following are the steps to calculate irregular deviation:

- The absolute difference of the plain image *p* and the cipher image *c* is calculated.

$$E = |p - c| \tag{16}$$

- The histogram of *E* is calculated.

$$HS = histogram(E) \tag{17}$$

- The amplitude of histogram hs_j is the value of number of occurrence of the gray value *j*. The mean value of *HS* is:

$$M = \frac{1}{256} \sum_{j=0}^{255} hs_j \tag{18}$$

- The absolute difference value between the histogram value for each gray value and the mean value is calculated as follows:

$$Hd_j = |HS_j - M| \tag{19}$$

- The irregular deviation IR is measured using Eqn. (20):

$$IR = \sum_{j=0}^{255} Hd_j \tag{20}$$

If the *IR* value is small, then the efficacy of encryption is better.

2.2.15 DEVIATION FROM UNIFORM HISTOGRAM

In Ref. [12], a new quality factor called deviation from uniform histogram is proposed. This factor measures the deviation of the cipher histogram from the ideal histogram. It is calculated using Eqn. (21).

$$D_h = \frac{\sum_{h=0}^{255} |H_h - I_h|}{M \times N} \tag{21}$$

$$I_h = \begin{cases} \frac{M \times N}{256}, 0 \le h \le 255 \\ 0, otherwise \end{cases} \tag{22}$$

where H_h is the histogram of the specific gray value between 0 and 255. The smaller the D_h value indicates, the better encryption quality. Small value indicates the less deviation from the ideal uniform histogram.

2.2.16 NPCR AND UACI

The small change in the original image must produce a significant change in the encryption image. Number of pixel changing rate (NPCR) [13] and Unified average changed intensity (UACI) are the two metrics to measure the change of pixel values in the encryption with respect to the original image. The NPCR and UACI are measured using Eqns. (23) and (24).

$$NP = \sum_{i,j} \frac{F(a,b)}{h \times l} \times 100\% \tag{23}$$

$$UA = \frac{1}{h \times l} \sum_{i,j} \frac{|cp_1(a,b) - cp_2(a,b)|}{256} \times 100\% \tag{24}$$

$$F(a,b)=\begin{cases}0\,if\;cp_1(a,b)\neq cp_2(a,b)\\1\,if\;cp_1(a,b)=cp_2(a,b)\end{cases} \tag{25}$$

Two cipher images cp_1 and cp_2 of plain images p_1 and p_2 with a single bit difference in only one pixel between p_1 and p_2.

2.2.17 STRUCTURAL SIMILARITY INDEX

Structural Similarity Index (SSIM) is a metric that measures the quality of the image in terms of similarity [14]. It tests the degradation of image caused by processing and loss during transmission. In image encryption, the image decrypted should be the same as the plain image. The SSIM value should be between –1 and 1 and value 1 is achieved for two identical images. This value should be nearer to 0 for dissimilar images. The similarity between original and encrypted image should be very low. This similarity can be calculated using Eqn. (26).

$$SS(k,l)=\frac{(2m_a m_b+d_1)(2v_{ab}+d_2)}{\left(m_a^2+m_b^2+d_1\right)\left(v_a^2+v_b^2+d_2\right)} \tag{26}$$

$$m_a=\frac{1}{N}\sum_{s=1}^{N} a_s \tag{27}$$

$$v_a=\frac{1}{N-1}\sum_{s=1}^{N} (a_s-m_a) \tag{28}$$

$$v_{ab}=\frac{1}{N-1}\sum_{s=1}^{N} (a_s-m_a)(b_s-m_b) \tag{29}$$

where; a and b are plain and decrypted images; m_a and m_b are the average of a and b; v_a^2 and v_b^2 are the variance of a and b. v_{ab} is the co-variance of a and b; $d_1=(s_1L)^2$ and $d_2=(s_2L)^2$ are the two values to stabilize the division in case of weak denominator. L is the range of pixel value, i.e., 256. s_1 and s_2 are 0.01 and 0.03, respectively.

2.2.18 UNIVERSAL IMAGE QUALITY (UIQ)

Another metric to measure the similarity between two images. The values of UIQ [14] ranges between –1 and 1, while the two identical images give the value 1. The similarity between the original and encrypted image should be as low as possible. It is expressed mathematically using Eqn. (30):

$$U(P,D) = \frac{4v_{PD}m_P m_D}{(m_P^2 + m_D^2)(v_P^2 + v_D^2)} \tag{30}$$

$$m_P = \frac{1}{A}\sum_{j=1}^{A} P_j \tag{31}$$

$$v_P = \frac{1}{A-1}\sum_{j=1}^{A}\left(P_j - m_P\right) \tag{32}$$

$$v_{PD} = \frac{1}{A-1}\sum_{j=1}^{A}\left(P_j - m_P\right)\left(D_j - m_D\right) \tag{33}$$

where m and v are the mean and variances, and v_{PD} is the covariance.

2.3 DIFFERENT SCAN PATTERN FOR PERMUTATION

There are different permutation techniques [15] available in the literature to permute the pixels. They are Raster scan, continuous raster scan, spiral scan, diagonal scan, z-order scan, U-index scan, Z-mirror scan, Gray scan and Hilbert scan. Figure 2.5 represents the various scan to permute the image. Figure 2.5(a)–(g) represents Raster scan, continuous raster scan, spiral scan, diagonal scan, z-order scan, U-index scan, Z-mirror scan, Gray scan and Hilbert scan.

2.4 DIFFERENT CHAOTIC MAPS

There are different chaotic maps available in literature to make the encryption or permutation of the pixels.

2.4.1 LORENZ SYSTEM

The Lorenz system [16] is given by:

$$\frac{da}{dt} = s\left(b-a\right) \tag{34}$$

$$\frac{db}{dt} = ta - ac - b \tag{35}$$

$$\frac{dc}{dt} = -uc + ab \tag{36}$$

where; a_0,b_0 and c_0 are the initial key values and control parameters s, t, and u are Prandtl number, Rayleigh number and geometric factor, respectively. It shows chaotic behavior when s = 3, t = 28, and u = 8/3.

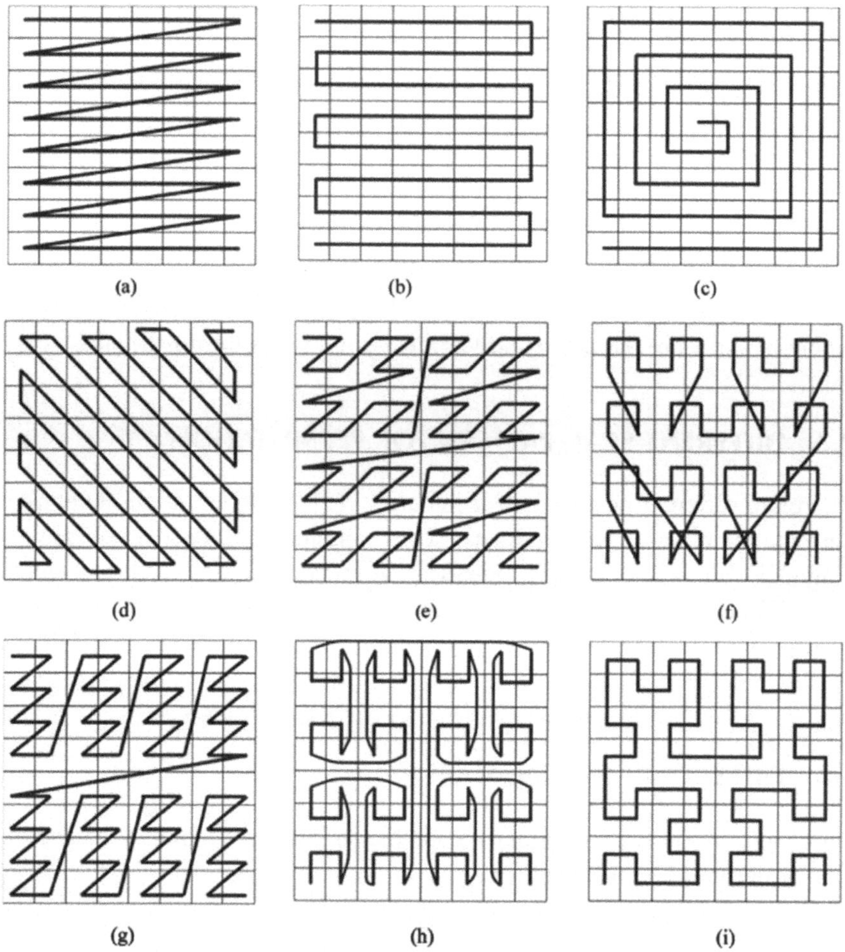

FIGURE 2.5 Different scan patterns for permutation.

2.4.2 ARNOLD'S CAT MAP

This is a discrete time dynamical system [17] and it is given by:

$$a_{n+1} = a_n + pb_n + qc_n \bmod \mathrm{N} \tag{37}$$

$$b_{n+1} = ra_n + prb_n + b_n + qrc_n \bmod \mathrm{N} \tag{38}$$

$$c_{n+1} = sa_n + pqrsb_n + rsc_n + c_n \bmod \mathrm{N} \tag{39}$$

where *p, q, r,* and *s* are the control parameters and a_0,b_0 and c_0 are the initial values.

2.4.3 BOGDANOV MAP

The Bogdanov [18] is a dynamical system, and it is given by:

$$s_{n+1} = s_n + t_{n+1} \tag{40}$$

$$t_{n+1} = t_n + et_n + ks_n(s_n - 1) + ms_n t_n \tag{41}$$

where s_0 and t_0 are the initial key values while *e, k*, and *m* are the control parameters.

2.4.4 DUFFING MAP

The duffing map [19] is a dynamical system based on discrete time. It is given by:

$$a_{n+1} = b_n \tag{42}$$

$$a_{n+1} = -qa_n + pb_n - b_n^3 \tag{43}$$

where; usually p = 0.275; q = 0.2 are used for control parameters. a_0, b_0 are the initial key values.

2.4.5 GAUSS MAP

It is a non-linear iterative map [20] which is given by:

$$a_{n+1} = \exp(-alp\ a_n^2) + beta \tag{44}$$

where; *alp* and *beta* are the real control parameters. a_0 is the initial key value.

2.4.6 GINGERBREADMAN MAP

It is a two dimensional map [21] and is expressed by:

$$s_{n+1} = 1 - t_n + |s_n| \tag{45}$$

$$t_{n+1} = s_n \tag{46}$$

where; a_0, b_0 are the initial key values.

2.4.7 TINKERBELL MAP

It is a dynamical system based on discrete time [22] and is given by:

$$a_{n+1} = a_n^2 - b_n^2 + pa_n + qb_n \quad (47)$$

$$b_{n+1} = 2a_n b_n + ra_n + sb_n \quad (48)$$

where; a_0, b_0 are the initial key values. *p, q, r,* and *s* are control parameters.

2.5 SIMPLE ENCRYPTION SCHEME (SES) – A CASE STUDY

This section introduces a simple example for the encryption, and all the discussed evaluation metrics are applied to the encrypted image, and results are tabulated.

2.5.1 ENCRYPTION PROCESS

In this section, the encryption and decryption process are explained in detail.

2.5.1.1 ENCRYPTION

The Encryption consists of three stages key generation, permutation, and diffusion. Figure 2.6 shows the block diagram of SES:

1. **Key Generation:** Initial values for chaotic maps are derived from pseudo-random generator as well as using the SHA algorithm. The plain image is given as the input to the SHA algorithm to get hash values, and these values are added together and multiplied by a random number. This hash-based value and the pseudo-random value are added to get the initial keys for the chaotic system.
2. **Permutation:** The image is divided into blocks, and inside the block, a Hilbert scan is applied. For the image as a whole, chaotic permutation is applied. The chaotic values are generated using initial values. These values are sorted, and an index of the chaotic values is used to permute the pixels.
3. **Diffusion:** The chaotic values obtained from the Bogdanov map are XORed to the pixel values. Finally, the pixel values are XORed with the Duffing map sequence to get the encrypted image.

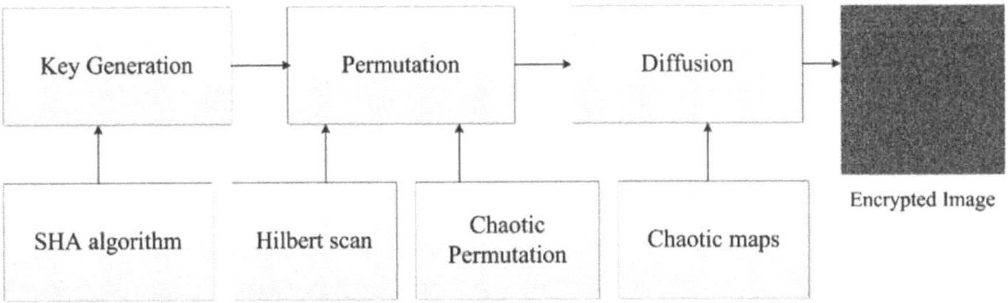

FIGURE 2.6 Block diagram of SES.

2.5.1.2 DECRYPTION

Decryption is the reverse process of encryption. The encrypted image is XORed with the duffing map sequence, and then the Bogdanov map sequence is XORed with the image. Then, the chaotic sequence is used to recover the original pixel positions using index values. Then, the image is divided into blocks, and Hilbert permutation recovery is performed.

2.5.2 ANALYSIS USING EVALUATION METRICS

The evaluation metrics discussed are used to analyze the efficiency of the SES encryption scheme. Plain images are taken from the USC SIPI database. MATLAB 2020a is used to program the encryption/decryption and also analyze the encrypted images, respectively.

2.5.2.1 KEY SPACE ANALYSIS

Two chaotic maps are used, and two initial keys should be assigned for each map. The precision of the values of initial key is 10^{-15} assuming the IEEE double-precision format. The size of the key is 10^{60}, which is higher than 2^{100} to resist the brute force attack.

2.5.2.2 CHI SQUARE ANALYSIS

The chi-square values for SES encryption are tabulated in Table 2.1. The significant value for chi-square is 293, and the values obtained are less than

293. Hence, the distribution of each number of pixels is uniform in SES encryption.

TABLE 2.1 Chi-Square Value

Image	Chi-Square Value
5.2.08	262.2422
5.2.09	242.2632
5.2.10	252.5195
7.1.01	288.3008
7.1.02	243.3125
7.1.03	251.4805
7.1.04	238.3867
7.1.05	252.0293
7.1.07	254.8730
Boat512	252.7168

2.5.2.3 HISTOGRAM VARIANCE ANALYSIS

The histogram variance analysis is a measure of uniformity in the pixel distribution. The results are tabulated in Table 2.2. The results are lesser, and hence, the distribution of pixels in the proposed SES encryption is uniform.

TABLE 2.2 Histogram Variance Values for Cipher Image

Image	Histogram Variance
5.2.08	1053.0823
5.2.09	972.8549
5.2.10	1014.0392
7.1.01	1165.7568
7.1.02	977.0667
7.1.03	1009.8666
7.1.04	957.2863
7.1.05	1012.0705
7.1.07	1023.4901
Boat512	1014.8313

2.5.2.4 CORRELATION COEFFICIENT ANALYSIS

The correlation coefficient value is nearer to 0 in Table 2.3, which shows it has less connectivity between the adjacent pixel values.

TABLE 2.3 Correlation Coefficient Values for Cipher Image

Image	Correlation Coefficient		
	H	V	D
5.2.08	0.0200	–0.0077	–0.0048
5.2.09	0.0043	0.0056	0.0114
5.2.10	–0.0139	–0.0104	0.0145
7.1.01	–0.0118	0.0053	–0.0045
7.1.02	–0.0153	–0.0228	0.0059
7.1.03	–0.0058	–0.0020	0.0027
7.1.04	0.0066	0.0087	–0.0068
7.1.05	–0.0148	0.00043	–0.00068
7.1.07	0.0028	–0.0115	–0.0150
Boat512	–0.0206	–0.0171	–0.0136

2.5.2.5 HISTOGRAM GRAPH ANALYSIS

The histogram graph analysis is a visual explanation for the uniformity in pixel distribution. The graph should be in the same range for every pixel and show a flattened appearance, while the plain image shows spikes for certain pixels having irregularity in the histogram image. Figure 2.7 shows the histogram graph of the cipher image, which is flattened and shows the distribution of pixels is uniform. Figures 2.7(a), (c), and (e) are the histogram of plain images 5.2.08, 5.2.09, and 5.2.10 where Figures 2.7(b), (d), and (f) are the corresponding histogram for cipher images, respectively.

2.5.2.6 PSNR AND MSE ANALYSIS

The PSNR and MSE values obtained using original and encrypted images should be as low as possible. These values show the similarity between the original and encrypted images. The obtained values are tabulated in Table 2.4, and it is lesser, which shows the dissimilarity between the plain and cipher images.

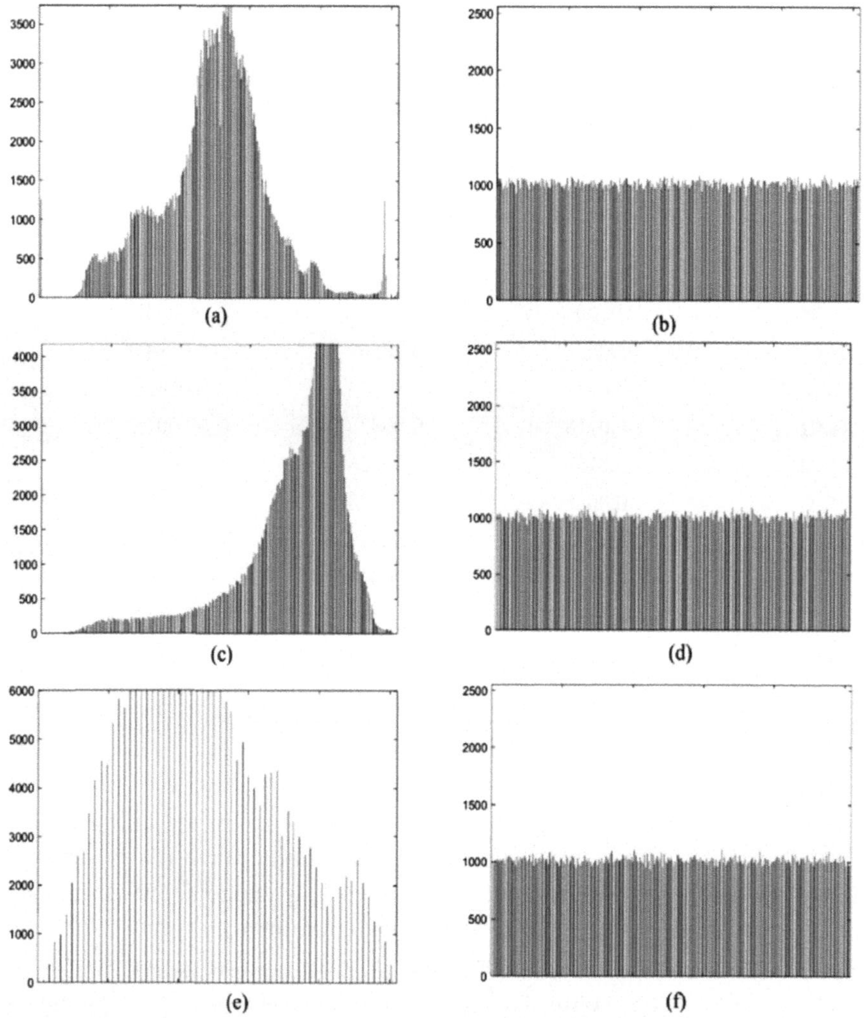

FIGURE 2.7 Histogram graph for different plain and cipher images.

2.5.2.7 GLOBAL INFORMATION ENTROPY ANALYSIS

The global information entropy gives the randomness in the pixel distribution. The obtained values are tabulated in Table 2.5, and it is nearer to 8, the ideal value for entropy, and shows the efficiency of the encryption algorithm.

TABLE 2.4 PSNR and MSE Values

Image	PSNR	MSE
5.2.08	9.6201	7096
5.2.09	8.2022	9837
5.2.10	8.7758	8619
7.1.01	9.9371	6597
7.1.02	8.9780	8227
7.1.03	10.2089	6197
7.1.04	9.8092	6794
7.1.05	9.5933	7140
7.1.07	10.0517	6425
Boat512	9.2959	7647

TABLE 2.5 Global Information Entropy for Cipher Image

Image	Global Entropy
5.2.08	7.9993
5.2.09	7.9993
5.2.10	7.9993
7.1.01	7.9992
7.1.02	7.9993
7.1.03	7.9993
7.1.04	7.9992
7.1.05	7.9993
7.1.07	7.9992
Boat512	7.9993

2.5.2.8 *LOCAL SHANNON ENTROPY ANALYSIS*

Local Shannon entropy gives the randomness towards pixel distribution for each block by dividing the cipher into blocks. Around 40% of cipher images show randomness in the local entropy. Local entropy values are tabulated in Table 2.6.

2.5.2.9 *CORRELATION COEFFICIENT GRAPH ANALYSIS*

The correlation coefficient graph of image boat512 is given in Figure 2.8. Figure 2.8(a), (c), and (e) shows the horizontal, vertical, and diagonal correlation coefficient graph of plain image and Figure 2.8(b), (d), and (f) show the horizontal, vertical, and diagonal correlation coefficient graph of cipher image. The Plain graph shows the correlation between the adjacent pixel values, whereas the cipher pixels show the values are spread all over the range and less correlation between the pixel values.

TABLE 2.6 Local Shannon Entropy for Cipher Image

Image	Local Entropy
5.2.08	7.9051
5.2.09	7.9027
5.2.10	7.9013
7.1.01	7.9004
7.1.02	7.9045
7.1.03	7.9049
7.1.04	7.9024
7.1.05	7.9008
7.1.07	7.9045
Boat512	7.9022

2.5.2.10 MAXIMUM DEVIATION ANALYSIS

This is another metric that shows the quality of encryption, and its values should be larger to ensure good encryption. The obtained results are tabulated in Table 2.7 and are large enough, and these values convey better quality of the encryption.

TABLE 2.7 Maximum Deviation Values

Image	Maximum Deviation
5.2.08	2,21,655
5.2.09	2,63,846
5.2.10	3,98,777
7.1.01	3,51,770
7.1.02	4,60,509
7.1.03	3,77,963
7.1.04	3,48,613
7.1.05	3,09,163
7.1.07	3,69,407
Boat512	2,20,342

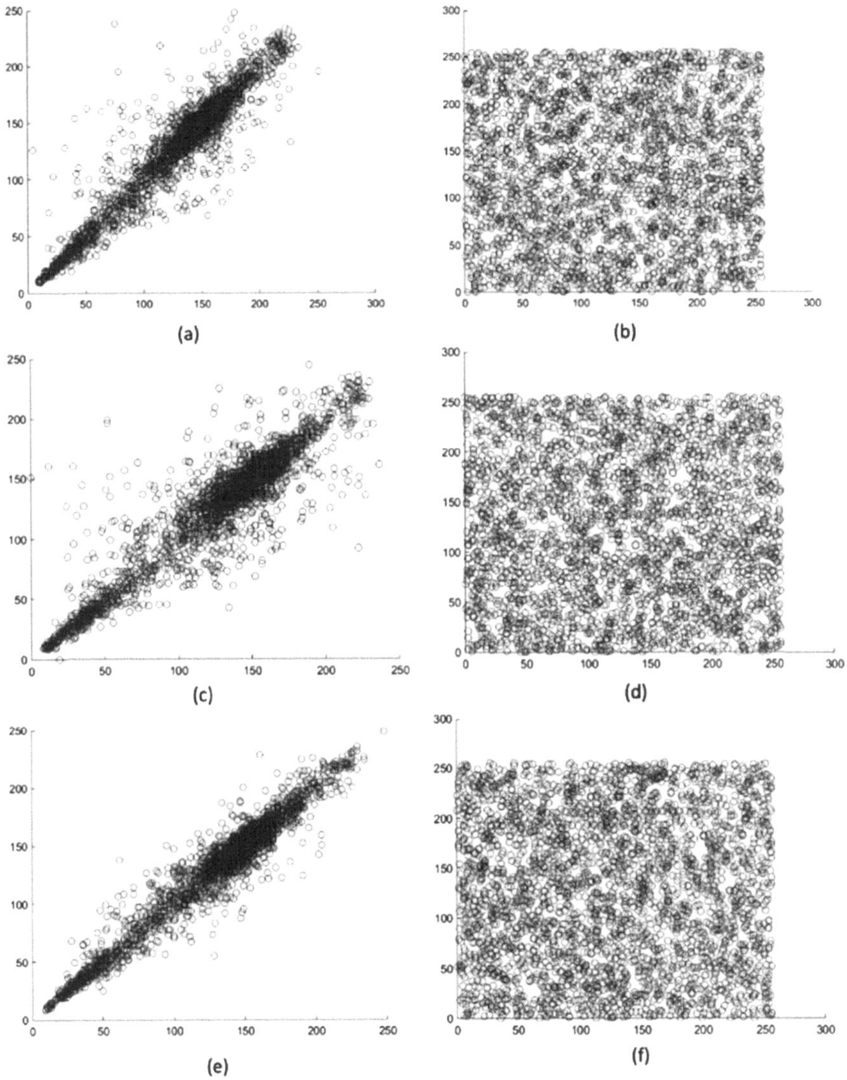

FIGURE 2.8 Correlation coefficient graph of horizontal, vertical, and diagonal pixels of plain and cipher image.

2.5.2.11 IRREGULAR DEVIATION ANALYSIS

The difference between the plain and cipher image should be uniform. The irregular deviation should be as small as possible to ensure good encryption. The values obtained are tabulated in Table 2.8.

TABLE 2.8 Irregular Deviation Values

Image	Irregular Deviation
5.2.08	1,99,818
5.2.09	1,36,662
5.2.10	1,64,258
7.1.01	2,12,258
7.1.02	1,56,718
7.1.03	2,15,788
7.1.04	2,09,288
7.1.05	1,94,730
7.1.07	2,13,184
Boat512	1,86,526

2.5.2.12 DEVIATION FROM UNIFORM HISTOGRAM ANALYSIS

This metric shows the uniformity in the distribution of pixel values. The obtained results are tabulated in Table 2.9 and the results are nearer to 0 shows that the less deviation from the ideal histogram value.

TABLE 2.9 Deviation Values

Image	Deviation Value
5.2.08	0.0259
5.2.09	0.0249
5.2.10	0.0251
7.1.01	0.0272
7.1.02	0.0239
7.1.03	0.0247
7.1.04	0.0239
7.1.05	0.0245
7.1.07	0.0247
Boat512	0.0250

2.5.2.13 NPCR AND UACI ANALYSIS

The possibility of differential attacks is analyzed using NPCR and UACI values. The obtained results are tabulated in Table 2.10. The obtained results for NPCR are larger than the threshold value determining the less possibility for differential attack.

TABLE 2.10 NPCR and UACI Values

Image	NPCR	UACI
5.2.08	99.6201	27.5943
5.2.09	99.6025	31.8303
5.2.10	99.6090	29.9156
7.1.01	99.6098	26.8418
7.1.02	99.5998	29.3323
7.1.03	99.6147	26.1980
7.1.04	99.6223	27.1403
7.1.05	99.6346	27.6816
7.1.07	99.6250	26.5824
Boat512	99.6185	28.4548

2.5.2.14 STRUCTURAL SIMILARITY INDEX ANALYSIS

This measures the similarity between the plain and cipher image. The results obtained are tabulated in Table 2.11 and show very less values and confirm that the two images are dissimilar.

TABLE 2.11 SSIM Values

Image	SSIM Value
5.2.08	0.0463
5.2.09	0.0616
5.2.10	0.0648
7.1.01	0.0410
7.1.02	0.0248
7.1.03	0.0417
7.1.04	0.0388
7.1.05	0.0534
7.1.07	0.0494
Boat512	0.0481

2.5.2.15 UNIVERSAL IMAGE QUALITY (UIQ) ANALYSIS

Universal image quality (UIQ) conveys the similarity between two images. The similarity between the plain and cipher image should be as low as possible. The obtained results are tabulated in Table 2.12, and the values are nearer to 0 by ensuring the dissimilarity.

TABLE 2.12 UIQ Values

Image	UIQ Values
5.2.08	–0.00054
5.2.09	–0.00046
5.2.10	0.00050
7.1.01	0.00005
7.1.02	–0.00057
7.1.03	0.00072
7.1.04	–0.00034
7.1.05	0.00021
7.1.07	–0.0010
Boat512	–0.00014

2.5.2.16 GRAY DIFFERENCE DEGREE (GDEG) ANALYSIS

The Gdeg conveys the efficiency of the permutation. The obtained results are tabulated in Table 2.13 and are nearer to 1 or sometimes 1, thereby showing the permutation quality is good.

TABLE 2.13 Gray Difference Degree Values

Image	Gdeg Values
5.2.08	0.8516
5.2.09	0.9799
5.2.10	1
7.1.01	0.9718
7.1.02	1
7.1.03	0.4613
7.1.04	0.9899
7.1.05	0.9711
7.1.07	0.9632
Boat512	0.9009

1.5.2.17 MEAN ABSOLUTE ERROR (MAE) ANALYSIS

Mean absolute error (MAE) measures the difference between the plain and cipher images and the values should be larger to ensure the quality of encryption. The results are tabulated in Table 2.14.

TABLE 2.14 MAE Values

Image	MAE Value
5.2.08	70.3655
5.2.09	81.1672
5.2.10	76.2849
7.1.01	68.4465
7.1.02	74.7973
7.1.03	66.8049
7.1.04	69.2079
7.1.05	70.5881
7.1.07	67.7851
Boat512	72.5598

2.6 CONCLUSION

The evaluation metrics to measure/analyze the efficiency of the encryption algorithm is studied. The key space is the metric, which should be high enough to resist the brute force attack. Key sensitivity should give vast change in encryption or decryption. The correlation coefficient, chi-square, and histogram are used to analyze whether the encryption does not reveal any features/pattern to the attacker. Global and local entropy values are used to analyze the randomness in the distribution of pixels. The Gdeg is used to analyze the quality of the permutation. PSNR, MSE, SSIM, and UIQ are used to analyze the similarity between the original and encrypted image. MAE, maximum deviation, irregular deviation and deviation from histogram are used to analyze the quality of the substitution. NPCR and UACI values are used to analyze the efficiency of encryption towards differential attacks. Different permutation techniques are discussed, and in addition to it, different chaotic systems are discussed. Using a permutation technique and a few chaotic systems, a simple encryption scheme (SES) is devised. The evaluation metrics studied are applied to the encrypted image, and the results are tabulated.

KEYWORDS

- **chaotic map**
- **chi square analysis**
- **diffusion**
- **evaluation metrics**
- **image encryption**
- **irregular deviation analysis**
- **mean absolute error**
- **permutation**

REFERENCES

1. Dong, H., Bai, E., Jiang, X. Q., & Wu, Y., (2020). Color image compression-encryption using fractional-order hyperchaotic system and DNA coding. *IEEE Access, 8*, 163524–163540.
2. Rehman, A. U., Wang, H., Shahid, M. M. A., Iqbal, S., Abbas, Z., & Firdous, A., (2019). A selective cross-substitution technique for encrypting color images using chaos, DNA rules and SHA-512. *IEEE Access, 7*, 162786–162802.
3. Zhou, S., Wang, B., Zheng, X., & Zhou, C., (2016). An image encryption scheme based on DNA computing and cellular automata. *Discrete Dynamics in Nature and Society.*
4. Chai, X., Chen, Y., & Broyde, L., (2017). A novel chaos-based image encryption algorithm using DNA sequence operations. *Optics and Lasers in Engineering, 88*, 197–213.
5. Belazi, A., Talha, M., Kharbech, S., & Xiang, W., (2019). Novel medical image encryption scheme based on chaos and DNA encoding. *IEEE Access, 7*, 36667–36681.
6. Wang, X., & Guan, N., (2020). A novel chaotic image encryption algorithm based on extended Zigzag confusion and RNA operation. *Optics & Laser Technology, 131*, 106366.
7. Swathika, E., Karthika, N., & Janet, B., (2019). Image encryption and decryption using chaotic system. In: *2019 IEEE 9th International Conference on Advanced Computing (IACC)* (pp. 30–37). IEEE.
8. Feng, W., He, Y., Li, H., & Li, C., (2019). A plain-image-related chaotic image encryption algorithm based on DNA sequence operation and discrete logarithm. *IEEE Access, 7*, 181589–181609.
9. Zhu, H., Zhang, X., Yu, H., Zhao, C., & Zhu, Z., (2016). A novel image encryption scheme using the composite discrete chaotic system. *Entropy, 18*(8), 276.
10. Dalhoum, A. L. A., Mahafzah, B. A., Awwad, A. A., Aldhamari, I., Ortega, A., & Alfonseca, M., (2012). Digital image scrambling using 2 D cellular automata. *IEEE Multimedia Magazine, 19*(4), 28–36.
11. Hamid, A., Ragab, M., Alla, O. S. F., & Noaman, A. Y., (2014). *Encryption Quality Analysis of the RCBC Block Cipher Compared with RC6 and RC5 Algorithms.*

12. Ahmad, J., & Ahmed, F., (2010). Efficiency analysis and security evaluation of image encryption schemes. *Computing, 23*, 25.
13. Wu, Y., Noonan, J. P., & Agaian, S., (2011). NPCR and UACI randomness tests for image encryption. *Cyber Journals: Multidisciplinary Journals in Science and Technology, Journal of Selected Areas in Telecommunications (JSAT), 1*(2), 31–38.
14. Abd EL-Latif, A. A., Abd-El-Atty, B., & Venegas-Andraca, S. E., (2019). A novel image steganography technique based on quantum substitution boxes. *Optics & Laser Technology, 116*, 92–102.
15. Shahna, K. U., & Mohamed, A., (2020). A novel image encryption scheme using both pixel level and bit level permutation with chaotic map. *Applied Soft Computing, 90*, 106162.
16. Al-Hazaimeh, O. M., Al-Jamal, M. F., Alhindawi, N., & Omari, A., (2019). Image encryption algorithm based on Lorenz chaotic map with dynamic secret keys. *Neural Computing and Applications, 31*(7), 2395–2405.
17. Pan, T. G., & Li, D. Y., (2011). A new algorithm of image encryption based on 3D Arnold cat. In: *Advanced Engineering Forum* (Vol. 1, pp. 183–187). Trans Tech Publications Ltd.
18. Janardhanan, S. V., & Sanjeeva, P., (2020). Bogdanov map-based permuted double image encryption. *Anais da Academia Brasileira de Ciências*, 92.
19. Stoyanov, B., & Kordov, K., (2014). Novel image encryption scheme based on CHEBYSHEV polynomial and duffing map. *The Scientific World Journal, 2014*.
20. Sahay, A., & Pradhan, C., (2017). Gauss iterated map based RGB image encryption approach. In: *2017 International Conference on Communication and Signal Processing (ICCSP)* (pp. 0015–0018). IEEE.
21. Khan, M., & Asghar, Z., (2018). A novel construction of substitution box for image encryption applications with Gingerbreadman chaotic map and S 8 permutation. *Neural Computing and Applications, 29*(4), 993–999.
22. Enayatifar, R., Sadaei, H. J., Abdullah, A. H., Lee, M., & Isnin, I. F., (2015). A novel chaotic based image encryption using a hybrid model of deoxyribonucleic acid and cellular automata. *Optics and Lasers in Engineering, 71*, 33–41.

CHAPTER 3

MACHINE LEARNING APPROACHES TO MALWARE DETECTION SYSTEMS

A. P. SHERLY[1] and P. ARUN RAJ KUMAR[2]

[1]*Graduate Student Member, IEEE; Research Scholar, NIT Calicut, Kerala, India*

[2]*Senior Member, IEEE; Department of Computer Science and Engineering, NIT Calicut, Kerala, India*

ABSTRACT

Due to the increasing popularity of digital media, steganography is a promising technique in critical applications such as military, healthcare, industry, space exploration, etc. For example, the malicious software sent in a stego video may affect life-or-death decisions, making fail-safe security a necessity. Such unknown or new malwares are not easy to detect. However, in the literature, to detect malware, various techniques such as static analysis, dynamic analysis, and reverse engineering exist. Static analysis analyzes the code or binary (signature based) without executing the malware. The static analysis is inefficient in the present scenario due to rapid increase in malware variants. Dynamic analysis executes the malware within a controlled environment. The dynamic analysis has limitations due to limited network access in a controlled environment. Reverse engineering involves disassembling or decompiling a software program to analyze malware. If the malicious program structure is complex, the reverse engineering framework is inefficient. Due to complex structure and hidden malicious instructions, there is a need to automate the detection process to find the malware instantly in real time. Therefore, artificial intelligence (AI) techniques such as machine learning (ML), deep learning, and fuzzy logic (FL) help

Multimedia Security: Tools, Techniques, and Applications. Bhaskar Mondal and Shyam Singh Rajput (Eds.)

especially in static and dynamic malware analysis for quick detection. To demonstrate the malware analysis in this chapter, a sample malware from the public repository is hidden in an image and analyzed using freely available Open Source Intelligence tools (OSINTs). From the experimental results, it is evident that the malware detection in stego image is indeed challenging due to its hidden behavior and malicious scripts. The primary objective of the chapter is to discuss and analyze the existing ML techniques used in detecting the presence of malware.

3.1 INTRODUCTION TO MALWARE AND ITS CURRENT THREATS

In the cyber world, the rapid increase of digital technologies has affected our daily activities. Malicious software (Malware) is one of the challenging threats. Malware is a computer program designed to invade and damage computers without the user's consent. Malware threats continue to expand by numbers, volumes, types, and functionality due to the technological advancement. The advancement in technologies allows creating smart and sophisticated malware. Recently Spyware, Ransomware, and Cryptomining are emerging as the most prolific types of malware. Pegasus [1], a project led by Amnesty International, came into possession of a list of 50,000 mobile phone numbers which it claims 'people of interest' potentially targeted by Niv, Shalev, and Omri (NSO) Group. The Pegasus Trojan (spyware) is capable of acquiring complete control of an infected device. The phone's camera and microphone are used for spying, data exfiltration, intercepting communications, and tracking location data. Ransomware program infects a host or network and demands a ransom from the victim. The Ransomware encrypts the files on the infected system or locks down the system. Further, the messages are sent in such a way that the victims will be forced to pay the ransom. Crypto mining uses the victim's computing power to challenge a complex cryptographic puzzle. From the attacker's perspective, the reasons for creating the malwares are as follows: (i) popularity; (ii) political; and (iii) personal revenge. Malware causes a significant loss of business intelligence and intellectual property (IP), brand reputation damage, loss of money, etc. Using malware, the attacker steals confidential data, demands a ransom amount, discloses the private information of a victim, etc. Depending on the purposes and proliferation systems, the malware is divided into the following categories:

i. **Adware:** Through advertisements.

ii. **Backdoor:** Bypass through open ports.
iii. **Bot:** Malicious script.
iv. **Downloader:** Downloads and installs without user's consent.
v. **Rootkit:** Conceals the malicious scripts.
vi. **Spyware:** Hiding and finding the secret information.
vii. **Trojan:** Illegitimate but behaves as a genuine software.
viii. **Virus:** Propagation from device to device.
ix. **Worms:** Spreading the infection.

Adware is designed to generate online advertisements automatically. The malware generates revenue by displaying advertisements on the user's interface or the screen. Backdoor is designed to bypass a system's security mechanisms and allows the attacker to access it. A Bot is malicious code to launch attacks such as denial of service (DoS), distributed denial of service (DDoS), etc. An infected BOT is called Zombie. The collection of Zombies is called Botnet, which are controlled by command and control (C&C) software. A downloader downloads and installs malicious programs. Rootkit conceals the existence of other malicious programs. Spyware spies and collects sensitive information without permission from a victim's computer. Examples of spyware such as key-loggers, password gravers, and sniffers. A Trojan disguise as a legitimate software, downloads, and installs malware on the systems.

A Virus propagate itself from device to device and Worm exploits vulnerabilities of the operating system to spread the infection. Malwares are created by trojan development kit (TDK), ransomware construction kit (RCK), and Open-source framework (such as Metasploit), etc. The anti-virus software uses signature-based detection, which is inefficient in the present scenario due to the rapid increase in the number and variants of malware. In the last decade, machine learning (ML) has triggered a radical shift in many sectors including cybersecurity. It has a variety of approaches to solve rather than a single method. The contributions in this chapter are as follows:

1. **Stegomalware Analysis:** How a malware is embedded in an image using open source framework and stego image analysis using open source tools are discussed.
2. **Static Malware Analysis:** A comprehensive survey on the existing static malware analysis using machine learning techniques.
3. **Dynamic Malware Analysis:** A comprehensive survey on the existing dynamic malware analysis using machine learning techniques.

3.2 EMBEDDING A MALWARE IN THE IMAGE USING METASPLOIT IN KALI/PARROT OS

Metasploit [2] is a pre-installed framework on Kali Linux machine that creates custom payloads and connects from the victim's computer. The system configuration for running the experiments is depicted in Table 3.1. *msfvenom* is a command line instance to generate and output all types of shell code in Metasploit. *msfvenom* is a combination of msfpayload and msfencode.

3.2.1 MALWARE ANALYSIS USING OSINT: VIRUS TOTAL

Kaushik et al. [3] created a malware in Parrot OS. The malware is downloaded and is hidden in an image using steganographic tool. The cover image and stego image are analyzed using freely available OSINTs such as VirusTotal. VirusTotal analyzes files and URLs for viruses, worms, Trojans, and the other malicious contents. Virus Total classifies the cover and stego image by differences. Before embedding the malware with an image, 59 out of 73 engines were able to scan the file using Virus Total. After embedding, malware with an image 36 out of 73 engines were able to scan and classified the file as malicious. The existing malware analysis tool is not sufficient to detect the presence of malware in image.

TABLE 3.1 System Configuration Used in Malware Creation

	Host Machine
Model	**Assembled**
Processor	AMD Ryzen 7 3700X, 8 Core @ 4.4 GHZ 36 MB cache
RAM	16.0 GB DDR4 3000 MHz
System type	64-bit operating system
Operating system	Windows 10
Tools/packages	OSINTs metasploit
	Hypervisor
Version	VMware® Workstation Pro15.5.0 build14665864
Processor	64-bit operating system
	Virtual Machine
Kali	v64-bit, 2 GB RAM, Hard Disk: 20 GB
Windows 7	v64-bit, 1 GB RAM, Hard Disk: 20 GB

3.3 LIMITATIONS OF THE STATIC ANALYSIS METHODS TO DETECT THE PRESENCE OF MALWARE

Static analysis is the process of examining the code (source, intermediate, or binary) and analyzing the behaviors of a program without executing. A wide range of malware analysis techniques fall under the category of static analysis. The static features to identify the malwares are hash value, N-grams, opcodes, strings, and portable executable (PE) header information, etc. Antivirus programs rely on a database of identifiable pieces of known suspicious code (file signatures), as well as behavioral and pattern-matching analysis (heuristics) to identify the malware. Websites such as VirusTotal (http://www.virustotal.com) allows uploading a file for scanning by multiple antivirus engines. VirusTotal generates a report that provides the details as follows:

- Total number of engines that marked the file as malicious;
- The malware name;
- Additional information about the malware (if retrieved).

Hashing is a common method to uniquely identify the malware. The malicious software is passed as an input to a hash function that produces a unique digest (fingerprint). The message-digest algorithm 5 (MD5) hash function is commonly used for malware analysis, though the secure hash algorithm 1 (SHA-1) is also popular. The md5deep and hashdeep is the freely available framework to generate the fingerprint. The existing static analysis techniques for malware detection are primarily classified as follows: (i) manual detection; and (ii) automated detection. Further, manual is classified into five subcategories, and automated is classified into three categories, as shown in Figure 3.1.

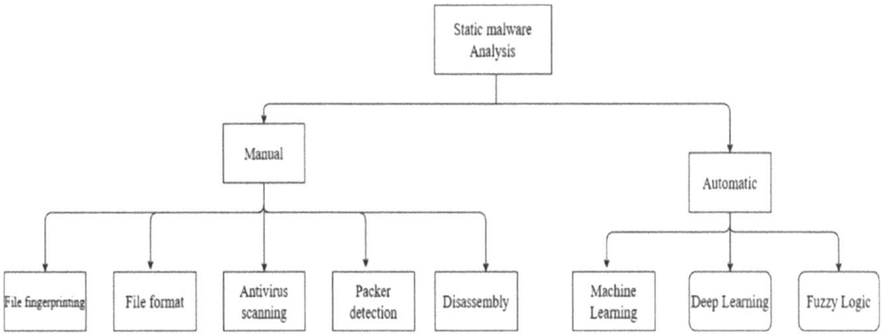

FIGURE 3.1 Static malware analysis classification.

3.3.1 MANUAL DETECTION TECHNIQUES

In manual detection, the binary file is inspected without execution. For example, if the source code is available, it contains information such as data structures, used functions, etc. The different techniques in static malware analysis are as follows:

1. **File Fingerprinting:** Besides examining the external features of the binary, it includes operations on the file level such as computation of a cryptographic hash (e.g., md5) of the binary to distinguish it from others and to verify that it has not been modified.
2. **File Format:** The metadata of a file format provides additional information about the file. In Windows binary, the PE format consists of useful information such as compilation time, imported and exported functions, strings, menus, icons, etc.
3. **Antivirus Scanning:** If the examined binary is a well-known malware, one or more Antivirus scanners detect it.
4. **Packer Detection:** Nowadays, malware is distributed in an obfuscated form, e.g., encrypted or compressed. The obfuscation is achieved using a packer. The program is completely different from a static analysis perspective and the logic and the metadata is hard to recover. Obfuscation is a major challenge in static malware analysis.
5. **Disassembly:** A major part of static analysis is the disassembly of a given binary. Disassembly is conducted using tools such as IDA Pro. It shows the assembly code of a binary and comes with advanced features that make understanding the assembly code as easy as possible. Based on the reconstructed assembly code, an analyst inspects the program logic.

3.3.2 AUTOMATED DETECTION TECHNIQUES

Moussas et al. [4] designed a two-level artificial neural network (ANN) for malware detection. The malware detection process is reduced to an image recognition problem. The classification is based on file and image features such as File size, Entropy, Contrast, Correlation, Energy, Homogeneity, Mean Image Intensity, etc. The malware is converted into grayscale images and the visual representation of malware is fed into ANN. The ANN architecture consists of three layers viz., input layer, output layer, and multiple hidden layers. The number of nodes in the input layer represents the number of features used. The number of nodes in the output layer represents the number of malware categories. Malware samples present significant visual similarities if converted to images. The accuracy for image classification is close to 99%. Damasevicius et al. [5] designed a malware detection technique by converting the complete binary file into a grayscale image. The convolutional neural network (CNN) algorithm was used for training the malware. The experiment was conducted on classification of malware with PE headers (ClaMP) dataset. The technique was evaluated using 10,568 binary files and produced an accuracy of 98.8%.

Wan et al. [6] used a ML method to detect and classify IoT malware programs. The detection and classification of malware is achieved by the information stored in the byte sequences at the entry points of executable programs. The entry point indicates the position that the processor enters a program or a code fragment and begins execution. Executable and link format (ELF) byte sequences obtained from executable files are the input features to the ML algorithms for IoT malware detection and malware classification. The experiment was conducted on a dataset from VirusTotal. The dataset contains malware samples belonging to eight variants including Mirai, Tsunami, Hajime, Dofloo, Bashlite, XORDDoS, Android, and Pnscan. The performance was evaluated using the various ML algorithms such as support vector machine (SVM), K-nearest neighbors (KNN), naive bayes (NB), and multi-layer perceptron (MLP). The result shows that the information in ELF headers contains information for malware detection, but that is not sufficient to differentiate the malware variants. The experimental results on malware detection showed that the optimal accuracy greater than 99.96%.

Bai et al. [7] developed graph isomorphism algorithms to identify known malware and their variants based on the function call graph (FCG). A disassembler IDA Pro was used to extract the sample files into assembly code. The relationship between functions in assembly code is represented as FCG. The graph isomorphism algorithm is employed to test if the FCG of testing sample is isomorphic with one of the FCG signature databases. If isomorphic, testing sample is identified as malware, otherwise it is identified as a benign program. The method achieved good experimental results. Wadkar et al. [8] developed a malware evolution-detecting system using SVM. Around 54 static features are extracted from the PE Header of the window executable. The SVM classifier was trained using a large dataset of 5,00,000 malware samples, which were collected from the Vxheaven [9] and VirusShare repositories. The technique was applied to 13 malware families. The experimental results on malware evolution detection showed that evolutionary changes had occurred within malware families.

Burnap et al. [10] and Huda et al. [11] used API calls but applied different ML approaches and produced better detection rate. But there are limitations of signature-based techniques. It is not possible to detect the new and obfuscated malware. Kumar et al. [12] implemented a learning model to detect maliciousness of PE using integrated feature set. In this method, raw malware and benign samples are collected from various sources. Raw samples are processed for training. The feature set is created by extracting values from all fields of three main headers (DOS header, File Header and Optional header, including standard and Windows-specific fields) present in every PE file and fed these features into machine-learning algorithms such as decision tree (DT), random forest (RF), KNN, logistic regression, linear discriminant analysis (LDA) and NB, etc. The experimental result showed that the classification accuracy is 98.4% for integrated feature set. Searles et al. [13] designed the conventional control flow graph (CFG)-based malware detection technique. The shortest path graph kernel (SPGK) was used to discover similarities between the CFG extracted from binary files. The SVM algorithm uses similarity matrix to train the classifier more accurately.

Different parallelization technique was evaluated to reduce the computational cost or to speed up the classifier. It showed greater accuracy as compared to the 2-gram and 3-gram model on 22,000 binary files. McLaughlin et al. [14] implemented a malware detection method that uses a deep CNN to process the raw dalvik byte code of an Android application. The application disassembled using Baksmali [15] to obtain the files that contain the human-readable dalvik bytecode of the application and fed into deep CNN. The system is capable of simultaneously learning to perform feature extraction and malware classification if the raw opcode sequences of a large number of labeled malware samples are alone provided. Niu et al. [16] designed an approach to classify Android malware based on deep learning and OpCode-level FCG. The FCG is obtained through static analysis of operation code (OpCode) from Android apps. The deep learning model, long short-term memory (LSTM), is used for training the data. The experiment was conducted on a dataset with 1,796 Android malware samples (obtained from VirusShare [17] and AndroZoo) and 1,000 benign Android apps. The experimental results showed that the accuracy is 97%.

Karbab et al. [18] developed MalDozer an automatic Android malware detection and attribution system that relies on deep learning techniques and raw sequences of API method calls in order to identify Android malware. The framework is based on an ANN. The input such as the raw sequences of API method calls are fed into the ANN. The MalDozer was evaluated on several small and large datasets, including Malgenome, Drebin, etc. MalDozer achieved a high accuracy in malware detection under multiple datasets. A disassembler Objdump (GNU binutils version 2.27.90) was used to extract the opcode sequence of each sample, and a dictionary of unique opcodes was generated and fed to Fuzzy pattern tree for malware classification. The experimental results showed that the accuracy is 95.74% for IoT and Vx-Heaven dataset. Vinayakumar et al. [19] implemented a DeepMalNet for static PE malware detection. The method evaluates the efficacy of shallow and deep networks to statically detect malicious Windows PE files. The technique used a recently released EMBER malware benchmark data set. The extracted PE features (Parsed and format agonostic features) fed into deep CNN for classification. The observed results of deep networks are high compared to the shallow networks. Dovom et al. [20] developed fuzzy pattern tree for edge malware detection and categorization in IoT. A disassembler Objdump (GNU binutils version 2.27.90) was used to extract the opcode sequence of each sample, and a dictionary of unique opcodes was generated and fed to Fuzzy pattern tree for malware classification.

The experimental results showed that the accuracy is 95.74% for IoT and Vx-Heaven dataset. Naik et al. [21] designed a static analysis technique fuzzy import hashing for malware detection. The technique is the integration of fuzzy hashing and import hashing. Fuzzy hashing attempts to find structural similarity between the two files. Import hashing attempts to find similarity between import address tables of files. The two techniques complement each other in finding a missed opportunity by one of the methods. The fuzzy import method was evaluated on four ransomware categories WannaCry, Locky, Cerber, CryptoWall, and goodware samples. The combined detection result increases the detection accuracy of the analysis process. The result showed that 80.05% detection accuracy for ransomware categories. Mohamad Arif et al. [22] implemented a mobile malware detection system based on risk evaluation using the fuzzy analytical hierarchy process (AHP) method. Permission-based features were selected to evaluate the mobile malware detection system. The experiment was conducted on a dataset from Drebin and AndroZoo, then decompile the APK file into the XML file. A total of 274 lists of permissions (known as features) were extracted from the AndroidManifest.xml file. The extracted features fed into ML techniques for classification. The results showed that the detection accuracy rate is 90.54% and the method classify the Android application into four different risk levels such as low, very low, medium, and high.

Li et al. [23] designed a malicious software classification with fuzzified features and boosted fuzzy RF algorithm. The method used fuzzy RF algorithm and SVM to perform the malware classification. The feature set is created by extracting the PE files in the Windows platform. The experiment was conducted on a dataset from virus total [24]. The results showed that the accuracy rate is 90.54% for large-scale malicious data set such as Virus, Backdoor, SoftwareBundler, Ransom, VirTool, Trojan, TrojanSpy, TrojanDownloader, etc. Shalaginov et al. [25] implemented a multinomial malware detection method using the Neuro-Fuzzy method. Neuro-Fuzzy is the Hybrid Intelligence method for malware detection. PE file features were selected to evaluate the malware detection system. Bootstrap aggregation is used to generalize the data regions for each of the fuzzy rules and to reduce overfitting. The experiment was conducted on largescale malware dataset that include variety of malware samples such as trojan, trojandownloader, worm, virtool, backdoor, virus, rogue, trojandropper, trojanspy, etc. The experimental results showed that the accuracy is s 83.56%. Table 3.2 summarizes the advantages and limitations of static malware analysis techniques.

The static-analysis using ML is fast and reliable mechanism to classify malicious and benign samples. Most of the static analysis techniques use signature-based detection, which is inefficient in the present scenario due to the rapid increase in the number and variants of malware. The signature-based detection approach requires frequent updates of the virus signature database. The challenging part is that malware creators are well aware of the limitations of static analysis. For example, the packed code cannot be statically analyzed because it is encrypted and compressed. The static analysis fails at polymorphic (complex file infectors capable to create modified versions of itself to avoid detection) and metamorphic (malware is capable of changing its code and signature patterns with each iteration) malwares. Various techniques, such as code obfuscation, dynamic code loading, encryption, and packing, are used by malware writers to evade static analysis. Dynamic analysis is robust to these techniques and provide an insight regarding the analyzed file. Dynamic analysis consequently provides better detection of malicious software.

3.4 DYNAMIC ANALYSIS METHODS USING VIRTUALIZATION/ SANDBOX

A malware sample executed within a controlled environment and its actions analyzed to detect the malicious behavior is called dynamic malware analysis. The objective of dynamic analysis is to expose the malicious activity

TABLE 3.2 Advantages and Limitations of Static Malware Analysis Techniques

SL. No.	Category	Technique	Features	Malware	Advantages	Limitation
1.	Machine learning	Artificial neural network (ANN) [4]	Grayscale image	Mailing dataset	Low false positive rate.	Less accuracy for obfuscation-based malware.
2.	Machine learning	Convolutional neural network (CNN) [5]	Binary sequences of portable executable (PE) files.	Data from Microsoft malware classification challenge.	Detect malware from various malware family.	Time consuming.
3.	Machine learning	SVM, KNN, NB, MLP [6]	Byte sequences from entry points.	Data sets from virus total [24].	High level accuracy for detecting malware.	Low level accuracy for identifying malware families.
4.	Machine learning	Function call graph (FCG) [7]	Byte sequence	Trojan-banker malware families.	Low false positive rate.	Only specific malware classes are considered for the approach evaluation.
5.	Machine learning	Support vector machine (SVM) [8].	Portable executable (PE) file features.	Data set from Vxheaven [9] and VirusShare [17].	Low false positive rate.	Limited to specific malware samples.
6.	Machine learning	Decision tree learning (DT), naive Bayes (NB) [10].	Functions calls.	Dataset from virus total [24].	Low false positive rate.	Limited to Android-based malware samples.
7.	Machine learning	Decision tree (DT), random forest (RF), KNN, and naive bayes (NB) [11].	Portable executable (PE) files features and strings.	Dataset from virus total [24] and VirusShare [17].	High level of accuracy in PE features.	Low level of accuracy in string features.
8.	Machine learning	Bagged decision tree classifier [12].	Structural information in PE header.	Dataset from virus total [24] and VirusShare [17].	The true positive rate is 99.8.	Time consuming
9.	Machine learning	SVM and the shortest path graph kernel (SPGK) [13].	Byte sequence	Dataset from VirusShare [17].	Low false positive rate.	Limited to Android-based malware samples.

TABLE 3.2 *(Continued)*

SL. No.	Category	Technique	Features	Malware	Advantages	Limitation
10.	Deep learning	Deep convolutional neural networks [14].	n-grams and opcode sequence.	Dataset from Android malware genome project.	Low false positive rate.	Time consuming
11.	Deep learning	Deep convolutional neural network (CNN) [16].	API method calls	Malgenome and drebin datasets.	Low false positive rate.	Fails to detect obfuscated malware.
12.	Deep learning	DeepMalNet for static PE malware detection [18].	Portable executable (PE) file features.	Ember dataset	Quick and effective.	Fails to detect obfuscated malware.
13.	Deep learning	Function call graph (FCG) [19].	Opcode sequence and function call sequence.	Dataset from VirusShare [17] and AndroZoo.	Automatically classifies the malware into their respective families.	Time consuming. The dataset is small.
14.	Fuzzy logic	Racing algorithm [20].	Opcode sequence	IoT, Vx-heaven, Kaggle, and ransomware dataset.	Accurately detecting malicious codes.	Obfuscation techniques reduce the effectiveness of the employed approach.
15.	Fuzzy logic	Fuzzy-import hashing technique [21].	Portable executable (PE) file features.	Ransomware dataset (WannaCry, Locky, Cerber, and CryptoWall).	Quick and effective	Similarity scores are dependent on the size of files, blocks, and order of functions.
16.	Fuzzy logic	Fuzzy analytical hierarchy process (AHP) [22]	Android application permission-based features.	Data set from Drebin and AndroZoo	Low false positive rate.	Only specific malware classes are considered for the approach evaluation.
17.	Fuzzy logic	Fuzzy decision tree and support vector machine [23].	Portable executable (PE) file features.	Dataset from VirusShare [17]	Good classification with interpretability.	Time consuming and fails to detect obfuscated malware.
18.	Fuzzy logic	Neuro-fuzzy method [25].	Portable executable (PE) file features.	Dataset from virus total [23] and VirusShare [17]	Low false positive rate.	Time consuming.

performed by the executable while it is running without compromising the security of analysis platform. The malware is classified by type, behavior, and privileges. Examples of malware by type are viruses, worms, adware, spyware, etc. The classification of malware by behavior is more relevant to the dynamic malware analysis than malware by type. After analyzing binary code, the information about its behavior is easier to classify to the correct malware type. The different malware behaviors are stealing information, creating vulnerability, denying service, executing commands from the command and control (C&C), deceiving the user, annoying the user, etc. A malware exhibits more than one behavior at a time. Identifying these behaviors is the first step towards the dynamic malware analysis process. Privilege-based malware is harder to detect and analyze. The CPU is built on protection rings such as user mode, kernel mode, hypervisor, etc. The code running at kernel mode has more privileges (e.g., read/write permissions) than the code running at other protection rings.

3.4.1 DYNAMIC MALWARE ANALYSIS FRAMEWORKS

Dynamic malware analysis has potential benefits over static malware analysis to detect malware effectively. Recently, machine learning (ML), deep learning, and FL are the focus of the security experts to detect malware and predict its variants. Dynamic malware analysis is classified into three categories, as shown in Figure 3.2.

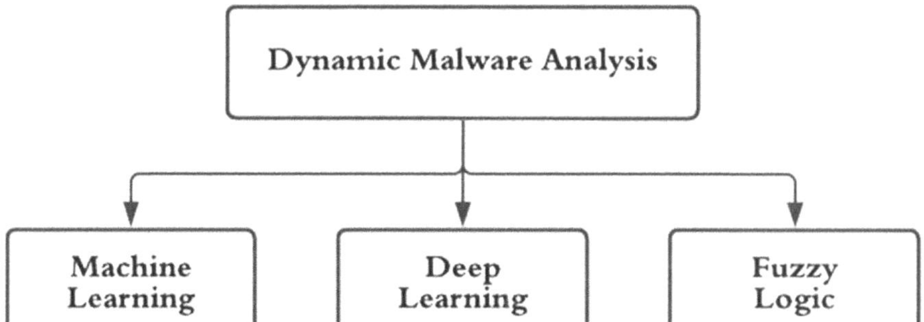

FIGURE 3.2 Dynamic malware analysis classification.

Usman et al. [26] designed a hybrid approach for dynamic malware analysis. Two honeypots were used in the pre-processing stage for the detection of malicious IPs and their attacks. During the activation period of an IP address, several features such as IP address, its associated malicious binary file, its activation time, its number of attempts, etc., are extracted. All these features compute the reputation of an IP address. Several ML techniques such as decision tree (DT), SVM, Mini Batch K-means (MBK) clustering, and NB are applied on the dataset from cuckoo sandbox [27]. Among these techniques, DT performs better while categorizing the unknown malware samples. After categorization, a kill chain methodology is applied to assign severity to a malware sample. The experimental results showed that the accuracy is 93.5%.

Hwang et al. [28] implemented a two-stage ransomware detection using machine-learning technique. The method observed the Windows API call sequences and reported abnormal behavior when the sequence differs from normal one. The two-stage ransomware detection model are as follows: (i) Markov chain detection at the first stage; and (ii) random forest (RF) method at the second stage. Two Markov chains used for malware detection, one for ransomware and the other for benign program. Markov chains decide whether incoming program is malign or benign by comparing likelihoods of an incoming samples. RF detects the remaining ransomwares. The results showed that two-stage mixed detection method achieved 97.3% accuracy with 4.8% FPR and 1.5% FNR.

Zhang et al. [29] designed an Android malware detection method based on method level correlation relationship of application's abstracted API calls. The Android application's source code is split into separate function methods. The confidence of association rules between the abstracted API calls is computed that forms behavioral semantics. The method used a combination of SVM and RF algorithms to identify the different behavioral patterns of malicious and benign apps to build the detection system. The experiment was conducted on the dataset Drebin (benign 5.9 K and malware 5.6 K) and AMD (benign 20.5 K and malware 20.8 K). The experimental results achieved 96% and 98% detection results both in accuracy and F-measure.

Escudero et al. [30] implemented an optimal feature configuration for dynamic malware detection. The data set included different file types such as .pdf, .docx, .exe, .html, .xlsx, etc. The method extracted different feature sets from Cuckoo sandbox and used them to distinguish between malicious and benign files. ML models were generated using the auto-sklearn library, and statistical tests were applied to the results to determine whether differences between different feature sets and hyperparameter optimization algorithms are significant. The datasets used in experiments were from Malshare [31]; and VirusShare [17]. The results showed that the accuracy is 93.6%.

Daniele et al. [32] developed a ML approach for dynamically analyzing and classifying ransomware. The method monitors a set of actions performed by applications in their first phases of installation, checking for the characteristic signs of ransomware. The ML component consists of two phases: feature selection and classification. Mutual Information criterion is used for feature selection that selects the most discriminating features for ransomware and goodware from the sandbox. Regularized Logistic Regression classifier was used for classification. The experiment was conducted on a dataset of 582 ransomware belonging to 11 families and 942 goodware applications. The experimental results showed that the accuracy is 93.6%. Zhang [33] developed a convolutional neural network-long short-term memory (CNN-LSTM) model for Malware Detection based on Dynamic Semantic Behaviors. The method integrated CNN and LSTM into a unified model. CNN extracts discriminative features from the raw API call sequences. LSTM is a type of recurrent neural network (RNN) capable of learning order dependence in sequence prediction. The experiment was conducted on a raw malware dataset, and the accuracy is 96%.

Vinayakumar et al. [34] designed a scalable framework ScaleMalNet to detect, classify, and categorize zero-day malwares. The framework applies deep learning models such as CNN and RNN on the malware from end-user hosts and follows a two-stage process for malware analysis. In the first stage, the executable files are classified as malware or legitimate. The malware executable files are categorized into corresponding malware families using image processing approaches in the second stage. The experiment was conducted on the publicly available benchmark datasets and privately collected datasets. The framework is capable of analyzing a large number of malware in real time.

Shibahara et al. [35] implemented a dynamic malware analysis technique based on HTTP requests using deep learning. The method analyzes a malware sample based on the network behaviors in their short-period analysis. Around 18 different features such as general features, hostname-based features, HTTP-based features are collected and fed in to Recursive Neural Network (RNN) for malware analysis. The experiment was conducted on 42,856 malware samples from virus total and achieved 98% of accuracy. Alzaylaee et al. [36] developed DL-Droid, a deep learning system to detect malicious Android applications through dynamic analysis using stateful input generation. DL-Droid used 31,125 Android applications, 420 static and dynamic features for malware detection. The features such as API calls and intents, Actions/Events are used as inputs for DL-Droid's deep learning-based classifier to detect Android malware. The experiments were performed with 30,000 applications (benign and malware) on real devices. DL-Droid achieved a 97.8% detection rate with dynamic features and 99.6% detection rate with dynamic and static features. Yuan et al. [37] designed a framework Droid Detector for android malware characterization and detection. The method extracted 192 features from Android apps and characterized the malware using a deep belief network (DBN)-based deep learning model. DroidDetector evaluated on 20,000 benign apps from the Google Play Store and 1,760 malwares from the Contagio Community and Genome Project. The results show that Droid Detector with a deep learning model achieved above 94.4% accuracy.

Mario et al. [38] designed a dynamic malware detection method using Process Mining (PM) and FL techniques. FL classifies the malware applications and verify its relations with the existing malware variants. The combination of two techniques allows to obtain a fingerprint of an application that is used to verify its maliciousness/trustfulness. The distances between the system call execution fingerprints (SEF) are used to characterise malware families. The experiment was conducted on a dataset of 3,000 trusted and malicious applications across 12 malware families. The method is highly effective in detecting malware with 0.821 Precision value. Mahbub et al. [39] implemented a Malware detection in edge devices with fuzzy oversampling and dynamic class weighting. The method used two approaches for malware detection. First, a fuzzy set theory-based oversampling technique and second, a hybrid technique with a loss function that dynamically assigns weights to the malware. Static features are extracted by parsing the application code, and dynamic features are extracted by executing the application and monitoring its run-time behavior. Extracted features from the data set fed in to fuzzy-synthetic minority oversampling technique (SMOT) for sampling and deep neural network (DNN) is used for classification. A dynamic minority class weighting scheme is used for malware detection. The experiment was conducted on 50,000 malwares from Drebin, AndroZoo, and VirusTotal dataset. The experimental results showed that the accuracy is 97%.

Shalaginov et al. [40] designed a deep neuro-fuzzy method for multi-label malware classification. The method analyzed multinomial malware categorization using static and dynamic behavioral features extracted from Windows PEs. A combination of DNN and FL is used for malware analysis. The experiment was conducted on 4,07,741 unique PE32 binaries from Virus total and achieved 69.44% accuracy. Naik et al. [41] implemented a method that utilizes fuzzy hashing and fuzzy rules to enhance YARA rules for malware detection. YARA rules are developed to detect malware by primarily matching its signatures/strings with the existing malware signatures/strings. Fuzzy Hash Matching is combined with the default String Matching condition of YARA rules for malware detection. The fuzzy hashing algorithm such as SSDEEP, SDHASH, and mvHASH-B methods were used to detect similarity for each ransomware category. The experiment was conducted on the data set from malshare and virus total and achieved 86% accuracy. Altaher [42] developed an Android malware detection scheme based on evolving hybrid neuro-fuzzy classifier (EHNFC) and permission-based features. The method partitioned selected permission-based features into clusters. EHNFC dynamically creates fuzzy rules to predict the newly generated malware, such as obfuscated Android malware, based on mining the fuzzy rules. The experimental results show that EHNFC achieved 99% accuracy and low false positive rate.

Table 3.3 summarizes the advantages and limitations of dynamic malware analysis. Dynamic malware analysis is a domain in which significant progress are made in recent years and it has enhanced performance, higher malware detection rates, and improved resilience against malware evasion techniques. Section 3.4.1 discusses on dynamic malware analysis based on ML, deep learning, and FL. Malware detects and evades analysis, and defenders constantly propose mitigations against evasion.

3.4.2 *DYNAMIC MALWARE ANALYSIS TOOLS*

A sandbox is an isolated environment where users safely test suspicious code without causing risk to the device/network. The sandbox extracts specific activity of malware such as behavior, signatures, version, id, etc. The malware analysis tools are as follows:

1. **GFI Sandbox:** The Sandbox analyze the behavior of suspected viruses, worms, trojans, and other malware by executing the program inside a controlled environment and recording all changes such as Windows API calls, network traffic, etc. The input to the GFI sandbox is files and URL and output from the Sandbox is the information related to file system changes, registry changes, network changes, memory changes, etc.
2. **Norman Sandbox:** The sandbox monitors memory and disassembles code, virtual hard disk, registers, and network activity. The sandbox analyzer enables users to analyze file behaviors, actions performed by the file, etc.

TABLE 3.3 Advantages and Limitations of Dynamic Malware Analysis

SL. No.	Category	Technique	Feature/Features	Malware	Advantages	Limitation
1.	Machine learning [26]	DT, SVM, MBK, and NB	IP address	Data sets from cuckoo sandbox [27].	Detect zero-day anomalies.	Time consuming
2.	Machine learning [28]	Markov model and random forest algorithm.	Windows API call.	Data sets from virus share [17].	Low false positive rate.	Limited to ransomware malware samples.
3.	Machine learning [29]	SVM, RF	API calls	Drebin [40] and AMD data set.	Low false positive rate.	Rare and new malware files cannot be detected.
4.	Machine learning [30]	Bayesian optimizer	n-grams, API calls, signatures, network activity.	Data sets from Malshare [31] and VirusShare [17].	Detect malware from various malware families.	Obfuscation techniques reduce the detection accuracy.
5.	Machine learning [32]	Logistic regression (LR)	Registry keys operations, API stats, strings, file extensions, files operations.	Data sets from virus total [24].	Detect malware from various ransomware families.	Time consuming
6.	Deep learning [33]	CNN, long short-term memory (LSTM)	n-grams, API calls.	Data sets from VX heavens.	Low false positive rate.	The dataset size is small.
7.	Deep learning [34]	CNN and deep neural network (DNN)	n-grams, portable executable (PE) file features.	Ember dataset, malimg data set, data set from cuckoo sandbox [27]. VirusSign, VirusShare [17].	Low false positive rate.	Time consuming.
8.	Deep learning [35]	Recursive neural network (RNN)	Hostname-based features, HTTP-based features.	Data set from virus total [24], AV class.	Low false positive rate.	Advanced packing techniques reduce detection accuracy.

TABLE 3.3 *(Continued)*

SL. No.	Category	Technique	Feature/Features	Malware	Advantages	Limitation
9.	Deep learning [36]	Permission features, actions/events features.	API/system calls	Data set from McAfee Labs.	Quick and effective.	Not able to detect bots with rootkit capabilities.
10.	Deep learning [37]	Deep belief networks (DBN)	N-gram features	Data set from Contagion community and genome project	Low false positive rate.	Not able to detect bots with rootkit capabilities.
11.	Fuzzy logic [38]	Dynamically expanded neuro-fuzzy (DENF) method	CPU load, memory usage, API calls-based features.	Data set from virus total [24].	Low false positive rate.	Time consuming.
12.	Fuzzy logic [39]	Synthetic minority oversampling technique (SMOT), DNN.	API /system calls	Data set from Drebin and Andro-Zoo.	Quick and effective.	Lose the accuracy when obfuscation or encrypt technique was applied.
13.	Fuzzy logic [40]	DNN, FL	API /system calls, portable executable (PE) file features.	Data set from virus total [24], virus share [17].	Zero-day vulnerabilities detection.	Time consuming.
14.	Fuzzy logic [41]	YARA rules, fuzzy hashing, fuzzy rules	Statistically improbable features (SIF)	Data set from hybrid-analysis, Malshare [31] and VirusTotal [24].	Quick and effective	Dependency of fuzzy rules.
15.	Fuzzy logic [42]	Evolving hybrid neuro-fuzzy classifier (EHNFC)	Permission-based features	GNOME project	Low false positive rate	The dataset used in the experimental evaluation is small.

3. **Joe Sandbox Document Analyzer:** It analyzes potential malicious files and URLs on Windows, Android, Mac OS, Linux, and iOS for suspicious activities. It performs deep malware analysis and generates comprehensive and detailed analysis reports.
4. **Cuckoo Sandbox:** It is an open-source dynamic malware analysis system. It executes the Malware in an isolated Windows Operating System and provides the consolidated report. It analyzes many different malicious files (executables, office documents, pdf files, emails, etc.) and malicious websites under Windows, Linux, macOS, and Android virtualized environments.

The dynamic malware analysis methods in ML, deep learning, and FL create adaptive malware detection systems that defend and protect the network and information of users and organizations.

3.5 FEATURES EXTRACTION USING REVERSE ENGINEERING

Reverse engineering disassembles or decomposes a software program and converts the binary instructions to code mnemonics or higher-level constructs. By knowing its details, a malware analyst analyzes the program's intended malicious effects. To analyze the threat associated with malicious software, its behavior is examined in a controlled environment. In reverse engineering, malware analysts run the malware samples in debuggers, disassemblers, or virtual environments to analyze the functionality and behavior. FLARE VM [43] is a reverse engineering framework on the Windows platform. It includes many tools such as IDA, radare, YARA, Ghidra, OllyDbg, PE Explorer, Wireshark, etc.

3.5.1 REVERSE ENGINEERING TOOLS

1. **Disassemblers:** Reverse engineering of malware involves disassembling or decompiling a software program to analyze malware. A disassembler (IDA Pro) converts machine language into assembly language. IDA Pro is a multi-platform, multi-processor disassembler for debugging and reverse engineering. IDA Pro displays the assembly code in a text view. It contains address, instruction, parameters, and comments arranged row by row or in graph view, that draws the assembly code in logic blocks. The division of blocks

is based on jumps, conditions, and loops. It is used as a local or as a remote debugger on various platforms. IDA Pro Plug-ins support a variety of executable formats for different processors and operating systems. IDA pro analyzes strings, functions, imports, exports, and the names of malicious programs.

i. **Strings:** These are ASCII and Unicode-printable sequences of characters embedded within a file. The string extraction contains the program functionality and indicators associated with a suspect binary file. For example, if malware creates a file, the filename is stored as a string in binary.
ii. **Exported Functions:** A list of functions that are offered from an executable to the external environment. Exported functions are called and executed by an external program.
iii. **Names:** A list of all entity names (library function, regular function, instruction, string literal, data, and imported name).
iv. **Imported Functions:** A list of API functions loaded from external libraries and used by an executable. From the list of imported functions, it is possible to identify how an executable interacts with the operating system and its resources (filesystem, registry, networking, encryption, etc.).
v. **Functions:** A list of all functions incorporated in the code of an executable. In addition, the Fast Library Identification and Recognition Technology (F.L.I.R.T.) feature allows the IDA to recognize standard library functions generated by compilers and improves the usability and readability of generated disassembles.

2. **Debuggers:** The debuggers monitor malicious code execution including insight into memory, registers, stack, and control elements. The debugger runs the code, controls the execution (instruction by instruction, breakpoints), and analyzes particular values in registers, parameters of functions, return values, etc. There are several debuggers for executables, e.g., WinDbg, x64dbg, Immunity Debugger, and OllyDbg.
3. **PE Viewer:** It is a tool for viewing Windows portable executable (PE) file structures. Features such as function call frequency and suspicious section counts from executables are extracted, e.g., CFF Explorer. CFF Explorer is a tool to modify the PE. It runs on Windows OS and has the capability of listing processes or dumping the process to a file. CFF Explorer extracts the compilation date and

architecture type from the analyzed malware sample based on the information inside the PE Header. By analyzing the section header, the analyst identifies if the malware is packed or not. Packers tend to change section names from the regular names (.text, .data, .rsrc, etc.) to other names.

4. **Network Analyzers:** Wireshark is a popular network analyzer tool. The network analyzer monitors the network connections and the data traffic used by an application with remote systems. The tool displays the list of network (ethernet/wireless) interfaces. From the captured traffic, the malware behavior is identified by eliminating the noises. It is easy to filter the packets using protocol options in the wireshark tool.

   The malware feature identification is significant in ML classification. The accuracy of ML-based malware analysis methods depends on the features present in malware samples. The malware analysis method focuses on determining the best features that help in detecting the malware accurately. Tools such as disassemblers, debuggers, PE viewer, network analyzers, etc., are freely available software to extract multiple features from the malware. A good selection of features improves the efficiency of the classifier. The feature selection methods are broadly classified as follows: (i) filter method; and (ii) wrapper method. In Filter Method, feature selection is independent of the learning algorithm. In Wrapper Method, a random subset of features is used to train the model.

3.6 MACHINE LEARNING (ML) TECHNIQUES IN AUTOMATING THE DETECTION AND IMPROVING THE PRECISION RATE

Social network service (SNS) messengers grow rapidly in the cyber world and cyber attackers are targeting messengers through various methods such as ransomware, steganography, etc. A steganography-based botnet (stego-botnet) is constructed and implemented in SNS messengers. Stego botnet C&C messages are secretly hidden into multimedia files such as images or video and shared in the SNS messenger. The botmaster in the stegobotnet uploads a stego-image containing bot commands in the chatroom, and victims read and download the stego-image. A bot malware automatically extracts the secret message embedded in the image.

Park et al. [44] developed an automated detection procedure of steganography Images. The detection procedure consists of the following steps:

- Defender reads and clicks all shared image files in the chatroom;
- Image files are download at Defender's device;
- Stored image copies are automatically and periodically transferred from Defender's device to Defender's inspection server;
- All collected image files are examined by the detection system and the presence/absence of malware is reported.

Automated detection procedure consists of two components: (i) an automated collection of stego images; and (ii) automated detection of stego malware. The defender reads and clicks images that are stored in the local storage of the defender's device. To move the images from the defender's smartphone to the inspection server, the automated detection component will automatically examine whether the collected image files consist of hidden steganography messages using ML-based malware detection approaches. Hex Editor is used to extract the feature from stegomalware. The experiment was conducted on 40 benign and 8 malware images and the experimental results showed that the accuracy is 100%.

The different ML algorithms for malware classifications are NB, K-Nearest Neighbors (K-NN), SVM, DT, RF, etc. NB, SVM, DT, and KNN are nonparametric supervised ML algorithms.

- Naive Bayes (NB) algorithm classifies the data samples using the Bayes theorem. Different types of NB algorithms are applied based on the distribution of data.
- K-Nearest Neighbors (KNN) is the instance-based algorithm. It stores all the available data and classifies a new data point based on the similarity. The user specifies the value of k-nearest neighbor's instance.
- SVM (support vector machine) is a supervised learning model that analyzes high-dimensional and sparse data and recognizes patterns. It produces the vectors which partition the feature space into multiple regions belonging to different classes. To produce the maximum marginal hyperplane, the kernel function and the C parameter (regularization parameter) are chosen. Kernel defines the processing of training datasets using some mathematical equations to produce the hyperplane. The three most commonly used kernel functions are Linear, Polynomial, and Radial Kernel.
- Decision tree (DT) classification is a non-parametric supervised ML algorithm. It performs regression, classification, and generates a model that predicts the value or a class of a target variable through decision rules derived from data features. Training features are partitioned

using splitting criteria (Gini index or info gain) to represent a DT. Leaf nodes of the tree specify the class labels.
- Random forest (RF) algorithm extends the basic idea of the DT. The difference is that multiple trees are created. It is a type of bagging (bootstrap aggregation) algorithm. A bootstrap aggregation algorithm produces multiple models from the training dataset and considers all models averages to create the final robust model.
- Adaboost combine multiple weak classifiers into a single strong classifier. In Adaboost, lower learning rate means the weight will be decreased or increased to a small extent. Updating the weights of model bit by bit produces better performance. It is a trade-off between n-estimator and learning rate.

Jagsir Singh et al. [45] developed a malware detection technique and surveyed the assessment of supervised ML algorithms. The runtime features are captured through Cuckoo Sandbox for malware classification. During execution, the malware performs different malicious activities. The API calls of the process classify malware (benign or not). The result shows that the data set from Virus Share, malware, Contagio, and virus total, NB algorithms achieved 85% accuracy, KNN algorithms achieved 98% accuracy, and SVM algorithms achieved 98% accuracy. RF and Adaboost algorithm achieved 99% accuracy. ML algorithms have many variations in training malware detection classifiers. An automatic system periodically examines the running behavior of malicious processes from executables. It detects hidden, dead, and dubious processes on predicting early infection of the malware symptoms.

3.7 CONCLUSION

We described the advancements in malware analysis techniques using ML, deep learning, and FL. The static analysis using ML is the fast and reliable mechanism, which is inefficient in the present scenario due to the rapid increase in the number and variants of malware. Dynamic malware analysis methods using ML, deep learning, and FL create adaptive malware detection systems that defend and protect the network and information of users and organizations. Dynamic malware analysis is an ever-evolving domain in which significant progress was made in recent years in enhanced performance, higher detection rates, and improved resilience against evasion techniques.

KEYWORDS

- **automated detection techniques**
- **dynamic analysis**
- **machine learning**
- **malware**
- **manual detection techniques**
- **reverse engineering**
- **virtualization**

REFERENCES

1. NSO Group in Trouble Again Over Spyware, (2021). *Network Security, 2021*(8), 2, 3. https://www.sciencedirect.com/science/article/pii/S1353485821000829 (accessed on 28 September 2023).
2. Metasploit-Framework, (2003). *Rapid7, Penetration Testing Tool*. https://tools.kali.org/exploitation-tools/metasploit-framework (accessed on 28 September 2023).
3. Kaushik, M., Malik, M., & Narwal, B., (2020). Developing malware and analyzing it afore after steganography with OSINTs. *IEEE International Conference for Innovation in Technology, INOCON 2020*, 20–23.
4. Moussas, V., & Andreatos, A., (2021). Malware detection based on code visualization and two-level classification. *Information (Switzerland), 12*(3), 1–14.
5. Damasevicius, et al., (2021). Ensemble-based classification using neural networks and machine learning models for windows pe malware detection. *Electronics (Switzerland)*, 1–26.
6. Wan, T. L., & Ban, T., (2020). Efficient detection and classification of internet-of-things malware based on byte sequences from executable files. *IEEE Open Journal of the Computer Society*, 262–275.
7. Bai, J., Shi, Q., & Mu, S., (2019). A malware and variant detection method using a function called graph isomorphism. *Security and Communication Networks*, 2–12.
8. Wadkar, M., Di Troia, F., & Stamp, M., (2020). Detecting malware evolution using support vector machines. *Expert Systems with Applications, 143*, 1–22.
9. *VX Heaven Virus Collection*, (2016). http://vxheaven.org/vl.php (accessed on 28 September 2023).
10. Burnap, P., French, R., Turner, F., & Jones, K., (2018). Malware classification using self-organizing feature maps and machine activity data. *Computers and Security, 73*, 399–410.
11. Huda, S., et al., (2016). Hybrids of support vector machine wrapper and filter-based framework for malware detection. *Future Generation Computer Systems, 55*, 376–390.
12. Kumar, A., Kuppusamy, K. S., & Aghila, G., (2019). A learning model to detect maliciousness of portable executable using integrated feature set. *Journal of King Saud University – Computer and Information Sciences*, 252–265.

13. Searles, R., et al., (2017). Parallelization of machine learning applied to call graphs of binaries for malware detection. *Proceedings – 25th Euromicro International Conference on Parallel, Distributed, and Network-Based Processing*, 69–77.
14. McLaughlin, N., Del, R. J. M., Kang, B. J., Yerima, S., Miller, P., Sezer, S., Safaei, Y., et al., (2017). Deep Android malware detection. *Proceedings of the 7th ACM Conference on Data and Application Security and Privacy*, 301–308.
15. Baksmali, (2010). Ben Gruver. *Penetration Testing Tool.* https://github.com/JesusFreke/smali (accessed on 28 September 2023).
16. Niu, W., Cao, R., Zhang, X., Ding, K., Zhang, K., & Li, T., (2020). Opcode-level function call graph based android malware classification using deep learning. *Sensors (Switzerland), 20*(13), 1–23.
17. Virus Share, (2012). *Nils Kuhnert, CERT-Bund.* Retrieved from: https://virusshare.com (accessed on 28 September 2023).
18. Karbab, E. B., Debbabi, M., Derhab, A., & Mouheb, D., (2017). *Android Malware Detection Using Deep Learning on API Method Sequences*. arXiv:1712.08996. [Online]. Available: https://arxiv.org/abs/1712.08996 (accessed on 28 September 2023).
19. Vinayakumar, R., Alazab, M., Soman, K. P., Poornachandran, P., & Venkatraman, S., (2019). Robust intelligent malware detection using deep learning. *IEEE Access, 7,* 46717–46738.
20. Dovom, E. M., Azmoodeh, A., Dehghantanha, A., Newton, D. E., Parizi, R. M., & Karimipour, H., (2019). Fuzzy pattern tree for edge malware detection and categorization in IoT. *Journal of Systems Architecture, 97*, 1–7.
21. Naik, N., Jenkins, P., Savage, N., Yang, L., Boongoen, T., & Iam-On, N., (2021). Fuzzy-import hashing: A static analysis technique for malware detection. *Forensic Science International: Digital Investigation, 37*, 301139.
22. Mohamad, A. J., Ab Razak, M. F., Tuan, M. S. R., Awang, S., Ismail, N. S. N., & Firdaus, A., (2021). Android mobile malware detection using fuzzy AHP. *Journal of Information Security and Applications, 61*, 102929.
23. Li, F.-Q., Wang, S.-L., Liew, A. W.-C., Ding, W., & Liu, G.-S. (2021). "Large-Scale Malicious Software Classification With Fuzzified Features and Boosted Fuzzy Random Forest," in *IEEE Transactions on Fuzzy Systems*, *29*(11), 3205–3218, doi: 10.1109/TFUZZ.2020.3016023.
24. Virus Total, (2019). https://www.virustotal.com/#/home/upload (accessed on 28 September 2023).
25. Shalaginov, A., & Franke, K., (2018). A deep neuro-fuzzy method for multi-label malware classification and fuzzy rules extraction. In: *2017 IEEE Symposium Series on Computational Intelligence Proceedings* (pp. 1–8).
26. Usman, N., Usman, S., Khan, F., Jan, M. A., Sajid, A., Alazab, M., & Watters, P., (2021). Intelligent dynamic malware detection using machine learning in ip reputation for forensics data analytics. *Future Generation Computer Systems*, 118.
27. Cuckoosandbox., (2010). *Claudio Guarnieri.* https://cuckoosandbox.org (accessed on 28 September 2023).
28. Hwang, J., Kim, J., Lee, S., & Kim, K., (2020). Two-stage ransomware detection using dynamic analysis and machine learning techniques. *Wireless Personal Communications, 112*(4), 2597–2609.
29. Zhang, H., Luo, S., Zhang, Y., & Pan, L. (2019). An efficient android malware detection system based on method-level behavioral semantic analysis. *IEEE Access, 7*, 69246–69256.

30. Escudero, D., & Decastro-Garcia, N., (2021). Optimal feature configuration for dynamic malware detection. *Computers & Security, 105*, ISSN: 0167–4048.
31. Malshare, (2012). *Malware Repository*. https://malshare.com (accessed on 28 September 2023).
32. Scudero, D., & Decastro-Garcia, N., (2021). *Automated Dynamic Analysis of Ransomware: Benefits, Limitations, and Use for Detection*. arXiv:1609.03020[Online].
33. Zhang, J., (2020). DeepMal: A CNN-LSTM model for malware detection based on dynamic semantic behaviors. *Proceedings–2020 International Conference on Computer Information and Big Data Applications*, 313–316.
34. Vinayakumar, R., Soman, K. P., Prabaharan, P., Vysakh, S. M., & Amara, D. K., (2018). ScaleNet: Scalable and hybrid framework for cyber threat situational awareness based on DNS, URL, and email data analysis. *Journal of Cyber Security and Mobility, 8*(2), 189–240.
35. Shibahara, T., Yagi, T., Akiyama, M., & Chiba, D., (2019). Efficient dynamic malware analysis for collecting HTTP requests using deep learning. *IEICE Trans. Inf. & Syst.,* (Vol. E102-D, NO.4, pp. 725–736).
36. Alzaylaee, M. K., Yerima, S. Y., & Sezer, S., (2020). DL-droid: Deep learning based android malware detection using real devices. *Computers and Security, 89*, ISSN 0167–4048.
37. Yuan, Z., Lu, Y., & Xue, Y., (2016). Droid detector: Android malware characterization and detection using deep learning. In: *Tsinghua Science and Technology* (Vol. 21, No. 1, pp. 114–123).
38. Mario, L. B., et al., (2017). A fuzzy-based process mining approach for dynamic malware detection, IEEE international conference on fuzzy systems. *IEEE International Conference on Fuzzy Systems (FUZZ-IEEE)* (pp. 1–8).
39. Mahbub, E. K., Joarder, K., Iqbal, G., & Tasadduq, I., (2017). Malware detection in edge devices with fuzzy oversampling and dynamic class weighting. *Applied Soft Computing, 112*. ISSN 1568-4946.
40. Shalaginov, A., (2017). A deep neuro-fuzzy method for multi-label malware classification and fuzzy rules extraction. *IEEE Symposium Series on Computational Intelligence (SSCI), 2017*, 1–9.
41. Naik, N., et al (2021). Embedded YARA rules: Strengthening YARA rules utilizing fuzzy hashing and fuzzy rules for malware analysis. *Complex & Intelligent Systems, 7*(2), 687–702.
42. Altaher, A., (2017). An improved Android malware detection scheme based on an evolving hybrid neuro-fuzzy classifier (EHNFC) and permission-based features. *Neural Computing and Applications, 28*(12), 4147–4157.
43. *FireEye Labs Advanced Reverse Engineering (FLARE VM)*, (2004). https://www.fireeye.com (accessed on 28 September 2023).
44. Park, J., & Cho, Y., (2020). Design and implementation of automated steganography image-detection system for the Kakao talk instant messenger. *Computers, 9*(4), 1–14.
45. Jagsir, S., & Jaswinder, S., (2020). Assessment of supervised machine learning algorithms using dynamic API calls for malware detection. *International Journal of Computers and Applications*, 1–8.

CHAPTER 4

SECURING IDENTITIES THROUGH BIOMETRIC TEMPLATE SECURITY

VIVEK SINGH BAGHEL and SURYA PRAKASH

Department of Computer Science and Engineering, Indian Institute of Technology Indore, Indore, Madhya Pradesh, India

ABSTRACT

Biometric authentication systems provide multiple advantages over conventional authentication systems, which are based on tokens and passwords. However, they still have some limitations, which primarily include the issues related to the security and privacy of the biometric template stored in the system. This poses a great challenge on the biometric systems as the user's biometric data is not changeable, unlike passwords, if an adversary steals it. Moreover, it has been discussed in the literature that the reconstruction of biometric data is feasible in many cases from an original biometric template. Hence, if it is compromised, it will be a permanent loss of a user's biometric data. In order to provide the security and privacy of the biometric data, various biometric template protection (BTP) techniques have been proposed. The critical aspect of these techniques is that the encrypted or transformed biometric template of a user is stored in the database instead of the original biometric template. In order to do that, the original biometric template is transformed in such a way that it is infeasible to reconstruct the original biometric data from the transformed template if it is compromised. The BTP techniques can be mainly classified into two categories, which are biometric cryptosystems and cancelable biometrics. The biometric cryptosystems combine the cryptographic key with the original biometric template or directly generate the key from the original template for securing the original biometric template. On the other side, the techniques based on cancelable

Multimedia Security: Tools, Techniques, and Applications. Bhaskar Mondal and Shyam Singh Rajput (Eds.)

biometrics introduce some distortions into the original biometric template before storing it into the database. These distortions are obtained with the help of a transformation function such that it is infeasible to reconstruct the original biometric data from the transformed template. Besides this, there are a few more techniques, such as hybrid approaches and Homomorphic encryption, which have been discussed relatively less in the literature. Hybrid approaches are the combination of cryptosystems and cancelable techniques or multi-modalities, whereas Homomorphic encryption relies on an encryption-based approach. There are some essential properties, such as revocability, diversity, security, and performance, which need to be followed to determine the overall effectiveness of a BTP technique. This chapter provides a detailed discussion on all these aspects of biometric template security (Figure 4.1).

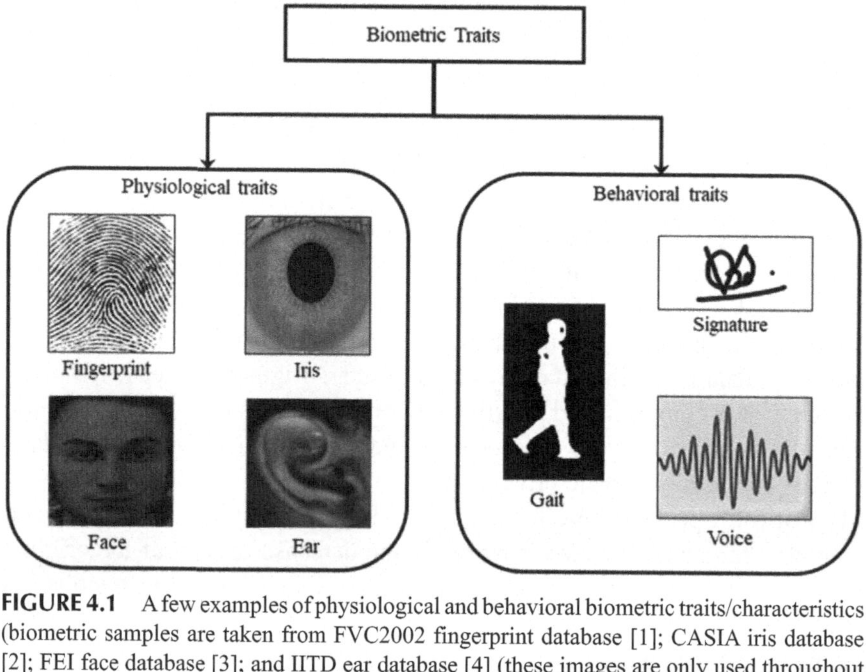

FIGURE 4.1 A few examples of physiological and behavioral biometric traits/characteristics (biometric samples are taken from FVC2002 fingerprint database [1]; CASIA iris database [2]; FEI face database [3]; and IITD ear database [4] (these images are only used throughout the chapter for representation purposes).

4.1 INTRODUCTION

Establishing a person's identity is a crucial task to provide security against the illegitimate access of different resources/applications. The identity recognition of a person can be done by utilizing the three different kinds of information as discussed in Refs. [5, 6], which are – (i) the knowledge, (ii) the possession, and (iii) the person's physiological and behavioral characteristics. In the first category, a person knows some confidential information, such as passwords, keys, PINs, etc., through which the identification process is done. In the second category, a person is identified with the help of a physical token such as a smart card, driver's license, etc. In the last category, the identification of a person is made using the physiological and behavioral characteristics/traits of a person, and these characteristics are known as biometrics. The recognition systems that are based on biometrics have many leverages over the knowledge-based and possession-based recognition systems. The biometric information of a person is non-changeable, non-shareable, and is impossible to steal, making it a more reliable way to identify a person than knowledge-based and possession-based recognition systems. Although biometrics-based recognition systems have several advantages over traditional systems (e.g., passwords/token-based), they face some challenges, such as identity theft, which causes the permanent loss of a person's biometric data. To prevent the identity theft of a person, different BTP techniques have been discussed in the literature (Figure 4.2).

The biometric authentication systems and biometric template security are discussed in detail in subsections.

4.1.1 BIOMETRIC AUTHENTICATION SYSTEM

Biometric authentication systems establish the identity of a person based on his/her physiological and behavioral biometric traits. The physiological biometric traits include the face, fingerprint, ear, iris, etc., and the behavioral biometric traits consist of the signature, gait, keystrokes, etc. Some of the examples of different biometric traits have been depicted pictorially in Figure 4.1. A biometric trait of a person exhibits some unique features/characteristics that are utilized to differentiate the identity of a person from others. In a biometric-based authentication system, these distinctive features are extracted from the biometric data of a user, which is also known as an original gallery template and is stored in the database during the enrollment stage. During the verification stage, the features are extracted from the biometric probe data and are matched with the stored gallery template in the database to compute the matching score, which is further used to decide for successful/unsuccessful verification, as shown in Figure 4.2. Similarly, a probe template is computed and matched with all stored templates in the database to recognize a person's identity during the identification stage, as depicted in Figure 4.2.

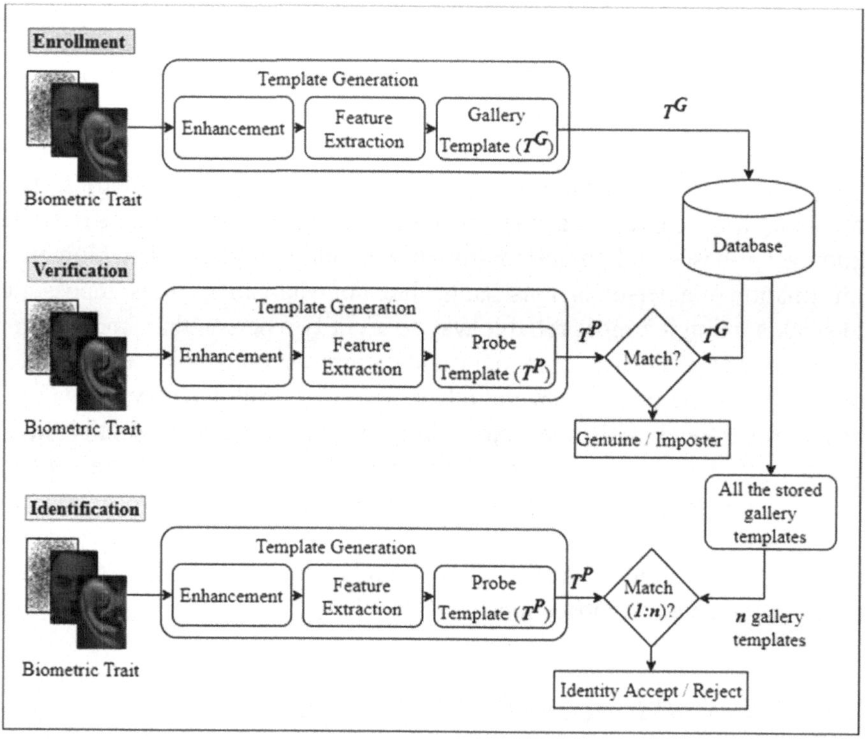

FIGURE 4.2 Overview of a biometric system.

4.1.2 BIOMETRIC TEMPLATE SECURITY

Biometric systems provide high-level security to the different resources and the applications of various fields. However, biometric systems themselves have some security issues as well, which have been discussed in Ref. [7]. The different kinds of attacks that are possible at the various points of a biometric system are depicted in Figure 4.3. In this chapter, we have discussed the security issue at point 6 (as shown in Figure 4.3), which is a stored biometric template in the database of a biometric system. A biometric system provides security to prevent illegitimate access; however, in the event of an attack, if an adversary inspects the database and gets the original biometric template, then it can cause a permanent identity loss of a person as the biometric data is immutable in nature. These acquired templates can be used to get illegitimate access to the biometric system by an adversary. Further, in the case of attack, these can also be modified, which will restrict the access of the legitimate user to the biometric system. In addition, it has been discussed in the literature [8–12] that the original biometric data can be reconstructed from the user template stored in the database.

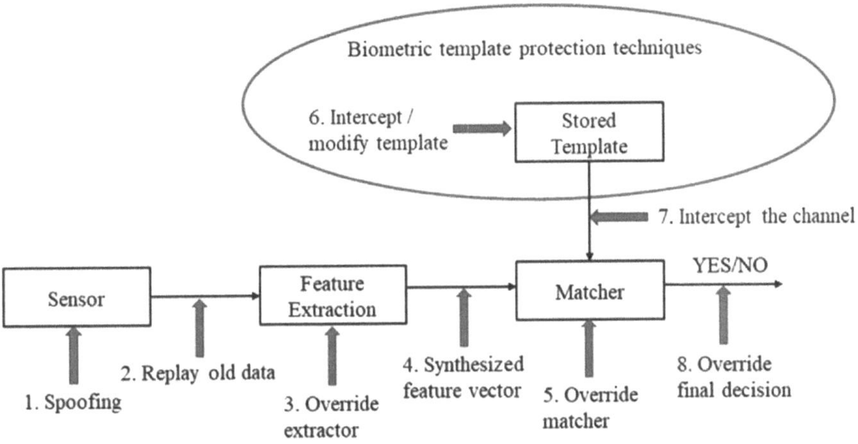

FIGURE 4.3 Different kinds of attacks at the various points of a biometric system [7].

In order to secure the stored template in the database, several techniques, known as BTP/security techniques, have been discussed in the literature [13–15]. The primary concept behind these techniques is to store a non-invertible transformed/encrypted biometric template instead of the original biometric template; hence, if an attack occurs in the database, an adversary will be unable to recover the original information of biometric data. According to the literature, the BTP techniques can be mainly categorized into two different categories which are biometric cryptosystems [16, 17], and cancelable biometrics [7, 18]. Further, there are a few more techniques discussed in the literature, which are basically homomorphic encryption and hybrid approaches (combination of cancelable and cryptosystems or different cryptosystems). In Figure 4.4, different categories of BTP techniques are shown in a hierarchical manner. Any of these mentioned techniques should follow the four essential requirements, which are revocability, unlinkability/diversity, security, and performance, according to Ref. [19]. These essential properties are explained in the next Section 4.2 and a thorough discussion of each BTP technique is provided in Sections 4.3–4.5.

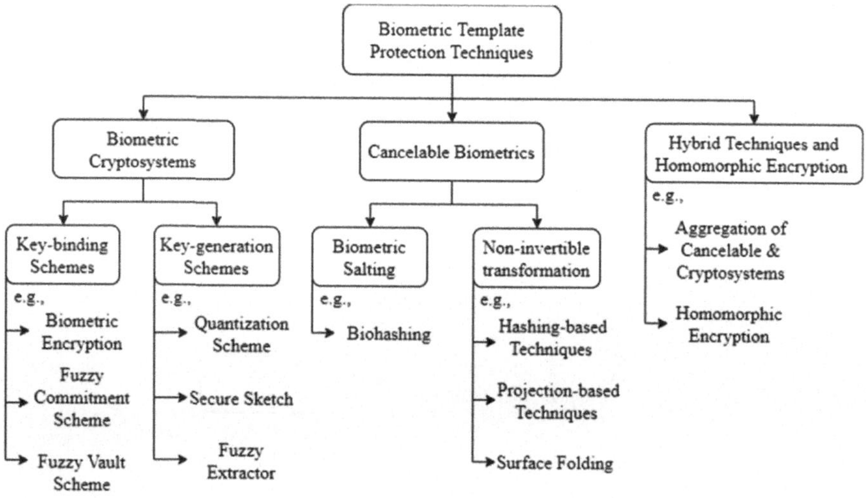

FIGURE 4.4 Classification of biometric template protection techniques.

4.1.3 CHAPTER ORGANIZATION

The rest of the chapter is organized as follows. In Section 4.2, different essential properties, which need to be followed by a BTP technique, have been discussed with a brief information regarding the analysis of these properties. In Section 4.3, biometric cryptosystems and a few techniques that are proposed in the literature have been discussed. Further, cancelable biometrics and some of the past work based on cancelable techniques are discussed in Section 4.4. A discussion about the hybrid and homomorphic encryption based template protection techniques have been provided in Section 4.5. The chapter is concluded in the last section.

4.2 ESSENTIAL PROPERTIES OF A TEMPLATE PROTECTION TECHNIQUE

A BTP technique should meet some essential requirements/properties, *viz.*, revocability, unlinkability, security, and performance [19]. Without meeting the aforementioned requirements, a template protection technique cannot be deemed robust and effective. The following is a discussion of these essential properties.

4.2.1 REVOCABILITY

The capacity to replace a stored secured template (compromised template) in the database with a totally new secured template that is generated using the same biometric data but different keys/parameters for altering the previous biometric template is termed as revocability. Although generated using the same biometric data, these two secured templates, the one stored in the database and another one newly generated, will not have any significant similarities with each other. In order to verify the notion that a template protection technique exhibits revocability property, some frameworks such as proposed in Refs. [20, 21] have been studied in the literature.

An attack scenario has been introduced in Ref. [20] to assess the revocability of the suggested approach in this study. The attack scenario is known as a revoked template attack, and it involves attacking a biometric system using a revoked template in two separate scenarios. In the first example, known as a Type-I attack, the revoked template is matched with the replaced new template, which is produced using the same fingerprint impression but different transformation keys/parameters. In the second situation, known as Type-II attack, a revoked template is matched with the replaced new template, which is constructed using another fingerprint impressions of the same finger but different transformation keys/parameters. The outcomes of these matching will now determine how effective a template protection strategy is in consideration of the revocability property. As a result, if the percentage of successful matches is close to zero, the secured templates are considered revocable, and the technique meets the revocability criteria overall.

A framework has been used in Ref. [21] to describe the revocable properties of the obtained secured templates of an iris biometric data. To do this, the mean and variance of genuine, imposter, and pseudo-imposter scores are calculated. The pseudo-imposter scores are calculated by matching the two secured templates that are constructed using the same impression but different transformation keys. According to Ref. [21], if the mean and variance values of pseudo-imposter scores are distinct from that of genuine scores and may be close to imposter scores, then the created templates are considered revocable, and the template protection technique adheres to the revocability property.

4.2.2 UNLINKABILITY

According to the property of unlinkability, the secured templates that are generated using a template protection technique for the same biometric impression and different keys/parameters of transformation should be unlikable to each other. In other words, there should be no similarity between the secured templates; thus, a template stored in the database can be easily replaced with a completely new and different template in the event of a database attack. To demonstrate that a template protection technique adheres to the unlinkability property, a few frameworks [22, 23] have been discussed in the literature.

The authors of Ref. [22] have introduced a new generalized framework to analyze template protection in the light of unlinkability. The framework computes two matching score distributions, namely, mated and non-mated matching score distributions. The mated score distribution is computed after matching the templates obtained from the same impression of a finger and different keys/parameters of the transformation function. Non-mated score distribution, on the other hand, is computed by matching the templates obtained from the different impressions of the different fingers and different keys/parameters of the transformation function. Now, if these two distributions overlap, as shown in Figure 4.5(a), it is safe to say that the template protection technique exhibits the unlinkability property and the templates are considered highly unlinkable.

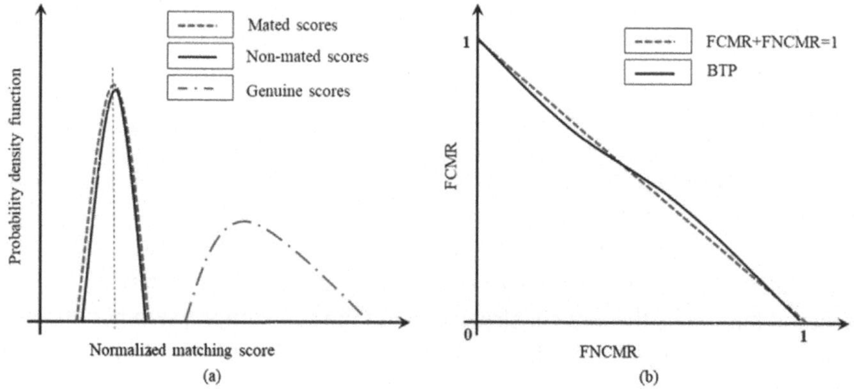

FIGURE 4.5 Representation of outcomes to assess the unlinkability of a biometric template protection (BTP) technique using two different frameworks [22, 23].

Another framework to analyze the unlinkability property of a given BTP technique has been discussed in Ref. [23]. By considering the given template protection technique, two different rates, i.e., false cross match rate (FCMR) and False Non-cross Match Rate (FNCMR) have been computed. In this case, FCMR denotes the rate of successful matching between non-mated samples (as discussed in Ref. [22]), whereas FNCMR denotes the rate of unsuccessful matching between mated samples. As shown in Figure 4.5(b), the values of FNCMR and FCMR for various threshold values are plotted against each other. According to the framework discussed in Ref. [23], if the plot of FNCMR-Vs-FCMR nearly overlaps with the plot of FCMR+FNCMR=1, then the given BTP technique adhere to unlinkability as depicted in Figure 4.5(b).

4.2.3 SECURITY

According to this criterion, the BTP technique must be capable of preventing data breaches and unauthorized access to the biometric system. To ensure the privacy of the original biometric data, the encrypted/transformed template must adhere to the non-invertibility property. This assures that an intruder cannot recover the original biometric data (fingerprint impression, face, iris, etc.) from a compromised modified template. Furthermore, an attacker might attempt to guess the original template from a compromised transformed template by attempting all conceivable permutations in order to get unauthorized access to the biometric system. This attack scenario is known as brute force attack. As a result, the transformed template should be constructed in such a way that an intruder will find it computationally impossible to predict the original biometric template.

Furthermore, if an intruder has prior knowledge of matching score calculation as well as the minimum necessary matching score to get access, then he/she may be able to get illegitimate access to the biometric system by applying the false accept attack. In this attack scenario, an intruder tries to guess the combination in such a way that the final matching score is close to the minimum necessary matching score, reducing the number of available combinations compared to a brute force attack. To avoid this attack scenario, a transformed template should be built in such a manner that guessing the original template is computationally impossible, even if the number of alternatives is lower than that in a brute force attack.

Now, assume an intruder obtains multiple templates that are calculated using the same biometric data; it is probable that the intruder may try to link these templates in order to get knowledge about the original biometric data, which may then be used to gain access to the biometric system. This attack scenario is also known as a record multiplicity attack (ARM) scenario. To avoid this attack, the created transformed templates should not have any links between themselves that may end up revealing information about the original biometric data. In addition to these attack scenarios, a few other attack scenarios, such as a birthday attack [24], a hill-climbing attack [25], and dictionary attack [26], have been described in the literature to assess the security level of a template protection technique.

4.2.4 PERFORMANCE

This is one of the most important requirements of a template protection technique. When an original biometric template is encrypted/transformed in order to protect the original biometric information, the overall recognition performance of a biometric system may suffer. However, encryption/transformation should be used without compromising the performance of a biometric system in order to develop a robust and effective template protection technique. There are a few measures which have been widely used in the literature to measure the performance of a biometric system. These performance measures are false rejection rate (FRR), false acceptance rate (FAR), genuine acceptance rate (GAR), equal error rate (EER), and accuracy.

1. **False Rejection Rate (FRR):** It is defined as the ratio of the falsely rejected genuine user and the total number of genuine comparisons. FRR is sometimes denoted as False Non-match Rate (FNMR).
2. **False Acceptance Rate (FAR):** It is defined as the ratio of the falsely accepted imposters and the total number of imposter comparisons. FAR is sometimes denoted as False Match Rate (FMR).
3. **Genuine Acceptance Rate (GAR):** It is defined as the ratio of the truly accepted genuine users and the total number of genuine comparisons. The relation between GAR and FRR is given as $GAR = 1 - FRR$.
4. **Equal Error Rate (EER):** It denotes the rate at which the value of FAR and FRR becomes equal as depicted in Figure 4.6. The value of EER should be always near to zero for a robust and effective biometric system.
5. **Accuracy:** The overall accuracy of correctly accepting the genuine and rejecting the imposter users is given as follows: $Accuracy = [1 - (FAR + FRR)/2] \times 100\%$

4.3 BIOMETRIC CRYPTOSYSTEMS

Biometric cryptosystems [17, 27, 28] provide a way to secure the template by means of combining the cryptographic key with biometric data or directly generating it from the biometric data during verification. In order to bind a key with biometric or retrieve it from biometric, a piece of public information has to be stored in the database, which is called helper data [13, 14]. Helper data is basically a piece of information, which is derived from the biometric data and doesn't disclose any fruitful information of the original biometric in the event of an attack on the database. Hence, the encrypted form of biometric is involved in the key-release process instead of the original biometric, which makes the biometric cryptosystem a secure BTP technique.

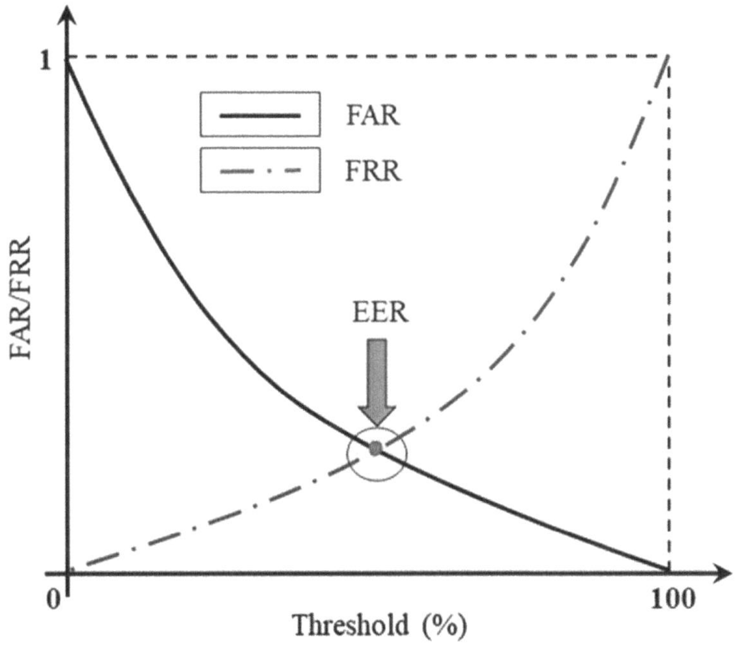

FIGURE 4.6 The threshold (%) vs. FAR/FRR plots where intersection of FAR and FRR represents the value of equal error rate (EER).

Biometric cryptosystems provide a replacement of conventional (password-based) key-release, where biometric data of an individual is utilized to release the key. This makes biometric cryptosystems more secure than the password-based key-release as biometric data is difficult to steal, forge, and share compared to passwords. In the conventional biometric systems, the decision of accepting or reject during the verification is taken based on the matching score, which is kind of a fuzzy decision depending on the threshold value. Whereas, In the biometric cryptosystems, the decision is always binary, which depends on the successful key-release or not. Further, biometric cryptosystems are mainly classified into two categories, i.e., key-binding schemes [16, 17, 27] and key-generation schemes [28]. In key-binding schemes, a key is combined with the original biometric of a user, and the constructed encrypted information is stored in the database, which is later used as a helper data to the key-release step during verification. In contrast, a key is generated directly from the biometric data itself instead of using a separate key in the key-generation schemes. These schemes are further discussed in the following sections.

4.3.1 KEY-BINDING SCHEMES

Key-binding schemes are one of the categories of biometric cryptosystems, which is defined in terms of the generation of helper data. In the key-binding scheme, the helper data is generated by combining a key with the original gallery biometric template, then stored into the database during the enrollment stage. The generated helper data would not reveal any information about the original biometric template if it is stolen through an attack. During the verification stage, the stored helper data and the probe biometric template is used to compute the key, and if the key is successfully released, then the probe template would be accepted; otherwise, it would be rejected. The working of key-binding scheme has been depicted in Figure 4.7. Further, the key-binding schemes can be mainly divided into three categories, which are Biometric Encryption™ [29], fuzzy commitment scheme [27], and fuzzy vault scheme [17]. These schemes have been discussed in the following sections, along with the different findings that are discussed in the literature.

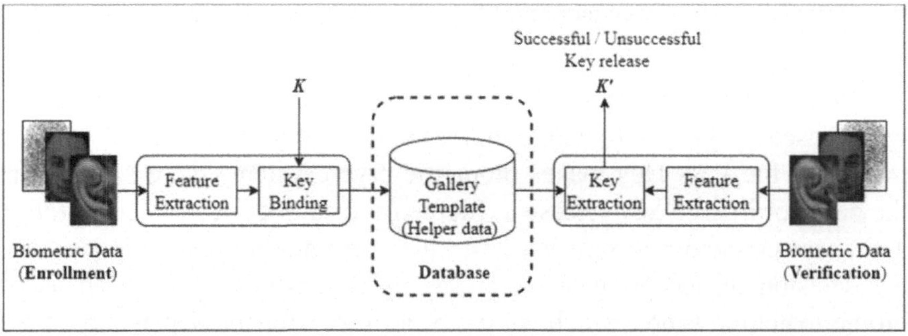

FIGURE 4.7 A schematic block diagram representing the working of the key-binding scheme.

4.3.1.1 BIOMETRIC ENCRYPTION™

Biometric Encryption™ [30, 31] is a procedure that has been developed to securely bind and regenerate a key from a biometric trait, e.g., fingerprint, by combining it with a piece of information, which is known as Bioscrypt™ (a filter function). Bioscrypt™ function is computed by utilizing an image processing approach and some additional information necessary to obtain and validate the key. Key is released employing a resultant pattern, which is generated by combining the biometric data with the Bioscrypt™ function. The first method that is based on this scheme is known as Mytec1 [32]. In this work, a public key cryptography system is implemented, which is controlled by fingerprint biometric. Mytec1 was the first algorithm, which was based on biometric cryptosystems. However, the overall performance and security of the proposed system were not good enough. The updated version of Mytec1 was Mytec2, which has been proposed by Soutar et al. [30, 31]. Biometric Encryption™ (Mytec1 and Mytec2) utilizes the correlation as a basis for implementation. A brief description of the working of both of these approaches has been discussed as follows.

At the enrollment stage, there are mainly three steps are involved, namely image processing, key linking, and identification code generation as discussed in Refs. [30, 31]. Filter function $H(u)$ is computed at the first step by utilizing the T training fingerprint samples that are represented as $\{f_0^1(x), f_0^2(x), \ldots, f_0^T(x)\}$. Here, the filter function $H(u)$ is a complex quantity, where $|H(u)|$ and $e^{i\phi}H(u)$ are representing the magnitude and phase of filter function. As filter function needs to be stored in the database, which is vulnerable to attack and possibly an attacker can regain the information of original fingerprint data $f(x)$ from $H(u)$. Hence, instead of storing $H(u)$ directly, $H_t(u)$ is stored in the database, which is computed by multiplying the phase component of $H(u)$ with a random array. Now, before linking a cryptographic key k_0, an output pattern $c_0(x)$ is computed using the following equation, $c_0(x) = F_0^t(u)H(u)$, where $F_0^t(u)$ is the Fourier transform of biometric data $f_0^t(x)$. Now, the output pattern $c_0(x)$ is linked with a cryptographic digital key k_0 to generate a lookup table by utilizing a linking algorithm. In the third step, the identification code is generated by computing the hash of an encrypted form of the key k_0 with the help of $H_t(u)$. Further, these three pieces of information, *viz.*, filter function $H_t(u)$, lookup table, and identification code id_0 are stored in the database, which is known as Bioscrypt and used during the verification stage.

At the verification stage, first an output pattern $c_1(x)$ is computed using the stored $H_t(u)$ and the query fingerprint samples that can be represented as

$c_1(x) = F_1(x)H_t(u)$. After computing the output pattern $c_1(x)$, the key retrieval algorithm is used to generate the key k_1 from the output pattern by means of a lookup table, which is stored in the database. Once the key k_1 is retrieved, the identification code id_1 is generated similar to the enrollment stage. Further, the stored identification code id_0 is compared with the id_1, and if $id_0=id_1$, then $k_0=k_1$ must be true and it shows the successful key release. In contrast, if $id_0 \neq id_1$, then $k_0 \neq k_1$ must be true, which is an unsuccessful key release. The block diagram depicting the complete enrollment and verification stages of Biometric Encryption™ is shown in Figure 4.8.

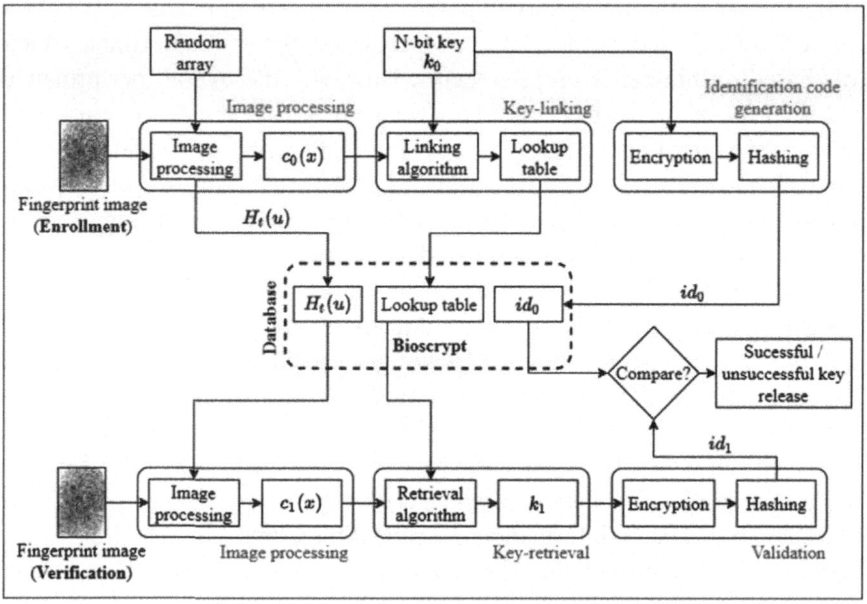

FIGURE 4.8 A schematic block diagram representing the working of biometric encryption™.

4.3.1.2 FUZZY COMMITMENT SCHEME

Fuzzy commitment scheme is one of the examples of biometric cryptosystems based on key-binding strategy. This scheme combines the error-correcting codeword with cryptography to secure the original biometric data of a user. It has been introduced by Juels and Wattenberg in Ref. [27]. In this work, authors have proposed the fuzzy commitment scheme, taking inspiration from the conventional bit commitment, and have shown the application for secure biometric authentication. In order to provide template security, the fuzzy commitment scheme utilizes the binary form of the user's biometric data. A brief description for working of fuzzy commitment scheme considering both enrollment and verification stage of a biometric system is given as follows.

In order to construct the fuzzy commitment scheme, commitment function F is computed in such a way that $F(c,x)=(h(c), \delta)$. Here, x is a n-bits binary string, which is basically the witness/biometric data. $c \in C$ is a codeword of length n-bits, which is generated using an error correction codeword. $h(c)$ represents the cryptographic hash (e.g., SHA-1) of codeword c, and δ represents the difference between witness x and c. Hence, the difference δ can be written as $\delta = x-c$. Now, the commitment $F(c,x)$ is stored in the database at the stage of enrollment. It is clear from the above discussion that just by knowing the $h(c)$ and δ it is infeasible to get back the original witness/biometric data x. Further, at the stage of verification, the value of a codeword c' is computed by taking the difference between a new query biometric data x' of length n-bits and the stored binary vector δ, which is given as $c' = x' - \delta$. Hence, if the query witness x' closely resembles enrolled x (or belong to the genuine subject), then the hash of codeword c,' i.e., ($h(c')$) would be equal to the stored hash value $h(c)$, which represents the successful verification otherwise the verification would be unsuccessful. A block diagram representing the working of fuzzy commitment scheme is shown in Figure 4.9.

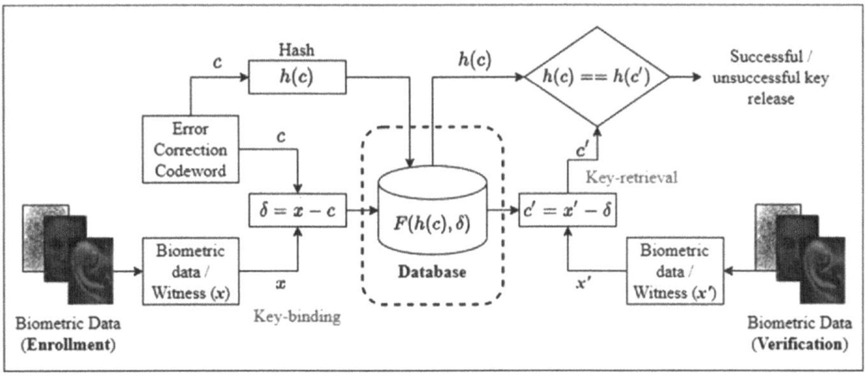

FIGURE 4.9 A schematic block diagram representing the general working of the fuzzy commitment scheme [27].

In the literature, so many techniques have been proposed to secure the original templates of different biometric traits, which are basically inspired from the general framework of fuzzy commitment scheme [27]. Further, in this section, we have included a brief description of as many as possible works related to the fuzzy commitment schemes by following the timeline of publications.

In Ref. [33], a fuzzy commitment scheme has been used to protect the iris biometric data. In this work, the authors have used a 2048-bits long iris code, which is combined (XOR) with a codeword that is generated from a two-layer error correction scheme. Two-layer error correction scheme utilizes Hadamard codeword at the outer layer and Reed-Solomon code at the inner layer. The outer layer corrects the error at the binary level, whereas the inner layer corrects the error at the block level. In this work, the helper data is stored in a token (smart card), which is used during the verification/identification stage. Zheng et al. [34] have utilized the concepts of fuzzy commitment scheme and k-nearest neighborhood classification to secure the iris biometric data. In this work, the biometric data is mapped from feature space to the lattice space by means of a codeword. After mapping, this lattice system is stored in the database, which does not reveal any information regarding the secret key and is used during the verification stage. Teoh & Kim [16] have proposed a new approach to generate the binary equivalent from the fingerprint features. This proposed approach is known as randomized dynamic quantization transformation (RDQT), which has used finger-code [35] as an input feature of a fingerprint biometric. RDQT consists of three steps, *viz.*, random projection, quantization, and condensation. First, features are randomly projected to a vector by using an under-determined linear system of equations, then quantization is performed to generate the binary equivalent of the feature vector. Condensation eliminates the perturbations in binary strings if they exist. Further, this binary vector is secured by using the fuzzy commitment scheme, where a concatenated codeword is used, *viz.*, Hadamard and Reed-Solomon codes. In Ref. [36], a theoretical analysis has been discussed for choosing the best suitable error correction codeword to secure the different types of biometric data by means of a fuzzy commitment scheme. In this work, the authors have discussed about the upper bound of the error-correcting capacity of a codeword along with the min-sum decoding algorithm for an error correction codeword. In the end, the authors have tested the theoretical findings on iris and fingerprint biometric database to establish the empirical proof of theoretical study. In Ref. [37], authors have proposed a signature template protection technique by using the fuzzy commitment scheme. In this work, BCH encoder and decoder have been used for the error correction purpose. In order to compute the binary equivalent of signatures, first, the feature vector has been extracted from the signature image, then converted into the binary form by using the inter-class and intra-class vector mean. Later, the reliable bits have only been selected from the binary vector, and the resultant binary vector is considered for computing the commitment. Further, the resultant binary vector is combined with the codeword, which is generated using a BCH encoder and a random key. During the verification, the key is retrieved by using the BCH decoder.

Ao & Li [38] have proposed a fuzzy commitment based cryptosystem to secure the user template that is constructed using infrared face images. In order to use an error correction codeword for infrared face images, authors have proposed an enhanced BioHash methodology by using an NXOR filter and then pass the resultant to the error correction codes as an input. BCH codes have been used to perform the key-binding and key-retrieval in the form of BCH encoder and decoder, respectively. In Ref. [39], a cryptosystem based on fuzzy commitment has been proposed to secure the fingerprint template. In order to accomplish it, first minutiae points are represented in the form of a delta function and then computed the Fourier spectrum corresponding to the delta function. Further, the Fourier spectrum is quantized using the polar logarithmic grid to compute the binary string. In this work, a method to extract reliable bits has also been proposed. This reliable binary vector is secured by using the fuzzy commitment scheme, which utilizes turbo codes as an error correction codeword. A new approach to calculate the binary vector from a fingerprint biometric has been discussed in Ref. [40]. In this work, minutiae triplet has been utilized to compute the binary vector. In the process of generating binary string from minutiae triplets, more than one fingerprint images are used for a single user. In addition, the generation of the binary vector is inspired by the concept of user-specific questions, which is discussed in Ref. [41]. Further, in order to secure the binary template using fuzzy commitment scheme, three different error correction codewords are used, *viz.*, BCH codes, Two-layer concatenated codes (BCH and Reed-Solomon codes), and LDPC codes. Yadigar et al. [42] have proposed a bio-cryptosystem by combining the texture-based features of fingerprint images. These features are namely finger code [35], local binary pattern (LBP) [43], and local directional pattern (LDP) [44]. All of these features provide a binary vector as output after using the reliability-based binarization method. Further, the binary vector has been secured by using the fuzzy commitment scheme with the BCH and LDPC codes as error correction codewords. In this work, the fusion of binary vectors generated from three different texture-based features has also been performed.

In Ref. [45], author has proposed a multi-modal (face and ear) template protection approach using fuzzy commitment scheme. In this work, the features have been extracted using principal component (PC) analysis from face and ear biometric images. These features are further fused and then converted into the binary vector using the non-invertible transformation. In the end, a fuzzy commitment scheme has been used to secure the binary vector by means of BCH codes. Wang et al. [46] have proposed a multi-biometric system by fusing both iris and face (visual and thermal) images. In order to secure the biometric data, a chaotic system and fuzzy commitment scheme have been used. The chaotic system is used to encrypt the feature vector of the iris (from both the eyes). Features of visual and thermal faces are fused in the form of a binary vector. The fusion between cross-modality is performed at the score level. In this work, BCH codes have been used as an error correction codeword to secure the biometric data. In Ref. [47], authors have proposed the feature level sequential fusion approach to compute the feature of the finger-vein biometric data. In this work, the optimality of the proposed fusion approach has been proved theoretically. Further, they have utilized a fuzzy commitment scheme to protect the finger-vein biometric template. BCH codes, shortened cyclic code, and LDPC codes have been utilized for the error correction purpose in the implemented biometric cryptosystem. A fuzzy commitment scheme based fingerprint cryptosystem is proposed by Bentahar et al. [48] to protect the information transmitted in the internet of things (IoT) applications. Discrete wavelet transform (DWT) is used to extract the binary vector from the fingerprint image. The length of the binary vector is 96-bits, which is generated after computing the DWT of 64×64 block of fingerprint image around the core point. For computing the commitment, a key of 48-bits length is randomly generated and converted into the 96-bits long codeword using Hamming codes, which is further used to combine with the binary biometric data. Due to this implementation, the transmission would take place through an encrypted channel between the smart connected things, owner, and remote server. In Ref. [49], the authors have proposed a technique to provide the security to multi-server-based E-healthcare systems by using biometrics. In order to protect the biometric data, a fuzzy commitment scheme is used, which provides the error correction capability as a slight intra-class variation exists in the biometrics. Also, the cancelable biometric templates are generated, which enables the revocable nature of biometric templates. Shi et al. [50] have proposed an approach to protect the fingerprint template, which is used in the secure transmission in cloud-based IoT systems. In order to provide privacy to user's fingerprint data, authors have used fuzzy commitment schemes and BCH codes as an error correction codeword. In order to extract the binary feature vector from a fingerprint image, sector encoding has been used around the singular (core) point of a fingerprint image.

4.3.1.3 FUZZY VAULT SCHEME

Fuzzy vault is one of the biometric cryptosystems based on the key-binding approach. It is also one of the popular cryptosystems to secure the original biometric template of a user. It is first introduced by Juels & Sudan [17]. In the fuzzy vault scheme, the biometric features, which can be in any order, are mapped to a polynomial, and these mapped points, along with a large number of chaff points, are stored in the database as a vault. The polynomial is generated by utilizing an error correction codeword and a cryptographic key. The query biometric features, which mostly overlap with the template biometric features, can only successfully unlock the vault; otherwise, it is infeasible to unlock the vault. Moreover, if the information stored in the vault is compromised, then the reconstruction of original biometric data is infeasible from the vault information. Hence, it ensures the privacy of user's biometric data. Further, according to Ref. [17], the working of fuzzy vault based biometric cryptosystem is discussed as follows.

At the enrollment stage, the polynomial p is generated by using the secret key k in such a way that the coefficients of the polynomial are derived from the key k, which can be represented as $p \leftarrow k$. Now, the biometric data A (can be unordered as well) is mapped to the polynomial p, i.e., $p(A)$. Further, chaff points will be added to hide the original points, and the number of chaff points is always greater than the number of points in A. Chaff points are the points in the two-dimensional space where the y-axis values should not lie on the polynomial p. So, collectively $\{A, p(A)\}$ and chaff points are stored into the database, which is called vault V_A. Here, the key idea is to generate a codeword for generic error correction codeword to hide the secret key k. The codeword is generated by utilizing the biometric data A. Thus, during the verification stage, if biometric data B sufficiently overlaps with A, then most of the genuine points can be filtered out from all the vault points V_A ($\{A, p(A)\}$ and chaff points). Now, the polynomial p' can be reconstructed from these filtered genuine points using the error correction codeword. Further, secret key k can be retrieved using the reconstructed polynomial p'; thus, the vault would be unlocked. In contrast, if set B does not largely overlap with set A (i.e., case of imposter), then the secret key k would not be released successfully, and verification/authentication would be unsuccessful. A block diagram representing the working of fuzzy vault scheme is shown in Figure 4.10. Some of the proposed work (regarding to fuzzy vault scheme) in the literature are discussed as follows.

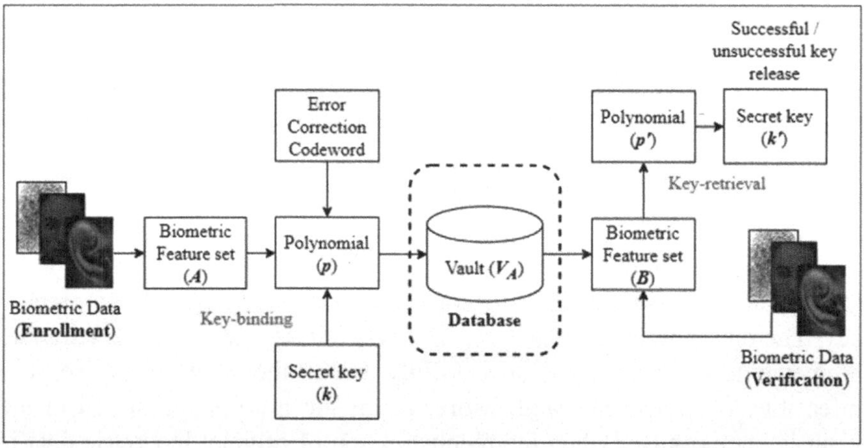

FIGURE 4.10 A schematic block diagram representing the general working of the fuzzy vault scheme [17].

In Ref. [51], a template protection technique based on fuzzy vault and cyclic redundancy check (CRC) has been proposed. In this technique, Lagrange's polynomial interpolation has been used to reconstruct the polynomial at the time of authentication. Nandkumar et al. have extended this work in Ref. [52] and have proposed a new approach to filter the minutiae point locations from the vault at the time of authentication, which they have called coarse-filtering. In this work, the locations of minutiae points and orientation values are utilized to map on the polynomial during the construction of the vault. In Ref. [53], Nandkumar et al. have addressed the issues with respect to a fuzzy vault scheme and developed the fuzzy vault by means of transformed minutiae information (by using a password), which makes the approach revocable in nature. They made use of CRC codes to check the error of extracted key during the authentication. Nyang & Lee [54] have presented a face template protection technique using the fuzzy vault scheme. They have used Reed-Solomon encoder and decoder for locking (enrollment) and unlocking (authentication) the weighted face features to a vault, respectively. In this work, an extra layer, which contains the genuine weight values of the enrolled face image, is incorporated to unlock the vault during authentication. In Ref. [55], the authors have developed an approach based on fuzzy vault to protect the iris template, which utilizes Reed-Solomon codeword for encoding and decoding purposes. In this approach, the clustering algorithm is employed on features of the different local regions on an iris image to compute the final feature vector (iris codes).

Zhou et al. [56] have presented a fingerprint fuzzy vault technique in which an extra user-specific password is used to randomly compute the artificial minutiae that are combined with the genuine minutiae to form the vault. In addition, the user-specific password has also been utilized to calculate the hash of the secret key, which is used to encode the polynomial. In Ref. [57], a classifier based on an artificial neural network (ANN) has been utilized to extract the features from an online signature biometric. The essential features have been selected by dividing the signature into time slices and further used to train the ANN using weighted backpropagation and AdaBoost algorithm. The final binary feature vector is secured by using the fuzzy vault scheme. Liu et al. [58] have presented a new multi-dimensional fuzzy vault scheme to protect the palmprint biometric. In this work, linear discriminant analysis (LDA) has been used to extract the features from a palmprint image. Before encoding of the fuzzy vault, extracted features have been projected to sub-vectors, which construct a linear space to eliminate the intra-subject variance. In Ref. [59], Yang et al. have proposed a technique based on bio-cryptosystem and the modified version of Voronoi neighborhood structure (VNS). The proposed technique is an alignment-free technique as it has utilized the VNS local structures of minutiae. In Ref. [60], the authors have presented an approach that transforms the pair-wise minutiae features using the user-specific password. Further, these features have been secured by utilizing the fuzzy vault scheme. In Ref. [61], a detailed security analysis of minutiae-based fuzzy vault has been given, and the minutiae locations centered on a hexagonal grid have been used to eliminate the chances of cross-matching attack. Li and Hu have proposed a fuzzy vault based template protection technique in Ref. [62] by utilizing the pair-polar structure of minutiae points. In this technique, the number of the fuzzy vault for a fingerprint impression is equal to the total number of minutiae because each minutia is considered as a reference point to compute the pair-polar minutiae structures.

Neu et al. [63], have presented an improvement for fuzzy vault scheme by means of minutia angles to eliminate the chances of correlation attack. In this work, authors have also addressed a new attack scenario, which is the angle correlation attack and is occurred due to the use of minutia angles. In Ref. [64], an approach based on fuzzy vault has been proposed for protecting the health monitoring data of a person, which is generated by a smart-wearable. In order to generate the secure key, an onboard accelerometer sensor and the unique behavioral biometric data, i.e., gait, have been utilized. Mai et al. [65] have discussed a new technique to protect the fused multi-modal biometric template based on biometric cryptosystem. Dependency reductive method for bit-grouping and discriminative within the group for a fusion of multi-modal biometrics have been utilized to construct the binary template, which is secured by a biometric cryptosystem. A secure tele-healthcare system has been proposed in Ref. [66], in which a fuzzy vault scheme is utilized to secure the time domain information of a physiological signal, i.e., Electrocardiogram (ECG) signal. In this technique, the time domain information of ECG data of a person is locked into a vault using a secret key and send to the receiver node, which unlocks the vault and uses the ECG data for the health monitoring purpose. In Ref. [67], Bobkowska et al. have discussed an approach for protecting the e-passport system by utilizing a fuzzy vault scheme. In order to do that, the authors have integrated the different biometric traits such as fingerprint, iris, and face biometric. In Ref. [68], the authors have presented a technique to provide a secure dynamic signature verification. In this technique, a total of 15 global features of signature biometric data are considered for computing the fixed-length signature template, and Reed-Solomon encoder and decoder have been used for the error-tolerance purpose.

4.3.2 KEY-GENERATION SCHEMES

Key-generation schemes are also part of the umbrella term biometric cryptosystems, which can be defined by way of generating the helper data and key. In the key-generation scheme, a separate secret key is not required as the key is directly generated from the biometric data itself or the helper data (which is also derived using the biometric data). The helper data does not need to be always stored in the database as some of the approaches do not use any kind of helper data to generate the key during authentication. However, if any attack happens then, there is no chance of revocability in such approaches. In addition, other approaches store the helper data in the database to generate the key during verification/identification. The generic working of the key-generation scheme has been depicted in Figure 4.11. Key-generation schemes have the advantage of generating keys directly from the biometric data; however, the performance of biometric system suffers due to the less discriminability. Nevertheless, this issue can be addressed by analyzing the key stability and key entropy of any proposed approach based on the key-generation scheme. Quantization schemes, fuzzy extractors, and secure sketches are examples of the key generation schemes, which are discussed in the literature.

Quantization is a procedure of converting a large set of input values to a smaller number of values, which contains concrete information about the input data. Similarly, quantization scheme [69] has been used to generate a key directly from biometric data. In the quantization scheme, multiple biometric samples of a subject are used to find out the intervals of each feature parameter of input biometric data. Further, this information is stored in the database as helper data to compute the key during the verification stage. At the time of verification, the key is released directly from the query biometric data by using the interval information stored in the database. Fuzzy extractor and secure sketch are also examples of the key-generation schemes. These techniques have been introduced by Dodis et al. [70]. The fuzzy extractor extracts the random secret key from an input biometric data. It is an error-tolerant primitive, which means that the same random key would be generated even if an input changes a little bit (different samples of the same subject). The secure sketch generates a public information P_{info} from an input biometric data I_1 and P_{info} does not reveal any idea about the input I_1; however, at the time of verification, I_1 can be retrieved from P_{info} if another input I_2 is available, which largely overlaps with the input I_1. Further, the brief descriptions of a few proposed approaches related to the key generation scheme are discussed as follows.

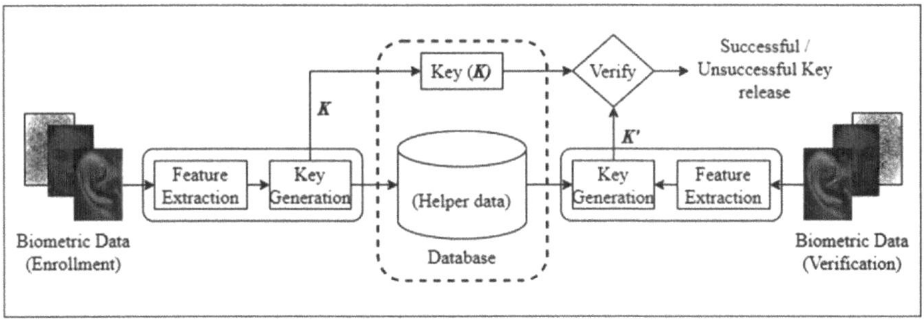

FIGURE 4.11 A schematic block diagram representing the working of the key-generation scheme.

In Ref. [69], a template protection approach based on quantization scheme has been proposed for online signature biometric system. In this work, an interval matrix is computed from more than one sample of online signature samples, and this matrix is used to compute the key/hash vector at the time of authentication. Hence, there is no need to store the original signature template in the database, which improves the security of the overall biometric system.

Chang et al. [71] have used the LDA to compute the transformed features of a genuine user's face images, which can easily distinguish them from the imposter users. These features are further used to generate the stable cryptographic key. Dodis et al. [70] have introduced the concepts of the fuzzy extractor and secure sketch primitives. These both the techniques are proposed to protect the any kind of biometric and noisy data. In this work, authors have also shown the closely optimal constructions of these primitive by means of Hamming distance, edit distance, and set difference closeness measures. In Ref. [72], two approaches are proposed to show the application of biometric data for mutual authentication or key exchange over an insecure channel. These approaches are inspired by the fuzzy extractor and secure sketch [70]. Sutcu et al. [73] have discussed about the issues related to the practical implementation of secure sketch's theory. A generalized framework has also been proposed in this work to implement and analyze the secure sketch for face biometric data. In Refs. [74, 75], authors have shown the implementation of the fuzzy extractor and secure sketch, respectively, for fingerprint biometric data. Whereas, Sutcu et al. [76] have shown the implementation of the secure sketch for the multi-model biometric system, which utilizes face and fingerprint biometrics. In Ref. [77], the definition of security and correctness for the fuzzy extractor has been extended. In this work, a generic construction of fuzzy extractor has been proposed, which works for both discrete and continuous noisy data. Tian et al. discussed about the solution for leakage-attack in remote authentication by using the fuzzy extractor in Ref. [78]. The generic framework that is proposed in this work allows a user to securely authenticate to the remote server by using his/her biometric data.

4.4 CANCELABLE BIOMETRICS

Cancelable biometrics is one of the extensively used template protection techniques. In cancelable biometrics, transformed biometric features are stored in the database instead of original features in order to cater the privacy of the user's biometric data. The transformation of biometric data is performed in such a way that if an attacker steals the transformed biometric features (stored in the database), then it is not feasible for an attacker to reconstruct the original biometric template from the transformed one. As the transformed features are stored in the database, the matching is also performed in the transformed domain during the verification/identification. A schematic diagram representing the generic working of cancelable biometrics is shown in Figure 4.12. In Ref. [7], different types of attacks on a biometric system have been discussed as depicted in Figure 4.3 and also the idea of cancelable biometrics have been introduced. Further, cancelable biometrics can be divided into two major categories, which are salting and non-invertible transformation [13, 14]. In the salting based techniques, the biometric features are transformed by means of a transformation function and the user-defined keys/parameters. These keys or parameters of the transformation function need to be kept securely by a user due to the invertible behavior of transformation for a large extent in the salting techniques. A person needs to provide the keys during the authentication stage. In contrast, the template protection techniques based on non-invertible transformation utilize a transformation function, which is non-invertible in nature. Hence, even if the transformation parameters/keys are compromised by an attacker, it is infeasible to reconstruct the original biometric feature from a transformed template. Further, we have incorporated a brief description of as many techniques as possible, which are discussed in the literature and related to the salting and non-invertible cancelable biometrics.

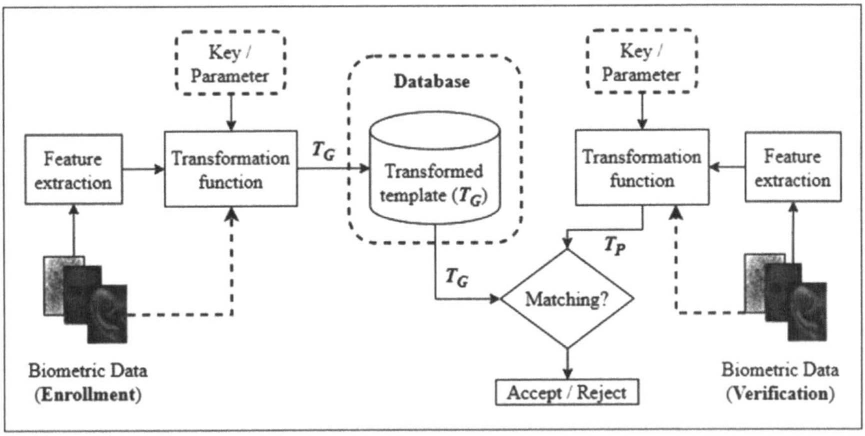

FIGURE 4.12 A schematic block diagram representing the generic working of the cancelable biometrics.

4.4.1 SALTING TECHNIQUES

In the literature, there are a smaller number of salting techniques have been proposed compared to the non-invertible transformation based techniques. A brief overview of some of the proposed techniques is as follows. Connie et al. have proposed a technique to compute the revocable template by utilizing the biometric salting in Ref. [79]. To compute the secured and revocable palmprint template, authors have used fisher discriminant analysis (FDA) for feature extraction, and binary discretization is used to compute the hash code, which is called palm hash. These palm hash codes are stored in the smart card or chip for verification purposes. Teoh et al. [80] have proposed a new technique to secure the fingerprint template, and they have called their formulation as BioPhasor. BioPhasor is a binary code, which is computed by combining the pseudo-random numbers (user-specific) with fingerprint features in an iterative manner. Further, in Ref. [81], Kuan et al. have utilized the BioPhasor and 2^N discretization to protect the hand signature template. They have also proposed a method to compute the compressed form of a dynamic hand signature by means of the DWT and discrete Fourier transform (DFT). In Ref. [82], a face template protection technique has been proposed by using biometric hashing, which provides non-invertibility, revocability, and effective performance. In order to do so, they make use of FDA for feature extraction and random multispace quantization (RMQ) to perform the mapping of extracted features.

Chin et al. [83] have proposed an iris encoding called S-Iris encoding to secure the original iris template. To perform the S-Iris encoding, authors have used inner-product between a user-specific pseudo-random number and iris features in an iterative manner. Thresholding has been used to exclude the weaker inner product, which eventually generates the stable and secured S-Iris binary codes. In Ref. [84], authors addressed the stolen-token attack issue of BioHash by considering a biometric system based on face biometric data. In this work, a quantized random projection ensemble based on the Johnson–Lindenstrauss lemma has been proposed to show the stability of BioHash mathematically. Ouda et al. [85] have proposed a technique based on single-factor authentication, and the generated secured iris codes are called BioCodes. According to the authors, this approach provides an edge on BioHashing as it does not utilize any kind of user-specific tokens to generate the secured codes. In this approach, results have been compared with the BioHashing considering the stolen-token scenario. In Ref. [86], a new approach which is called minutia vicinity decomposition (MVD), has been proposed. In this technique, the template has been computed using salting on geometrically invariant features, which are extracted from minutiae triplets. In this technique, the decomposition of minutia vicinity into minutiae triplets has been performed, and geometrically invariant features are extracted using these triplets. The template has been generated using these features and salting (random offset).

4.4.2 NON-INVERTIBLE TRANSFORMATION-BASED TECHNIQUES

In the non-invertible transformation, the biometric template is highly secure as it is infeasible to reconstruct the original biometric data from a transformed template even if an intruder acquires the transformation keys/parameters, unlike salting techniques. In the literature, several techniques based on non-invertible transformation have been discussed to protect the biometric template of an individual. We have tried to include a concise summary of some of the works as follows. In Ref. [18], different approaches to generate a cancelable fingerprint template have been proposed. In this technique, the template has been constructed using three different ways, which include the transformation of the locations of minutiae points in Cartesian and polar space, and surface folding is performed in the third one. In this technique, authors have computed the transformation of minutiae points using three different ways, which are Cartesian, polar, and functional transformations. In the case of Cartesian, minutiae points' locations are transformed in the Cartesian space, whereas in polar, the transformation has been performed on the polar coordinates of minutiae points. In the case of functional, surface folding has been used to transform the fingerprint features. An alignment-free template protection technique based on the patches of minutiae points has been proposed in Ref. [87]. The technique has generated the signature from patches of minutiae points, and these signatures are transformed by using a key. In Ref. [88], a fingerprint template protection technique has been proposed based on non-invertible transformation in which pairs of minutiae points have been projected on a circle around them, and by using bin-based quantization, a user template is generated. Lee et al. have proposed an alignment-free technique to protect a fingerprint based user template in Ref. [89]. In this work, a binary string has been computed by mapping the transformed locations of minutiae points on a 3D array. Further, this is used as a secure template by permuted using user keys. In this work, a binary string has been computed by using the location of minutiae points which are calculated after rotating the minutiae points with respect to a reference minutia point. These locations are mapped onto a three-dimensional array to generate the binary string, and the permutation of this binary string using a user key is used as a secured user template. In Ref. [90], Ahmed et al. have discussed a technique, which is based on pair-polar coordinates of minutiae. In this work, many-to-one mapping is used to map the pair-polar coordinates of minutiae into the different sectors around the reference minutia point, which is used to compute the pair-polar coordinate. Similarly, in Ref. [91], authors have made use of an infinite-to-one mapping scheme to secure the biometric template. Also, the proposed template protection technique is alignment-free in this work. The infinite-to-one mapping technique has been called densely infinite-to-one mapping (DITOM) in Ref. [91].

In Ref. [92], Boult et al. have proposed a new technique in which they have used public-key encryption for both invertible and non-invertible transformations. In this technique, the transformed user template is called bio-token. These bio-token is further stored in the database as a secured user template. Symmetric polynomials of different degrees have been used in Ref. [93] to compute the hash of minutiae points of a fingerprint image which is used as a user template. The intra-subject variations present in the fingerprint images make a significant difference between the hash values of fingerprint impressions of the same subject and hence poses a challenge in maintaining a stable performance in this technique. Ahn et al. [94] have proposed a non-invertible transformation based template protection technique that is alignment-free and utilizes the geometrical information of minutiae triplet to generate a secure user template. Kumar et al. [95] have proposed a technique to compute the secure template as an aggregation of symmetric hash functions calculated for multiple *k*-plets of minutiae points of a fingerprint image. Though this technique is capable of protecting a fingerprint based user template from different attacks, the performance of the system is seen to be reduced in this technique. Cappelli et al. [96] have introduced a new alignment-free minutiae representation to compute a cancelable fingerprint template, which is called minutia cylinder-code (MCC). MCC is basically computed with respect to the spatial and directional contribution of minutiae points around a reference minutia point in a fingerprint image. Although the performance is significantly better compared to other techniques, it suffers from the invertibility issue. Further, Ferrara et al. [97] have extended the technique, which was proposed in MCC [96]. In this work [97], the generated templates are completely secure and non-invertible in nature. Although this technique has depicted better performance, it lacks in terms of revocability property. Ferrara et al. have proposed another technique based on MCC [96] in Ref. [20], which is a two-factor scheme to protect the template and also eliminate the major problems of MCC [96, 97] such as invertibility and non-revocability.

In Ref. [98], 3-tuple quantization based approach has been proposed to generate a bit string from the set of minutiae points. In this technique, a set of minutiae points have been computed by means of dividing the overall fingerprint image in the form of a polar grid. In Ref. [99], multi-line code (MLC) has been utilized to compute the secure fingerprint template. The accuracy of the system is reduced due to the binary code in this technique, and in order to overcome it, a new similarity measure has also been proposed in Ref. [99]. In Ref. [100], authors have incorporated the randomized Hamming Embedding, which makes use of graphs to generate the secured fingerprint template. In order to compute the fingerprint features, authors have adapted MVD along with the randomized projection approach. In Ref. [101], authors have discussed a fingerprint template protection technique by using the *k*-nearest neighborhood structure (*k*-NNS) of minutiae. To compute the cancelable template, *k*-NNS structures are computed by considering each and every minutia as a reference point and are mapped on a 2D grid. This information is further used to form a binary string. These binary strings are then converted into the final cancelable template using DFT and non-invertible transformation. In Ref. [102], in order to compute the secured fingerprint template, features are generated using Delaunay triangulation composed by minutiae points. These Delaunay triangulation based features are further transformed by using the non-invertible transformation in order to compute the final cancelable template.

In Ref. [103], Jin et al. have used a kernelized PCA based approach to construct a fixed-length structure utilizing the minutiae points in a fingerprint image and have used multiple samples of a subject at the time of enrollment. In Ref. [104], in order to compute the cancelable fingerprint template, a blind system identification technique has been discussed. In this work, the authors have made use of a quantized minutiae pair vector in order to build the blind system identification. Sandhya et al. [105], have proposed a technique to construct the non-invertible fingerprint template. To compute the non-invertible template, authors have utilized the different set of features from the Delaunay triangulation structure of minutiae points. These sets of features are mapped into a 3D grid to construct a binary vector, which is eventually used as a template after computing the transformation. In Ref. [106], Khodadoust & Khodadoust have implemented indexing of fingerprints by utilizing extended Delaunay triangulation along with the k-means clustering algorithm for retrieval of fingerprints. According to this work, the proposed indexing approach can be encouraging to improve the efficiency of automatic fingerprint identification systems (AFISs). Sandhya & Prasad [107] have utilized fused feature structures to secure the fingerprint template. In this work, the authors have computed two types of features from minutiae points, namely local and distant features. These features are further converted into bit-strings, and the fusion of bit-strings is used to compute the cancelable template.

In Ref. [108], Wang et al. have used partial Hadamard transform to obtain a secure use template. In their technique, the transformation has been performed on the DFT of binary string computed from the fingerprint features to construct the cancelable user template. A new approach has been proposed in Ref. [109], which is called dense registration of a fingerprint, to eliminate the intra-subject variations like distortions that arise due to the elasticity of the skin. To design a secure partial fingerprint authentication system, Lee et al. [110] have proposed a technique based on minutiae points along with a new feature, called ridge shape feature. Wang et al. [111] have utilized the local minutiae structure by extraction of zoned minutiae pairs and have performed non-invertible transform based on DFT to construct a cancelable user template. Moujahdi et al. [112] have introduced a fingerprint template protection technique, which is known as Fingerprint Shell. In this work, the authors have utilized the distances of minutiae points from a singular point to compute the spiral curve, and Hausdorff distance has been employed to perform the matching between spiral curves. Jain & Prasad [113] have further used fingerprint shell along with clustering technique to index fingerprints. There is a major limitation of the template constructed using fingerprint shell [112]. In the case of an attack on the database, the distances which are used to compute the user template (spiral curve) can be easily computed from the spiral curve. To eliminate this shortcoming of fingerprint shell, Ali & Prakash have proposed an enhanced form of fingerprint shell in Refs. [114, 115], which is further improved in Refs. [116, 117]. In Ref. [118], a translation/rotation independent technique has been proposed. In this technique, locations of minutiae points are transformed by using a non-invertible transformation and user-specific key/parameters to secure the original fingerprint template. This technique has been further improved in Ref. [119]. In Ref. [120], Ali et al. have proposed a framework based on polynomial curves, which is known as Polynomial Vault. In this work, the polynomial curve is generated using the transformed distances of minutiae points from a singular point. In Ref. [121], a Delaunay triangulation based template protection approach has been discussed. This approach utilizes the internal angles of Delaunay triangles and orientation values of minutiae to construct the feature vector, which is transformed by means of user-specific keys.

4.5 HYBRID AND HOMOMORPHIC ENCRYPTION-BASED TECHNIQUES

In the literature, a few more hybrid and Homomorphic encryption-based techniques have been proposed other than the biometric cryptosystem and cancelable biometrics. Hybrid techniques basically combine the functionality of different template protection techniques such as fuzzy vault and fuzzy commitment or cancelable and biometric cryptosystems. By using these kinds of combinations, the overall security of the biometric system increases significantly along with the higher recognition rate. Nagar et al. [122, 123] have developed a hybrid technique to secure the fingerprint template. In this work, a fuzzy commitment scheme is utilized to encrypt the polynomial, which is evaluated using a fuzzy vault scheme. They have used a minutiae descriptor, which constitutes orientation values and ridge count around a minutia's neighborhood. This combination of two different biometric cryptosystems improves the security significantly. In Ref. [124], a hybrid technique has been presented to secure the palmprint biometric. In this work, 2D PalmHash code, i.e., a cancelable template, is used to secure the palmprint. Further, this template is stored in the database by using the fuzzy vault scheme. In the recent work, Ouda et al. [125] have presented a hybrid approach, which is called cancelable biometric vault. This work combines the functionality of cancelable biometrics in order to compute the secure features and encode them by means of cryptographic key.

Homomorphic encryption is an encryption scheme, which allows performing the computation on encrypted data. This encryption scheme is adapted in some of the work for encrypting the biometric data to protect the original template. The concept of Homomorphic encryption resembles with the cancelable biometric as the matching is performed in the encrypted domain in the case of Homomorphic encryption. Similarly, the transformed domain is used in the latter category. Some of the work related to the application of Homomorphic encryption for biometric security are discussed as follows. In Ref. [126], a protocol has been proposed to provide secure authentication using biometrics (hand geometry). The biometric data is encrypted by means of the asymmetric encryption approach. The generic SVM is used for verification in a client-server based architecture. Marta et al. [127] have proposed a multi-modal biometric system (for fingerprint and online signature) based on a Homomorphic probabilistic encryption approach. In this work, the proposed encryption based technique has been analyzed for three different types of fusions, *viz.*, feature-level fusion, score-level fusion, and decision-level fusion. In Ref. [128], a fully Homomorphic encryption based technique has been proposed to protect the iris biometric data. In this work, hamming distance measure has been used during the verification stage.

4.6 CONCLUSION

Biometrics provides a way to authenticate a person by using the physiological and behavioral characteristics. It has numerous advantages over traditional authentication systems such as password/token-based authentication. However, if the original biometric characteristics of an individual are stolen by an adversary, then it would be a permanent loss of biometric characteristics or an identity of a person due to the immutable characteristics of biometric data, and it can also cause illegitimate access to the biometric system. Moreover, it has been found in the literature that the biometric data can be reconstructed from the stored template, and the biometric data cannot be easily replaced like passwords or PINs in the event of an attack on the database. Therefore, in order to protect the original biometric data of an individual, BTP techniques have been developed. These techniques essentially suggest storing the helper data (computed after cryptographic key-binding or generation) or non-invertible transformed template in the database instead of storing the original template without compromising the recognition performance of the biometric system. These implementations enable the non-invertibility, revocability, unlinkability (diversity), and immune the biometric system against so many attacks (e.g., brute force attack, hill-climbing attack, false accept attack, dictionary attack, etc.). In this chapter, we have tried to include a brief discussion about the biometric template security along with the concise review of as many as possible techniques, which are published in the literature to secure the original template of different biometric traits (e.g., face, fingerprint, iris, signature, ear, etc.). After reviewing the different approaches, which are discussed in the literature, some of the future research directions are listed as follows:

- The work can be done to build more multi-modal biometric systems, which are more secure owing to the fusion of characteristics from several biometric data.
- Behavioral biometric characteristics have received less attention in the literature than physiological biometric characteristics. As a result, more biometric systems based on behavioral biometric features may be created in the future because a person's behavioral characteristics are likewise unique and hard to fake.
- In addition, the use of behavioral biometrics traits in conjunction with physiological features for active authentication on mobile devices has been discussed in the literature; hence, they can be further investigated.

- There is less work has been published related to hybrid and Homomorphic encryption; thus, in the future, more approaches can be developed based on hybrid and Homomorphic encryption.
- Benchmark databases have been used in the majority of template protection techniques; nevertheless, the size of these databases is still insufficient to verify the effectiveness of any suggested solution. As a result, large benchmark datasets may be created.
- Biometrics are now being utilized to offer secure authentication in IoT-based systems, e-healthcare systems, and other applications. However, a few approaches for securing the biometric templates used in such systems have only been presented. As a result, in the future, new approaches can be addressed.

KEYWORDS

- **biometric authentication system**
- **biometric cryptosystems**
- **biometric template security**
- **biometrics**
- **cancelable biometrics**
- **homomorphic encryption**

REFERENCES

1. *Fingerprint Verification Competition–2002 Database*. http://Bias.Csr.Unibo.It/Fvc2002/Default.Asp (accessed on 28 September 2023).
2. *CASIA Iris Image Database Version 1.0*. http://English.Ia.Cas.Cn/Db/201610/T20161026_169399.Html (accessed on 28 September 2023).
3. *FEI Face Database*. https://Fei.Edu.Br/~cet/Facedatabase.Html (accessed on 28 September 2023).
4. Kumar, A., & Wu, C., (2012). Automated human identification using ear imaging. *Pattern Recognition.*, *45*(3), 956–968.
5. Jain, A. K., Ross, A., & Prabhakar, S., (2004). An introduction to biometric recognition. *IEEE Trans. Circuits Syst. Video Technol.*, *14*(1), 4–20.
6. Jain, A. K., Ross, A. A., & Nandakumar, K., (2011). *Introduction to Biometrics*. Springer Publishing Company, Incorporated.

7. Ratha, N. K., Connell, J. H., & Bolle, R. M., (2001). Enhancing security and privacy in biometrics-based authentication systems. *IBM Syst. J.*, *40*(3), 614–634.
8. Ross, A. A., Shah, J., & Jain, A. K., (2005). Toward reconstructing fingerprints from minutiae points. In: *Proc. of SPIE Conf. on Biometric Technology for Human Identification* (pp. 68–80).
9. Cappelli, R., Maio, D., Lumini, A., & Maltoni, D., (2007). Fingerprint image reconstruction from standard templates. *IEEE Trans. Pattern Anal. Mach. Intell.*, *29*(9), 1489–1503.
10. Chen, F., Zhou, J., & Yang, C., (2009). Reconstructing orientation field from fingerprint minutiae to improve minutiae-matching accuracy. *IEEE Trans. Image Process.*, *18*(7), 1665–1670.
11. Feng, J., & Jain, A. K., (2011). Fingerprint reconstruction: From minutiae to phase. *IEEE Trans. Pattern Anal. Mach. Intell.*, *33*(2), 209–223.
12. Ross, A., Shah, J., & Jain, K. A., (2007). From template to image: Reconstructing fingerprints from minutiae points. *IEEE Trans. Pattern Anal. Mach. Intell.*, *29*(4), 544–560.
13. Jain, A. K., Nandakumar, K., & Nagar, A., (2008). Biometric template security. *EURASIP J. Adv. Signal Process, 2008.*
14. Uhl, R., & Andreas, C., (2011). A survey on biometric cryptosystems and cancelable biometrics. *EURASIP J. Inf. Secur.*, *3.*
15. Sandhya, M., & Prasad, M. V. N. K., (2017). Biometric template protection: A systematic literature review of approaches and modalities. In: Jiang, R., Al-maadeed, S., Bouridane, A., Crookes, P. D., & Beghdadi, A., (eds.), *Biometric Security and Privacy: Opportunities & Challenges in The Big Data Era* (pp. 323–370). Springer International Publishing: Cham.
16. Teoh, A. B. J., & Kim, J., (2007). Secure biometric template protection in fuzzy commitment scheme. *IEICE Electron. Express*, *4*(23), 724–730.
17. Juels, A., & Sudan, M., (2002). A fuzzy vault scheme. In: *Proc. IEEE International Symposium on Information Theory* (p. 408).
18. Ratha, N. K., Chikkerur, S., Connell, J. H., & Bolle, R. M., (2007). Generating cancelable fingerprint templates. *IEEE Trans. Pattern Anal. Mach. Intell.*, *29*(4), 561–572.
19. Maltoni, D., Maio, D., Jain, A. K., & Prabhakar, S., (2009). *Handbook of Fingerprint Recognition* (2nd edn.). Springer Publishing Company, Incorporated.
20. Ferrara, M., Maltoni, D., & Cappelli, R., (2014). A two-factor protection scheme for MCC fingerprint templates. In: *Proc. of BIOSIG 2014* (pp. 1–8).
21. Dwivedi, R., & Dey, S., (2018). Score-level fusion for cancelable multi-biometric verification. *Patt. Recogn. Lett.*
22. Gomez-Barrero, M., Galbally, J., Rathgeb, C., & Busch, C., (2018). General framework to evaluate unlinkability in biometric template protection systems. *IEEE Trans. Inf. Forensics Secur.*, *13*(6), 1406–1420.
23. Simoens, K., Yang, B., Zhou, X., Beato, F., Busch, C., Newton, E. M., & Preneel, B., (2012). Criteria towards metrics for benchmarking template protection algorithms. In: *2012 5th IAPR International Conference on Biometrics (ICB)* (pp. 498–505).
24. Jin, Z., Hwang, J. Y., Lai, Y. L., Kim, S., & Teoh, A. B. J., (2018). Ranking-based locality sensitive hashing-enabled cancelable biometrics: Index-of-max hashing. *IEEE Trans. Inf. Forensics Secur.*, *13*(2), 393–407.
25. Martinez-Diaz, M., Fierrez-Aguilar, J., Alonso-Fernandez, F., Ortega-Garcia, J., & Siguenza, J. A., (2006). Hill-climbing and brute-force attacks on biometric systems: A case study in match-on-card fingerprint verification. In: *Proceedings 40th Annual 2006 International Carnahan Conference on Security Technology* (pp. 151–159).

26. Shin, S. W., Lee, M. K., Moon, D., & Moon, K., (2009). Dictionary attack on functional transform-based cancelable fingerprint templates. *ETRI J.*, *31*(5), 628–630.
27. Juels, A., & Wattenberg, M., (1999). A fuzzy commitment scheme. In: *Proc. Computer and Communications Security, Singapore* (pp. 28–36).
28. Qiming, L., Muchuan, G., & Ee-Chien, C., (2008). Fuzzy extractors for asymmetric biometric representations. In: *Proc. IEEE Computer Society Conference on Computer Vision and Pattern Recognition Workshops* (pp. 1–6).
29. Soutar, C., Roberge, D., Stoianov, A., Gilroy, R., & Kumar, B. V., (1999). *Biometric Encryption*. ICSA Guide to Cryptography.
30. Soutar, C., Roberge, D., Stoianov, A., Gilroy, R., & Kumar, B. V., (1998). Biometric encryption: Enrollment and verification procedures. In: *Aerospace/Defense Sensing and Controls* (pp. 24–35).
31. Soutar, C., Roberge, D., Stoianov, A., Gilroy, R., & Kumar, B. V., (1998). Biometric encryption using image processing. In: *Photonics West'98 Electronic Imaging* (pp. 178–188).
32. Soutar, C., Tomko, G. J., & Schmidt, G. J., (1996). *Fingerprint Controlled Public Key Cryptographic System*. US Pat., 5541994.
33. Hao, F., Anderson, R., & Daugman, J., (2006). Combining crypto with biometrics effectively. *IEEE Trans. Comput.*, *55*(9), 1081–1088.
34. Li, W., Zhan, C., & Zheng, G., (2006). Cryptographic key generation from biometric data using lattice mapping. In: *Pattern Recognition, International Conference* (Vol. 4, pp. 513–516). IEEE Computer Society: Los Alamitos, CA, USA.
35. Jain, A. K., Prabhakar, S., Hong, L., & Pankanti, S., (2000). Filter bank-based fingerprint matching. *IEEE Trans. Image Process.*, *9*(5), 846–859. https://doi.org/10.1109/83.841531.
36. Bringer, J., Chabanne, H., Cohen, G., Kindarji, B., & Zemor, G., (2008). Theoretical and practical boundaries of binary secure sketches. *IEEE Trans. Inf. Forensics Secur.*, *3*(4), 673–683.
37. Maiorana, E., Campisi, P., & Neri, A., (2008). User adaptive fuzzy commitment for signature template protection and renewability. *J. Electron. Imaging*, *17*(1), 1–12.
38. Ao, M., & Li, S. Z., (2009). Near infrared face based biometric key binding. In: Tistarelli, M., & Nixon, M. S., (eds.), *Advances in Biometrics* (pp. 376–385). Springer Berlin Heidelberg: Berlin, Heidelberg.
39. Nandakumar, K., (2010). A fingerprint cryptosystem based on minutiae phase spectrum. In: *Proc. IEEE International Workshop on Information Forensics and Security* (pp. 1–6).
40. Li, P., Yang, X., Qiao, H., Cao, K., Liu, E., & Tian, J., (2012). An effective biometric cryptosystem combining fingerprints with error correction codes. *Expert Syst. Appl.*, *39*(7), 6562–6574.
41. Sutcu, Y., Rane, S., Yedidia, J. S., Draper, S. C., & Vetro, A., (2008). Feature transformation of biometric templates for secure biometric systems based on error correcting codes. In: *2008 IEEE Computer Society Conference on Computer Vision and Pattern Recognition Workshops* (pp. 1–6).
42. Imamverdiyev, Y., Teoh, A. B. J., & Kim, J., (2013). Biometric cryptosystem based on discretized fingerprint texture descriptors. *Expert Syst. Appl.*, *40*(5), 1888–1901.
43. Ahonen, T., Hadid, A., & Pietikainen, M., (2006). Face description with local binary patterns: Application to face recognition. *IEEE Trans. Pattern Anal. Mach. Intell.*, *28*(12), 2037–2041.
44. Jabid, T., Kabir, M. H., & Chae, O., (2010). Local directional pattern (LDP) – a robust image descriptor for object recognition. In: *Proc. 7th IEEE International Conference on Advanced Video and Signal Based Surveillance* (pp. 482–487).

45. Yuan, L., (2014). multimodal cryptosystem based on fuzzy commitment. In: *2014 IEEE 17th International Conference on Computational Science and Engineering* (pp. 1545–1549).
46. Wang, N., Li, Q., Abd El-Latif, A. A., Peng, J., Yan, X., & Niu, X., (2015). A novel template protection scheme for multibiometrics based on fuzzy commitment and chaotic system. *Signal, Image Video Process.*, *9*, 99–109.
47. Murakami, T., Ohki, T., & Takahashi, K., (2016). Optimal sequential fusion for multibiometric cryptosystems. *Inf. Fusion*, *32*, 93–108.
48. Bentahar, A., Meraoumia, A., Bendjenna, H., Chitroub, S., & Zeroual, A., (2018). Biometric cryptosystem scheme for internet of things using fuzzy commitment principle. In: *2018 International Conference on Signal, Image, Vision, and Their Applications (SIVA)* (pp. 1–6).
49. Barman, S., Shum, H. P. H., Chattopadhyay, S., & Samanta, D., (2019). A secure authentication protocol for multi-server-based e-healthcare using a fuzzy commitment scheme. *IEEE Access*, *7*, 12557–12574.
50. Shi, S., Cui, J., Zhang, X. L., Liu, Y., Gao, J. L., & Wang, Y. J., (2019). Fingerprint recognition strategies based on a fuzzy commitment for cloud-assisted IoT: A minutiae-based sector coding approach. *IEEE Access*, *7*, 44803–44812.
51. Uludag, U., & Anil, J., (2006). Securing fingerprint template: Fuzzy vault with helper data. In: *2006 Conference on Computer Vision and Pattern Recognition Workshop (CVPRW'06)* (p. 163).
52. Nandakumar, K., Jain, A. K., & Pankanti, S., (2007). Fingerprint-based fuzzy vault: Implementation and performance. *IEEE Trans. Inf. Forensics Secur.*, *2*(4), 744–757.
53. Nandakumar, K., Nagar, A., & Jain, A. K., (2007). Hardening fingerprint fuzzy vault using password. In: *Advances in Biometrics* (pp. 927–937). Springer Berlin Heidelberg: Berlin, Heidelberg.
54. Nyang, D., & Lee, K., (2007). Fuzzy face vault: How to implement fuzzy vault with weighted features. In: *Universal Access in Human Computer Interaction. Coping with Diversity* (pp. 491–496). Springer Berlin Heidelberg: Berlin, Heidelberg.
55. Lee, Y. J., Park, K. R., Lee, S. J., Bae, K., & Kim, J., (2008). A new method for generating an invariant iris private key based on the fuzzy vault system. *IEEE Trans. Syst. Man, Cybern. Part B*, *38*(5), 1302–1313.
56. Zhou, X., Opel, A., Merkle, J., Korte, U., & Busch, C., (2011). Enhanced template protection with passwords for fingerprint recognition. *2011 Third Int. Work. Secur. Commun. Networks*, 67–74.
57. Bhateja, A. K., Chaudhury, S., & Saxena, P. K., (2014). A robust online signature based cryptosystem. In: *2014 14th International Conference on Frontiers in Handwriting Recognition* (pp. 79–84).
58. Liu, H., Sun, D., Xiong, K., & Qiu, Z., (2014). Palmprint based multidimensional fuzzy vault scheme. *Sci. World Journal, Hindawi Publ. Corp.*, *2014*, 819031–819039.
59. Yang, W., Hu, J., Wang, S., & Stojmenovic, M., (2014). An alignment-free fingerprint bio-cryptosystem based on modified Voronoi neighbor structures. *Patt. Recogn.*, *47*(3), 1309–1320.
60. Benhammadi, F., & Beghdad, B. K., (2014). Password hardened fuzzy vault for fingerprint authentication system. *Image Vis. Comput.*, *32*(8), 487–496.
61. Tams, B., Mihăilescu, P., & Munk, A., (2015). Security considerations in minutiae-based fuzzy vaults. *IEEE Trans. Inf. Forensics Secur.*, *10*(5), 985–998.

62. Li, C., & Hu, J., (2016). A security-enhanced alignment-free fuzzy vault-based fingerprint cryptosystem using pair-polar minutiae structures. *IEEE Trans. Inf. Forensics Secur.*, *11*(3), 543–555.
63. Neu, M., Korte, U., & Ullmann, M., (2016). Improvement of fuzzy vault for multiple fingerprints with angles. In: *2016 International Conference of the Biometrics Special Interest Group (BIOSIG)* (pp. 1–8).
64. Revadigar, G., Javali, C., Xu, W., Vasilakos, A. V., Hu, W., & Jha, S., (2017). Accelerometer and fuzzy vault-based secure group key generation and sharing protocol for smart wearables. *IEEE Trans. Inf. Forensics Secur.*, *12*(10), 2467–2482.
65. Mai, G., Lim, M. H., & Yuen, P. C., (2017). Binary feature fusion for discriminative and secure multi-biometric cryptosystems. *Image Vis. Comput.*, *58*, 254–265.
66. Pirbhulal, S., Shang, P., Wu, W., Sangaiah, A. K., Samuel, O. W., & Li, G., (2018). Fuzzy vault-based biometric security method for tele-health monitoring systems. *Comput. & Electr. Eng.*, *71*, 546–557.
67. Bobkowska, K., Nagaty, K., & Przyborski, M., (2019). Incorporating iris, fingerprint, and face biometric for fraud prevention in e-passports using fuzzy vault. *IET Image Process*, *13*(13), 2516–2528.
68. Ponce-Hernandez, W., Blanco-Gonzalo, R., Liu-Jimenez, J., & Sanchez-Reillo, R., (2020). Fuzzy vault scheme based on fixed-length templates applied to dynamic signature verification. *IEEE Access*, *8*, 11152–11164.
69. Vielhauer, C., Steinmetz, R., & Mayerhofer, A., (2002). Biometric hash based on statistical features of online signatures. In: *Object Recognition Supported by User Interaction for Service Robots* (Vol. 1, pp. 123–126).
70. Dodis, Y., Reyzin, L., & Smith, A., (2004). Fuzzy extractors: How to generate strong keys from biometrics and other noisy data. In: Cachin, C., & Camenisch, J. L., (eds.), *Advances in Cryptology–EUROCRYPT 2004* (pp. 523–540). Springer Berlin Heidelberg: Berlin, Heidelberg.
71. Chang, Y. J., Zhang, W., & Chen, T., (2004). Biometrics-based cryptographic key generation. In: *2004 IEEE International Conference on Multimedia and Expo (ICME) (IEEE Cat. No. 04TH8763)* (Vol. 3, pp. 2203–2206).
72. Boyen, X., Dodis, Y., Katz, J., Ostrovsky, R., & Smith, A., (2005). Secure remote authentication using biometric data. In: Cramer, R., (ed.), *Advances in Cryptology–EUROCRYPT 2005* (pp. 147–163). Springer Berlin Heidelberg: Berlin, Heidelberg.
73. Sutcu, Y., Li, Q., & Memon, N., (2007). Protecting biometric templates with sketch: Theory and practice. *IEEE Trans. Inf. Forensics Secur.*, *2*(3), 503–512.
74. Arakala, A., Jeffers, J., & Horadam, K. J., (2007). Fuzzy extractors for minutiae-based fingerprint authentication. In: Lee, S. W., & Li, S. Z., (eds.), *Advances in Biometrics* (pp. 760–769). Springer Berlin Heidelberg: Berlin, Heidelberg.
75. Chang, E. C., & Roy, S., (2007). Robust extraction of secret bits from minutiae. In: Lee, S. W., & Li, S. Z., (eds.), *Advances in Biometrics* (pp. 750–759). Springer Berlin Heidelberg: Berlin, Heidelberg.
76. Sutcu, Y., Li, Q., & Memon, N., (2007). Secure biometric templates from fingerprint-face features. In: *2007 IEEE Conference on Computer Vision and Pattern Recognition* (pp. 1–6).
77. Verbitskiy, E. A., Tuyls, P., Obi, C., Schoenmakers, B., & Skoric, B., (2010). Key extraction from general nondiscrete signals. *IEEE Trans. Inf. Forensics Secur.*, *5*(2), 269–279. https://doi.org/10.1109/TIFS.2010.2046965.

78. Tian, Y., Li, Y., Sengupta, B., Li, N., & Su, C., (2020). Leakage-resilient biometric-based remote user authentication with fuzzy extractors. *Theor. Comput. Sci.*, *814*, 223–233.
79. Connie, T., Teoh, A., Goh, M., & Ngo, D., (2005). Palm hashing: A novel approach for cancelable biometrics. *Inf. Process. Lett.*, *93*(1), 1–5.
80. Teoh, A. B., & Ngo, D. C., (2006). Biophasor: Token supplemented cancellable biometrics. In: *2006 9th International Conference on Control, Automation, Robotics, and Vision* (pp. 1–5).
81. Wai, K. Y., Teoh, A. B. J., & Ngo, D. C. L., (2006). Secure hashing of dynamic hand signatures using wavelet-fourier compression with BioPhasor mixing and discretization. *EURASIP J. Adv. Signal Process.*, *2007*(1), 59125–59132.
82. Teoh, A. B. J., Goh, A., & Ngo, D. C. L., (2006). Random multispace quantization as an analytic mechanism for bio-hashing of biometric and random identity inputs. *IEEE Trans. Pattern Anal. Mach. Intell.*, *28*(12), 1892–1901.
83. Siew, C. C., Beng, J. A. T., & Chek, L. D. N., (2006). High security iris verification system based on random secret integration. *Comput. Vis. Image Underst.*, *102*(2), 169–177.
84. Teoh, A. B. J., Kuan, Y. W., & Lee, S., (2008). Cancellable biometrics and annotations on bio-hash. *Pattern Recognit.*, *41*(6), 2034–2044.
85. Ouda, O., Tsumura, N., & Nakaguchi, T., (2010). BioEncoding: A reliable tokenless cancelable biometrics scheme for protecting iris codes. *IEICE Trans. Inf. Syst.*, *E93.D*(7), 1878–1888.
86. Jin, Z., & Jin, A. T. B., (2011). Fingerprint template protection with minutia vicinity decomposition. In: *2011 International Joint Conference on Biometrics (IJCB)* (pp. 1–7).
87. Chikkerur, S., Ratha, N. K., Connell, J. H., & Bolle, R. M., (2008). Generating registration-free cancelable fingerprint templates. In: *2008 IEEE Second International Conference on Biometrics: Theory, Applications, and Systems* (pp. 1–6).
88. Huijuan, Y., Jiang, X., & Kot, A. C., (2009). Generating secure cancelable fingerprint templates using local and global features. In: *2009 2nd IEEE International Conference on Computer Science and Information Technology* (pp. 645–649).
89. Lee, C., & Kim, J., (2010). Cancelable fingerprint templates using minutiae-based bit-strings. *J. Netw. Comput. Appl.*, *33*(3), 236–246.
90. Ahmad, T., Hu, J., & Wang, S., (2011). Pair-polar coordinate-based cancelable fingerprint templates. *Patt. Recogn.*, *44*(10), 2555–2564.
91. Wang, S., & Hu, J., (2012). Alignment-free cancelable fingerprint template design: A densely infinite-to-one mapping (DITOM) approach. *Patt. Recogn.*, *45*(12), 4129–4137.
92. Boult, T. E., Scheirer, W. J., & Woodworth, R., (2007). Revocable fingerprint biotokens: Accuracy and security analysis. In: *Proc. of CVPR 2007* (pp. 1–8).
93. Tulyakov, S., Farooq, F., Mansukhani, P., & Govindaraju, V., (2007). Symmetric hash functions for secure fingerprint biometric systems. *Patt. Recogn. Lett.*, *28*(16), 2427–2436.
94. Ahn, D., Kong, S. G., Chung, Y. S., & Moon, K. Y., (2008). Matching with secure fingerprint templates using non-invertible transform. In: *Proc. of CISP 2008* (Vol. 2, pp. 29–33).
95. Kumar, G., Tulyakov, S., & Govindaraju, V., (2010). Combination of symmetric hash functions for secure fingerprint matching. In: *Proc. of ICPR 2010* (pp. 890–893).
96. Cappelli, R., Ferrara, M., & Maltoni, D., (2010). Minutia cylinder-code: A new representation and matching technique for fingerprint recognition. *IEEE Trans. PAMI*, *32*(12), 2128–2141.
97. Ferrara, M., Maltoni, D., & Cappelli, R., (2012). Noninvertible minutia cylinder-code representation. *IEEE Trans. IFS*, *7*(6), 1727–1737.

98. Jin, Z., Teoh, A. B. J., Ong, T. S., & Tee, C., (2012). Fingerprint template protection with minutiae-based bit-string for security and privacy preserving. *Expert Syst. Appl.*, *39*(6), 6157–6167.
99. Wong, W. J., Teoh, A. B. J., Wong, M. L. D., & Kho, Y. H., (2013). Enhanced multi-line code for minutiae-based fingerprint template protection. *Patt. Recogn. Lett.*, *34*(11), 1221–1229.
100. Jin, Z., Lim, M. H., Teoh, A. B. J., & Goi, B. M., (2014). A non-invertible randomized graph-based hamming embedding for generating cancelable fingerprint template. *Patt. Recogn. Lett.*, *42*, 137–147.
101. Sandhya, M., & Prasad, M. V. N. K., (2015). K-nearest neighborhood structure (k-NNS) based alignment-free method for fingerprint template protection. In: *Proc. of ICB 2015* (pp. 386–393).
102. Yang, W., Hu, J., Wang, S., & Yang, J., (2013). Cancelable fingerprint templates with delaunay triangle-based local structures. In: *Cyberspace Safety and Security* (pp. 81–91). Springer International Publishing.
103. Jin, Z., Lim, M. H., Teoh, A. B. J., Goi, B. M., & Tay, Y. H., (2016). Generating fixed-length representation from minutiae using kernel methods for fingerprint authentication. *IEEE Trans. Syst. Man, Cybern. Syst.*, *46*(10), 1415–1428.
104. Wang, S., & Hu, J., (2016). A blind system identification approach to cancelable fingerprint templates. *Patt. Recogn.*, *54*, 14–22.
105. Sandhya, M., Prasad, M. V. N. K., & Chillarige, R. R., (2016). Generating cancellable fingerprint templates based on Delaunay triangle feature set construction. *IET Biometrics*, *5*(2), 131–139.
106. Khodadoust, J., & Khodadoust, A. M., (2017). Fingerprint indexing based on expanded delaunay triangulation. *Expert Syst. Appl.*, *81*, 251–267.
107. Sandhya, M., & Prasad, M. V. N. K., (2017). Securing fingerprint templates using fused structures. *IET Biometrics*, *6*(3), 173–182.
108. Wang, S., Deng, G., & Hu, J., (2017). A partial Hadamard transform approach to the design of cancelable fingerprint templates containing binary biometric representations. *Patt. Recogn.*, *61*, 447–458.
109. Si, X., Feng, J., Yuan, B., & Zhou, J., (2017). Dense registration of fingerprints. *Patt. Recogn.*, *63*, 87–101.
110. Lee, W., Cho, S., Choi, H., & Kim, J., (2017). Partial fingerprint matching using minutiae and ridge shape features for small fingerprint scanners. *Expert Syst. Appl.*, *87*, 183–198.
111. Wang, S., Yang, W., & Hu, J., (2017). Design of alignment-free cancelable fingerprint templates with zoned minutia pairs. *Patt. Recogn.*, *66*, 295–301.
112. Moujahdi, C., Bebis, G., Ghouzali, S., & Rziza, M., (2014). Fingerprint shell: Secure representation of fingerprint template. *Patt. Recogn. Lett.*, *45*, 189–196.
113. Jain, A., & Prasad, M. V. N. K., (2016). A novel fingerprint indexing scheme using dynamic clustering. *J. Reliab. Intell. Environ.*, *2*, 159–171.
114. Ali, S. S., & Prakash, S., (2015). Enhanced fingerprint shell. In: *Proc. of SPIN 2015* (pp. 801–805.
115. Ali, S. S., & Prakash, S., (2017). Fingerprint shell construction with prominent minutiae points. In: *Proc. of COMPUTE 2017* (pp. 91–98). ACM.
116. Iyappan, G. I., Prakash, S., Dave, I. R., Joshi, P., Ali, S. S., & Srivastava, A. M., (2018). Ear recognition in 3D using 2D curvilinear features. *IET Biometrics*, *7*(6), 519–529.

117. Ali, S. S., & Prakash, S., (2019). 3-dimensional secured fingerprint shell. *Patt. Recogn. Lett., 126*, 68–77.
118. Ali, S. S., Iyappan, G. I., & Prakash, S., (2018). Robust technique for fingerprint template protection. *IET Biometrics, 7*(6), 536–549.
119. Ali, S. S., Iyappan, G. I., Prakash, S., Consul, P., & Mahyo, S., (2020). Securing biometric user template using modified minutiae attributes. *Patt. Recogn. Lett., 129*, 263–270.
120. Ali, S. S., Iyappan, G. I., Mahyo, S., & Prakash, S., (2019). Polynomial vault: A secure and robust fingerprint-based authentication. *IEEE Trans. Emerg. Top. Comput.*
121. Trivedi, A. K., Thounaojam, D. M., & Pal, S., (2020). Non-invertible cancellable fingerprint template for fingerprint biometric. *Comput. & Secur., 90*, 101690.
122. Nagar, A., Nandakumar, K., & Jain, A. K., (2008). Securing fingerprint template: Fuzzy vault with minutiae descriptors. In: *2008 19th International Conference on Pattern Recognition* (pp. 1–4).
123. Nagar, A., Nandakumar, K., & Jain, A. K., (2010). A hybrid biometric cryptosystem for securing fingerprint minutiae templates. *Pattern Recognit. Lett., 31*(8), 733–741.
124. Leng, L., & Teoh, A. B. J., (2015). Alignment-free row-co-occurrence cancelable palmprint fuzzy vault. *Pattern Recognit., 48*(7), 2290–2303.
125. Ouda, O., Nandakumar, K., & Ross, A., (2021). Cancelable biometrics vault: A secure key-binding biometric cryptosystem based on chaffing and winnowing. In: *2020 25th International Conference on Pattern Recognition (ICPR)* (pp. 8735–8742).
126. Upmanyu, M., Namboodiri, A. M., Srinathan, K., & Jawahar, C. V., (2009). Efficient biometric verification in encrypted domain. In: Tistarelli, M., & Nixon, M. S., (eds.), *Advances in Biometrics* (pp. 899–908). Springer Berlin Heidelberg: Berlin, Heidelberg.
127. Gomez-Barrero, M., Maiorana, E., Galbally, J., Campisi, P., & Fierrez, J., (2017). Multi-biometric template protection based on homomorphic encryption. *Pattern Recognit., 67*, 149–163.
128. Morampudi, M. K., Prasad, M. V. N. K., & Raju, U. S. N., (2020). Privacy-preserving iris authentication using fully homomorphic encryption. *Multimed. Tools Appl., 79*(27), 19215–19237.

CHAPTER 5

WATERMARKING AND COPYRIGHT PROTECTION OF REMOTELY SENSED IMAGES

ROHIT THANKI,[1] SUREKHA BORRA,[2] and PURVA JOSHI[3]

[1] *CTO & Senior IEEE Member, DetMedX, Germany*

[2] *Department of Computer Science and Engineering (IoT, Cybersecurity and Blockchain Technology), K. S. Institute of Technology, Bengaluru, India*

[3] *Institute for Communications and Navigation, Deutsches Zentrum für Luft und Raumfahrt, Weßling, Germany*

ABSTRACT

With the rise in technology and sensors, a lot of remotely sensed images are generated every day by satellites and unmanned aerial vehicles (UAV) and are wirelessly transmitted to the receiving stations. The security concern of remotely sensed images is the need of the hour as they are widely available on the internet for different applications such as research, weather, and agriculture monitoring to name a few. Digital watermarking is one of the solutions for protecting such expensive and valuable images. This chapter presents an overview of different watermarking schemes for copyright protection of remotely sensed images. Two types of watermarking techniques, in the spatial domain and transform domain for the remotely sensed image, are discussed in detail. Further, a watermarking scheme using a redundant discrete wavelet transform (RDWT) and Arnold scrambling is proposed and analyzed for remotely sensed images. The experimental results show that the

Multimedia Security: Tools, Techniques, and Applications. Bhaskar Mondal and Shyam Singh Rajput (Eds.)

proposed watermarking scheme performs better than existing watermarking schemes for remotely sensed image protection.

5.1 INTRODUCTION

Recently, the interest of researchers, scientists, and many industries is in satellite communication-related applications, due to the inexpensive availability of a variety of numerous remotely sensed data and images. In satellite communication, a satellite collects data from the earth and transmits the data to an earth station. The earth station creates several digital images as per the collected data using digital imaging algorithms for easy processing, analyzing, and interpretation. These images are used for various scientific or commercial applications such as weather monitoring, area observations, agriculture, and so on. These images are stored in DVD's, CD-ROMs, and hard disks, and transmitted via the internet.

In remote sensing, information in the image plays an important role because, based on this information, the interpretation and decision making is done in various applications. Similar to multimedia data, unauthorized modification or use of the remotely sensed images is possible. The problems with copyright protection or ownership authentication are rising for remotely sensed images and have been solved by watermarking technology, where secure marks or information such as copyright information or creator information is embedded in the original image to protect the owner. The availability of a watermark in the original image is also used to prove ownership and to protect the intellectual property (IP) rights of the creator against the creation of an unauthorized copy or use of data. This chapter gives various requirements of watermarking schemes for remotely sensed image protection. The various types of related watermarking schemes are presented, and also possible extensions are proposed in this chapter.

The word "watermark" comes from the German word 'watermarke.' The meaning of this word is that it gives identity information, like the effects of water on paper. Nearly 700 years ago, the Fabriano in Italy used paper watermarks in the art of papermaking [1, 2]. Emil Hembrooke gives the first used technology which is similar to the digital watermarking scheme in 1954 for the identification of music [3]. The first use of the word "digital watermarking" is by Komatsu & Tominaga in 1988 [4]. The research in the area of digital watermarking technology began to mushroom after 1995.

In digital watermarking schemes a digital mark is embedded into cover digital data, in such a way that the marks can identify the owner of IP. The cover digital data could be text information, digital image, digital audio signal, digital video or any digital data which is created by and wish to be protected by the owner [5]. The watermarking system has three parts such as an embedder or encoder, communication channel (which is optional) and detector or decoder. The embedder or encoder embeds the watermark into the cover digital data. The detector or decoder extracts watermark data from the watermarked digital data. The input of the decoding model is watermarked digital data and secret keys, the output being extracted watermark data. The watermarking schemes are divided into four categories according to the type of cover digital data to be watermarked such as text watermarking, image watermarking, audio watermarking and video watermarking [5, 6].

There are two major types based on where digital data are watermarked: spatial domain watermarking and transform domain watermarking. The spatial domain watermarking is easy to implement and provides high perceptibility to watermarked data. The transform domain watermarking is complex but far better and provides more robustness compared to spatial domain watermarking. For spatial domain watermarking, the cover digital data is directly modified in the spatial domain. For transform domain watermarking, the cover digital data is converted into the frequency domain using various image processing transforms such as DFT, discrete cosine transform (DCT), discrete wavelet transform (DWT), and so on. In spatial domain watermarking schemes, the cover digital data is modified by changing the least significant bits (LSBs) or adding various noise sequences. In the transform domain watermarking scheme, the cover digital data is modified by modification of efficient coefficients or/and by coefficients block similarities [4, 6–10]. But when these schemes are designed for remotely sensed images protection then some requirements are needed. The requirements of any watermarking scheme for remotely sensed image protection is given in next Section 5.2. The rest of the chapter is organized as follows.

In Section 5.2, the requirements of the digital watermarking scheme for the remotely sensed images are given. Section 5.3 introduces existing watermarking schemes for remotely sensed images in the spatial domain and transform domain. Section 5.4 provides an extended watermarking scheme for remotely sensed image protection. Finally, the conclusion of the chapter is given in Section 5.5.

5.2 REQUIREMENTS OF WATERMARKING FOR PROTECTION OF REMOTELY SENSED IMAGES

Satellite based remote sensing system collects information and observations of particular portions of the earth, transmits to earth stations, generates images using data processing algorithms, store, and makes images available on the website. The remotely sensed images are mainly used for weather monitoring and agriculture applications. So, every fine piece of information that lies in the image is very important. Therefore, the watermarking scheme should not degrade the information in the remotely sensed image while inserting a secure watermark into it. There are mainly three requirements for watermarking remotely sensed images: imperceptibility, robustness, and security [7–11].

5.2.1 IMPERCEPTIBILITY

The watermarking scheme must embed watermark information without affecting the perceptual quality of the remotely sensed image. This is the basic requirement for remotely sensed image protection.

5.2.2 ROBUSTNESS

In many situations, lossy or noisy image processing operations are involved while storing images in the database. For example, JPEG compression is a very common operation before the storage or transmission of the image. Similarly, when images are transmitted from one station to another station via a noisy transmission medium, they get distorted in several forms. The digital image may also be manipulated intentionally using various software packages. These manipulations may damage the quality of the cover remotely sensed an image and also damage the embedded watermark information. Furthermore, the impostor may manipulate the cover remotely sensed image in detecting the watermark information. Therefore, researchers and designers of the watermarking scheme must consider several types of image manipulations/attacks which eventually affect the quality of extracted watermark from the cover image. Also, Software's like StirMark, UnZign2 are available on the internet for testing robustness and extraction of watermark data from watermarked images [11]. Few manipulations which are modified watermarked content mentioned below:

- Additive noise, e.g., gaussian noise, speckle noise, and salt and pepper noise.
- Different types of filtering, e.g., low-pass filtering, median filtering, and mean filtering.
- Image compression, e.g., JPEG, JPEG 2000.
- Geometrical modification, e.g., rotation, cropping.

5.2.3 SECURITY

In remotely sensed image protection using watermarking, the embedding process must be secured so that an impostor is unable to extract inserted watermark information without the knowledge of the secret key. The security of the watermark embedding process is analogous to the security of the encryption method. The security in the watermarking scheme can be achieved in two ways such as by using some encryption method on watermark data before embedding into cover data and by using secret keys. Nowadays, the encryption methods like Arnold scrambling, advanced encryption standard (AES) and many are involved in improving the security of the watermarking scheme.

5.3 EXISTING WATERMARKING SCHEMES FOR REMOTELY SENSED IMAGES

In the last decade, researchers designed various watermarking schemes for remotely sensed image protection in the spatial domain and transform domain. A brief review of existing watermarking schemes for remotely sensed images is given in Subsections 5.3.1 and 5.3.2.

5.3.1 EXISTING WATERMARKING SCHEMES IN SPATIAL DOMAIN

This section discusses the existing watermarking schemes in the spatial domain. There are few watermarking schemes are implemented for the protection of remotely sensed images in the spatial domain. These schemes were designed using LSB based approach and XOR operation.

Caldelli et al. [12] proposed watermarking along with compression for remotely sensed image protection. Authors used a fragile watermark for authentication of the remotely sensed images. Authors modified Fridrich

[13] semi-fragile watermarking algorithm that embeds the watermark in the LSBs of the remotely sensed image. Authors provided authentication and tamper localization for remotely sensed images by JPEG-LS standard and semi-fragile watermarking. In this scheme, the image is divided into 8–16 blocks and the LSBs of each block are modified by watermark bits. The seven most significant bits (MSBs) of the pixels in the block are converted into a hash value by using a hash function. Then, a binary logo is inserted into the hash values by XOR operation, and the result of XOR is encrypted by a secret key encryption function and embedded into the LSBs of the same block. A watermark detection, watermarked image is again divided into 8–16 blocks. Then seven MSBs of each pixel are hashed, and LSBs are decrypted by the secret key. The XOR operation between hashed value and decrypted LSBs gives the extracted binary logo.

Li et al. [14] proposed LSBs based watermarking scheme for the remotely sensed images. In this scheme, the first LSB of cover remotely sensed image is extracted. The land reorganization planning map data are taken as watermark information and is encoded using encryption algorithms. The watermarked image is generated performing logical XOR between LSB of cover remotely sensed an image and encoded watermark data. The authors added XOR operation in the traditional LSB scheme in such a way that the watermark information cannot be detected easily.

5.3.2 EXISTING WATERMARKING SCHEMES IN TRANSFORM DOMAIN

This section provides the details of existing watermarking schemes in the transform domain. They are mostly based on DCT, DWT, singular value decomposition (SVD), fast Hadamard transform (FHT), Arnold Scrambling Transform and Karhunen Loeve Transform (KLT).

Yu [15] has proposed a watermarking scheme for the color remotely sensed images using Arnold scrambling transform and Morphological Component Analysis (MCA) theory. In this scheme, first, the watermark image is encrypted into binary sequences using Arnold scrambling transform. This encrypted watermark bit is inserted into the remotely sensed image using the MCA theory process [16]. Nazmudeen & Farsana [17] have proposed a DCT, DWT, and AES based watermarking scheme for remotely sensed image protection. In this scheme, noise is removed from the noisy image using a modified decision-based unsymmetrical trimmed median filter (MDBUTMF). The DWT is applied to the denoised image to get various wavelet subband of the image. Then, DCT is applied to the HH band to get DCT coefficients. Encoded watermark data is inserted into the DCT coefficients of the HH band of the denoised image to generate a watermarked image. In this scheme, the AES algorithm is used for the generation of encoded watermark data. Al Mansoori & Kunhu [18] have proposed three algorithms based on DCT for the security of remotely sensed images. In the first algorithm, two bits of watermark data are embedded into each 8×8 DCT block in the green channel of color remotely sensed image based on a secret key. In the second algorithm, two copies of two bits of watermark data are embedded into each 8×8 DCT block in the green channel of color remotely sensed image based on a secret key. In the third algorithm, three copies of two bits of watermark data are embedded into each 8×8 DCT block in the green channel of color remotely sensed image based on a secret key.

Hsu & Chen [19] have proposed a DWT and feature-based watermarking scheme for the security of remotely sensed image. In this scheme, scale-invariant feature transform (SIFT) algorithm is used to detect the scale-space feature points for watermark embedding in the remotely sensed image. Then three-layer wavelet packet transform is applied on feature points to get 64 subbands. The subbands with higher texture information are selected for embedding watermark data. These selected subbands are divided into 2×2 blocks. These blocks are modified according to watermark bits and then applied to inverse three-layer wavelet packets transform to get modified feature points of the remotely sensed image. The inverse SIFT algorithm is then applied on modified feature points to get watermarked remotely sensed image. Jiang & Xu [20] have proposed a DCT and AES based watermarking scheme. In this scheme, watermarked image is generated using DCT based scheme which is proposed by Miyazaki & Okamoto [21]. The AES encryption is applied on watermarked image to generate encrypted watermarked remotely sensed image. Chen et al. [22] have proposed a block-wise DCT based watermarking scheme for remotely sensed image. This scheme is robust against various watermarking attacks.

Hemalatha et al. [23] have proposed an SVD and DWT based watermarking scheme for remotely sensed image. In this scheme, the singular value of the approximation subband of the remotely sensed image is modified by watermark bits to generate the watermarked image. While Kaarna et al. [24] chose the 3D DWT domain, Kbaier & Belhadj [25] proposed DWT based scheme for the remotely sensed image. Ma et al. [26] have proposed

SVD and principal component analysis (PCA) based watermarking scheme for the remotely sensed image. In this scheme, the singular value of PCA transform coefficients of the remotely sensed image is modified according to the watermark bit. Panyavaraporn & Rangsanseri [27] have proposed KLT based watermarking scheme for remotely sensed image. In this scheme, KLT is applied to the cover image to get a principal coefficient of the image. Then two level of DWT is applied on most principal coefficients to get various subbands. Finally, the LH2 wavelet subband of the image is modified by a pseudo-random sequence according to the watermark bit to generate a watermarked image. Piva et al. [28] have proposed DWT-DCT based watermarking scheme for authentication of the remotely sensed image.

Kaarna & Toivanen [29] have proposed a PCA and DWT based watermarking scheme for protecting remotely sensed image. In this scheme, the wavelet subbands of PCA transform coefficients are modified by wavelet subbands of watermark data. Ziegeler et al. [30] have proposed a DWT based watermarking scheme for the remotely sensed image. Ho et al. [31] have proposed a FHT based watermarking scheme for the remotely sensed image. Barni et al. [32] have proposed two watermarking schemes based on DFT and DWT for the remotely sensed image.

The study of all the above-mentioned techniques shows that most of the existing watermarking schemes are designed using DFT, DCT, DWT, SVD, and PCA. Most of these watermarking schemes applied some kind of encryption method to watermark, before embedding it into the image. Also, many existing watermarking schemes did not perform the robustness test to verify the performance of their algorithms for various watermarking attacks. Further, these schemes have less imperceptibility. In other words, the quality of the remotely sensed image is degraded after watermarking. Hence, there is a need to design watermarking techniques which overcome the above-mentioned problems.

In the next section, a non-blind watermarking scheme is proposed as a solution to the above-mentioned problems. The scheme is designed using redundant discrete wavelet transform (RDWT) and Arnold scrambling transform. The scheme shows robustness against geometric attacks, signal processing attacks, and JPEG compression attacks. To test for copyright protection of remotely sensed image, a binary logo is embedded into the remotely sensed image. Implementation of this scheme includes the application of RDWT on the remotely sensed images first. The LL subband of RDWT is then modified by encoded watermark data. On the extraction side, non-blind watermark extraction is performed using the LL subband of the original preserved remotely sensed image.

5.4 PROPOSED WATERMARKING SCHEME FOR REMOTELY SENSED IMAGES

In this section, the complete information on the proposed watermarking scheme is given for remotely sensed image protection. The preliminaries for understanding this scheme are given in Subsection 5.4.1. The proposed watermark embedding and watermark extraction are given in Subsection 5.4.2. The results of the proposed scheme are presented in Subsection 5.4.3. The comparison of the proposed scheme with the existing scheme is given in Subsection 5.4.4.

5.4.1 PRELIMINARIES

5.4.1.1 REDUNDANT DISCRETE WAVELET TRANSFORM (RDWT)

So far, the most common transform used in watermarking is DWT, which has a limitation of restricted payload capacity due to downsampling of its subbands [33–35]. Further, the DWT is shift variance, which may cause a problem in the extraction of watermark information. So, to overcome these limitations of DWT in watermarking, RDWT is used in the proposed scheme. The RDWT provides shift-invariant, which is better for extraction of watermark information at the detector side. The RDWT eliminates downsampling and the upsampling process of the DWT. This transform provides a more robust process than DWT. The DWT is decomposed image with a size of $N \times N$ into its wavelet coefficients with the size of $N/2 \times N/2$ while RDWT decomposed into its wavelet coefficients with the size of $N \times N$. The difference between DWT and RDWT for an image with a size of 16×16 is shown in Figure 5.1 [33, 34]. For the proposed robust watermarking scheme, RDWT is applied on the remotely sensed image to obtain different wavelet subbands such as LL, LH, HL, and HH.

5.4.1.2 ARNOLD SCRAMBLING TRANSFORM [33, 34]

In the proposed watermarking scheme, Arnold scrambling is used to encrypt the watermark before embedding it into the cover remotely sensed image so that attacker cannot understand and extract watermark information from the remotely sensed image. The resultant chaotic image is secure and can't be decrypted without the knowledge of the scrambling algorithm and secure key. The 2D forward Arnold scrambling is defined by the below equation:

$$\begin{vmatrix} i' \\ j' \end{vmatrix} = \begin{vmatrix} 1 & 1 \\ 1 & 2 \end{vmatrix} \begin{vmatrix} i \\ j \end{vmatrix} \bmod N; i, j, i', j' = \{0, 1, 2, \ldots, N-1\} \quad (1)$$

where; i, j are the pixel coordinates of the original space; i,' j' are the pixel coordinates after iterative computation scrambling; N is the size of the watermark image. The original watermark image is obtained back by inverse Arnold scrambling as follows:

$$\begin{vmatrix} i \\ j \end{vmatrix} = \left(\begin{vmatrix} 2 & -1 \\ -1 & 1 \end{vmatrix} \begin{vmatrix} i' \\ j' \end{vmatrix} + \begin{vmatrix} N \\ N \end{vmatrix} \right) \bmod N \quad (2)$$

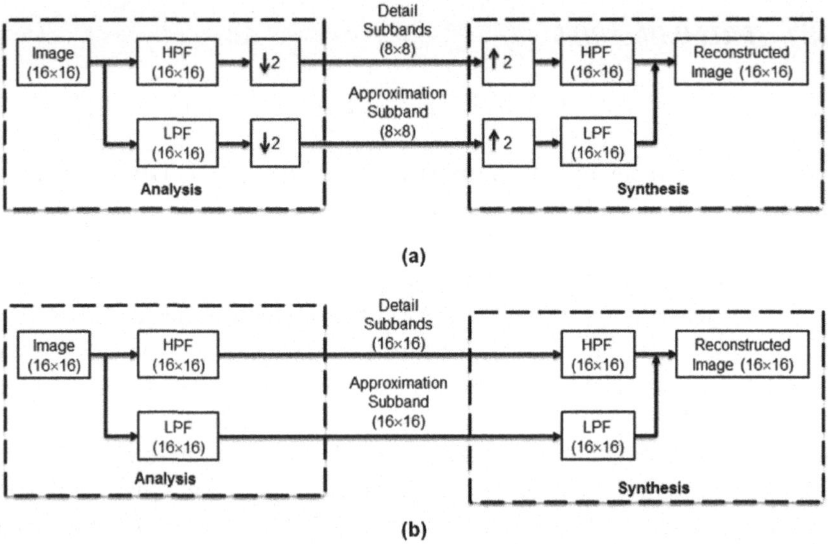

FIGURE 5.1 (a) DWT analysis and synthesis for image; and (b) RDWT analysis and synthesis for image.

5.4.2 PROPOSED WATERMARKING EMBEDDING AND EXTRACTION PROCESS

In this section, a non-blind watermarking scheme based on RDWT is proposed. The scheme is designed to be more imperceptible and robust against different watermarking attacks. In this scheme, RDWT is applied on cover remotely sensed image to get different wavelet subbands such as LL, LH, HL, and HH. The reason behind choosing the LL wavelet subband is that it is equal in size to the cover image and less affected by the attack.

The Arnold scrambling is used for encryption of watermark data before embedding it into the cover image. The steps for watermark embedding and extraction are given in the following subsections.

5.4.2.1 WATERMARK EMBEDDING PROCESS

The block diagram of a proposed watermark embedding process is given in Figure 5.2, and the algorithm for watermark embedding is given below:

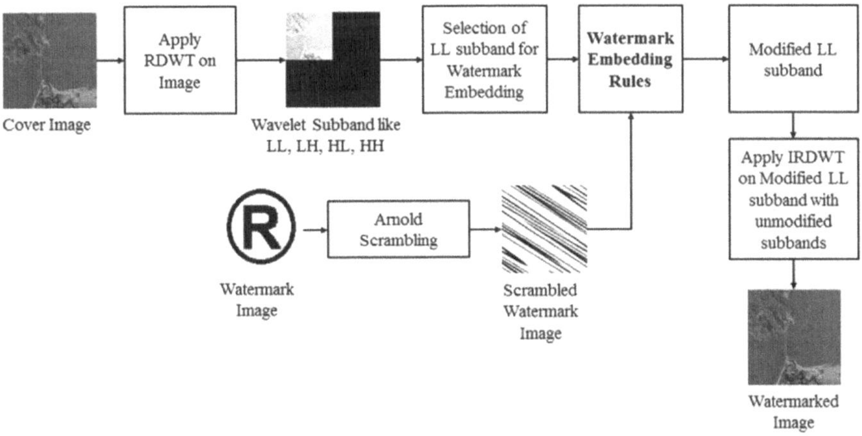

FIGURE 5.2 Block diagram of proposed watermark embedding process.

- **Step 1:** Calculate the size of watermark image W.
- **Step 2:** Apply Arnold scrambling with secret key K on the watermark image to generate scrambled watermark information EW in terms of the binary sequence.
- **Step 3:** Calculate the size of the cover remotely sensed image.
- **Step 4:** Apply RDWT to get LL, LH, HL, and HH subband of cover remotely sensed image I.
- **Step 5:** Choose LL subband coefficients for watermark bit embedding. Embed each bit of scrambled watermark in LL subband of cover remotely sensed image based on the following condition:

$$WI_{LL} = I_{LL} + \alpha \times EW \tag{3}$$

where WI_{LL} is the modified LL subband of the cover image, I_{LL} is the original subband of the cover image; EW is the encrypted watermark, and α is a gain factor.

- **Step 6:** Apply inverse RDWT on modified LL subband along with other unmodified subbands to get watermarked remotely sensed image *WI*.

5.4.2.2 WATERMARK EXTRACTION PROCESS

A watermark is extracted non-blindly using the original cover remotely sensed image. The block diagram of the proposed extraction process is given in Figure 5.3. The algorithm for watermark extraction is given below:

- **Step 1:** Calculate the size of watermarked remotely sensed image.
- **Step 2:** Apply RDWT to get LL, LH, HL, and HH subband of watermarked remotely sensed image. Choose modified LL subband coefficients for watermark bit extraction.
- **Step 3:** Apply RDWT to get LL, LH, HL, and HH subband of cover remotely sensed image. Choose LL subband coefficients for watermark bit extraction.

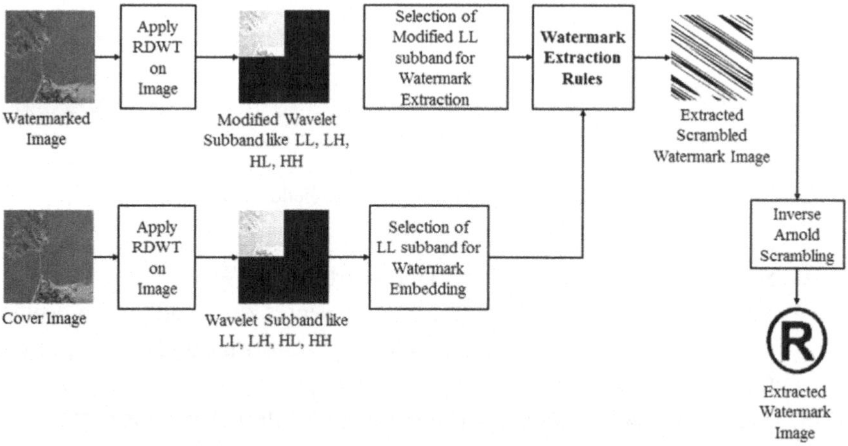

FIGURE 5.3 Block diagram of proposed watermark extraction process.

- **Step 4:** The scrambled watermark bit is extracted from the watermarked image as follows:

$$EW' = (WI_{LL} - I_{LL})/\alpha \quad (4)$$

where EW is the extracted encrypted watermark, WI_{LL} is the LL subband of the watermarked image, I_{LL} is the corresponding subband of the original cover image, and α is the gain factor.

- **Step 5:** Apply inverse Arnold scrambling using secret key K to get actual watermark information at the detector side.

5.4.3 *SIMULATION RESULTS OF PROPOSED WATERMARKING SCHEME*

The proposed scheme is tested and analyzed by grayscale remotely sensed image. The size of this image is 256×256. The test remotely sensed image is taken from the standard aerial Image Database [36] and is shown in Figure 5.4(a). A watermark image (logo) of size 256 × 256 shown in Figure 5.4(b) is used for testing.

(a) (b)

FIGURE 5.4 (a) Test cover remotely sensed image; and (b) watermark image.

In this technique, first, a watermark image is encrypted using Arnold scrambling transform with secret key K = 2. The RDWT is applied to remotely sense images to get LL, LH, HL, and HH subbands with the size of 256×256. The encrypted watermark logo with the size of 256×256 is inserted into the LL subband of a remote sensing image using the additive watermarking approach. Here LL subband is chosen for watermark embedding because it provides more robustness against watermarking attacks. On the extraction side, an encrypted watermark image with a size of 256×256 is extracted using the reverse process of watermark embedding. The inverse Arnold scrambling transform with secret key K = 2 is applied to extracted encrypted watermark to get the actual watermark image.

5.4.3.1 QUALITY MEASURES

Peak signal to noise ratio (PSNR) is used to measure imperceptibility between original and watermarked remotely sensed images. The PSNR given in Eqn. (5) depends on mean square error (MSE) which is an error between the original and processed image. The MSE is calculated using Eqn. (6). The MSE is measured in real values while PSNR is measured in dB. A high value of PSNR indicates more imperceptibility of the scheme.

$$PSNR = 10 \times \log\left(\frac{255^2}{MSE}\right) \tag{5}$$

$$MSE = \frac{1}{M \times N} \sum_{x=0}^{M-1} \sum_{y=0}^{N-1} (I(x,y) - WI(x,y))^2 \tag{6}$$

where I is the original remotely sensed image and WI is the watermarked remotely sensed image, respectively.

The robustness of watermarking scheme can be measured by structural similarity index measure (SSIM) [37]. The SSIM is calculated using Eqn. (7). SSIM measures the similarity between original and extracted watermark images. The SSIM lies in interval [0, 1]. The robustness of any watermarking scheme is high if the SSIM value is close to one.

$$SSIM(W, W^*) = \frac{(2\overline{W}\overline{W^*} + C1)(2\sigma_{WW^*} + C2)}{(\overline{W}^2 + \overline{W^*}^2 + 1)(\sigma_W^2 + \sigma_{W^*}^2 + 1)} \tag{7}$$

where W is the original watermark image, W^* is the extracted watermark image $\overline{W}$, and $\overline{W^*}$ are the corresponding mean of W and W^*, respectively, σ_W^2 and σ_{W*}^2 are corresponding variances of W and W^*, respectively, C1 and C2 are positive constants.

Using the above measures, the imperceptibility and robustness test of the proposed scheme is performed on a remotely sensed image for performance evaluation.

5.4.3.2 IMPERCEPTIBILITY TEST

An imperceptibility test is performed to check how degraded the remotely sensed image is after watermark embedding into it. The watermarked image appears as close as to the original remotely sensed image in the case of better imperceptibility. For the imperceptibility test, PSNR is calculated between the original and watermarked remotely sensed image and is compared with the Nazmudeen Scheme [17]. Nazmudeen Scheme is chosen for comparison because it is used for the protection of remotely sensed images and is more related to the proposed scheme. Figure 5.5 shows the quantitative results of the proposed scheme with the results of the Nazmudeen Scheme [17].

Watermarking Scheme	Nazmudeen Scheme (Nazmudeen and Farsana, 2014)	Proposed Scheme
Watermarked image (after embedding Watermark)	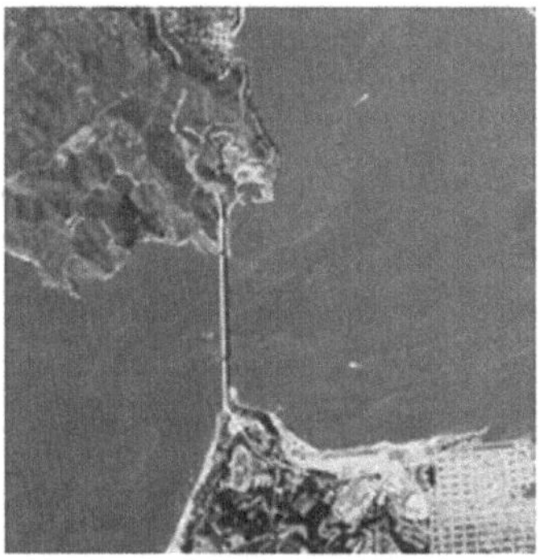	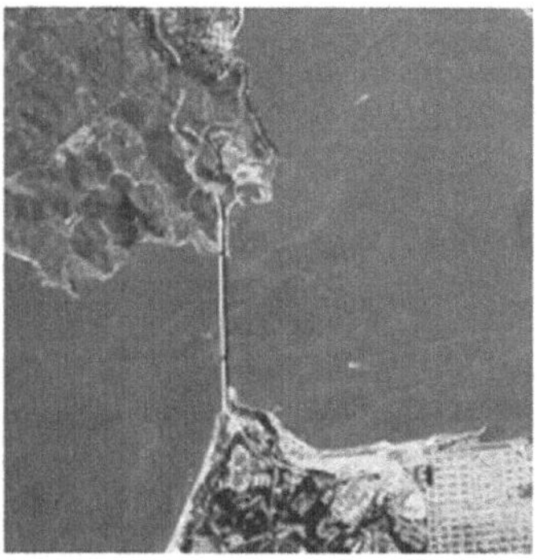
Extracted Watermark Images		

FIGURE 5.5 Comparison of watermarked image and extracted watermark image with Nazmudeen scheme [17].

The results include watermarked remotely sensed images and extracted watermark images. The corresponding PSNR values are tabulated and compared in Table 5.1. The PSNR of watermarked remotely sensed image of the proposed scheme is around 52.99 dB, while the PSNR of watermarked remotely sensed image of the Nazmudeen Scheme [17] is 52.04 dB. This indicates a slight improvement in the quality of watermarked remotely sensed images and, thereby, imperceptibility in the proposed scheme. The

SSIM values of extracted watermark images are also tabulated and compared in Table 5.1. An SSIM value for the proposed scheme is 0.9527, while the SSIM value for Nazmudeen Scheme [17] is 0.7540. This indicates a significant improvement in the quality of extracted watermarks with the proposed scheme than the Nazmudeen Scheme [17].

TABLE 5.1 Imperceptibility Test Comparison of Proposed Scheme and Nazmudeen Scheme [17]

Test Image	Nazmudeen Scheme [17]		Proposed Scheme	
	PSNR (dB)	SSIM	PSNR (dB)	SSIM
Remotely sensed image	52.04	0.7540	52.99	0.9527

5.4.3.3 ROBUSTNESS TEST

Various watermarking attacks such as JPEG compression, filtering such as median, mean, and Gaussian low pass filter, histogram equalization, sharpening, Gaussian noise, salt and pepper noise, speckle noise and geometric attack such as cropping are applied on watermarked remotely sensed image to verify the robustness of the proposed algorithm. The robustness is measured by calculating the SSIM. Figure 5.6 and Table 5.2 shows the objective and subjective robustness test results of the proposed scheme, along with their comparison with the Nazmudeen scheme [17] under various watermarking attacks. It is to be noted that the proposed scheme performs better than the Nazmudeen scheme [17] in resisting the attacks.

Watermarking Attacks	Nazmudeen Scheme [17]	Proposed Scheme
JPEG compression (Q = 90)		
Sharpening		

(Continued)

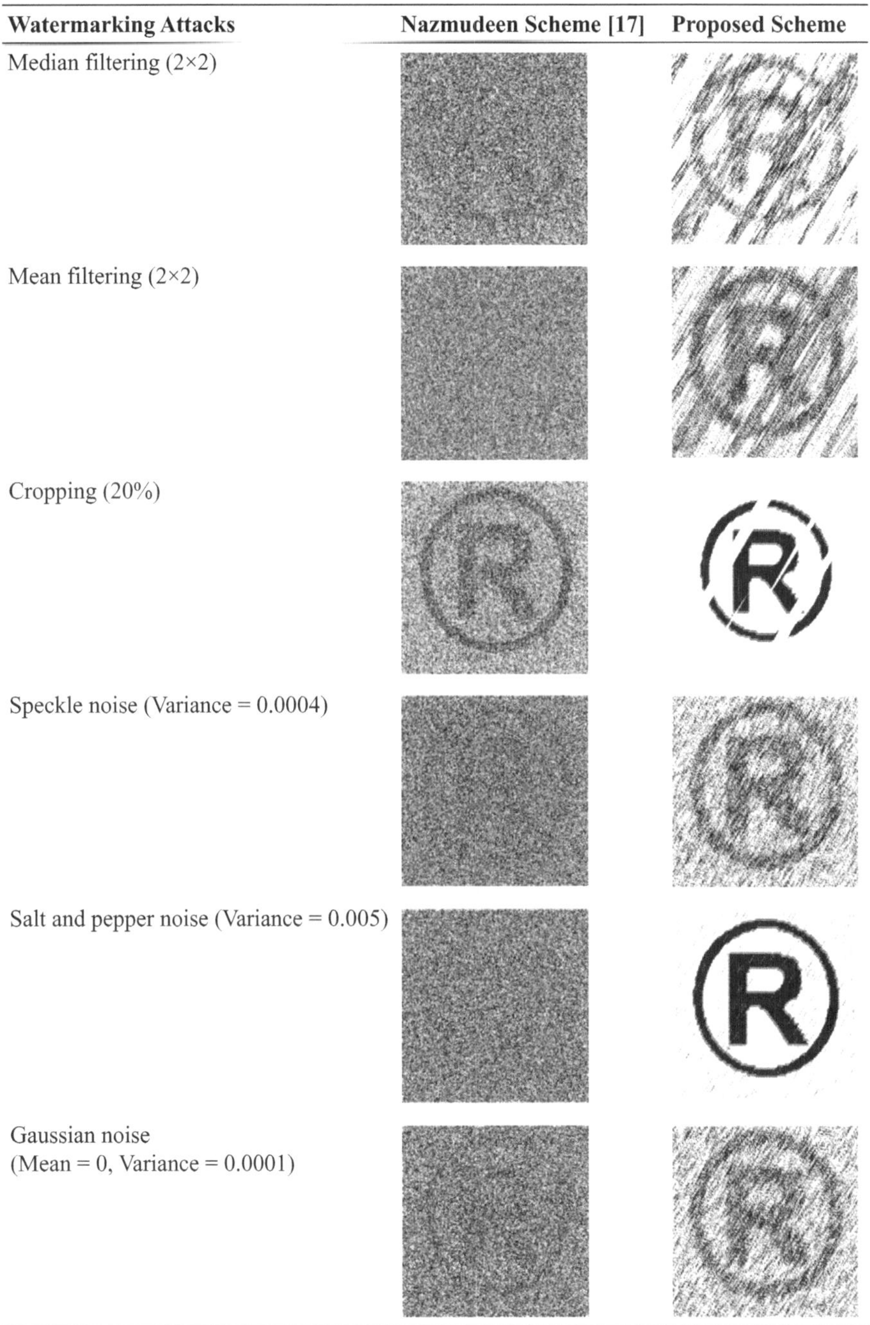

FIGURE 5.6 Extracted watermark images of proposed scheme and Nazmudeen scheme [17] under different watermarking attacks.

TABLE 5.2 Performance Comparison of Proposed Scheme with Nazmudeen Scheme [17] for under Various Watermarking Attacks

Watermarking Attacks	SSIM for Watermark Image	
	Nazmudeen Scheme [17]	Proposed Scheme
JPEG compression (Q = 90)	0.6091	0.9392
Sharpening	0.7415	0.7695
Median filtering (2×2)	0.6334	0.6501
Mean filtering (2×2)	0.6120	0.6345
Cropping (20%)	0.7283	0.8798
Speckle noise (Variance = 0.0004)	0.6410	0.8895
Salt and pepper noise (Variance = 0.005)	0.6124	0.8942
Gaussian noise (Mean = 0, Variance = 0.0001)	0.6394	0.7807

5.4.3.4 FALSE POSITIVE TEST

A false-positive test is a must for any watermarking scheme. In this test, a set of non-watermarked images is applied as an input to the watermark extraction algorithm. If a watermark is detected from the image, then it is called a false positive. For the false-positive test of the proposed scheme, 30 remotely sensed images are taken from the database and then the watermark extraction algorithm is applied. After obtaining the results of 30 images, it is found that the false positive rate of the proposed scheme is zero.

5.4.4 PERFORMANCE COMPARISON

In Table 5.3, the PSNR value of the proposed scheme is compared with the PSNR value achieved in some recently watermarking schemes developed for remotely sensed image protection. The comparison indicated that the imperceptibility of the proposed scheme is better than recently existing watermarking schemes.

The proposed scheme is compared with the Yu scheme [15] and the Nazmudeen scheme [17] by various features in Table 5.4. The comparison of the schemes is performed by the same remotely sensed image which is used in the proposed scheme. The watermarking in the Yu scheme (Yu scheme) is performed in the spatial domain, watermarking in the Nazmudeen scheme

[17] is performed in the DCT+ DWT domain, and watermarking is proposed is performed in the RDWT domain. The encryption of the watermark image is performed using Arnold scrambling in the Yu scheme [15] and the proposed scheme, while the encryption of the watermark image is performed using the Advanced Encryption method in the Nazmudeen scheme [17]. The average PSNR value of the Yu scheme [15] is 23.36 dB, the Nazmudeen scheme [17] is 52.04 dB, and the proposed scheme is 52.99 dB. The average similarity value of the Yu scheme [15] is 0.9975, Nazmudeen Scheme [17] is 0.7540, and the proposed scheme is 0.9527. The results indicate that the performance of the proposed scheme is better than the existing schemes [15, 17] in term of imperceptibility and security expect robustness.

TABLE 5.3 Comparison of PSNR (dB) Values of Proposed Scheme with Existing Schemes

SL. No.	Watermarking Scheme	PSNR (dB)
1.	Yu scheme [15]	23.36
2.	Nazmudeen scheme [17]	52.04
3.	Al-Mansoori scheme [18]	52.48
4.	Hsu scheme [19]	47.59
5.	Jiang scheme [20]	52.00
6.	Ma scheme [26]	48.74
7.	Panyavaraporn scheme [27]	41.42
8.	Kaarna scheme [29]	38.95
9.	Proposed scheme	52.99

TABLE 5.4 Comparison of Proposed Scheme with Existing Schemes [15, 17]

Features	Nazmudeen Scheme [17]	Yu Scheme [15]	Proposed Scheme
Type of watermarking	Robust	Robust	Robust
Number of image processing transform	2	0	1
Used transform for watermark embedding	Discrete wavelet transform (DWT), discrete cosine transform (DCT)	–	Redundant discrete wavelet transform (RDWT)
Used encryption method for security of watermark image	Advanced encryption method (AES)	Arnold scrambling transform	Arnold scrambling transform
Average PSNR (dB)	52.04	23.36	52.99
SSIM	0.7540	0.9975	0.9527

5.5 CONCLUSIONS

Since the last decade, a significant number of papers have been published for copyright protection of remotely sensed images using watermarking. In this chapter, various image watermarking schemes for copyright protection of remotely sensed image are discussed. This chapter also proposed a novel non-blind RDWT and Arnold scrambling based remotely sensed image watermarking scheme. The proposed scheme has shown improved performance in imperceptibility and security when compared with the existing schemes [15, 17]. The PSNR value is 52.99 dB and the SSIM value is 0.9527 for the remotely sensed image in the proposed scheme. This scheme may be used for copyright protection of remotely sensed images.

KEYWORDS

- **Arnold scrambling transform**
- **false positive test**
- **imperceptibility**
- **redundant discrete wavelet transform**
- **robustness test**
- **watermarking scheme**

REFERENCES

1. Kim, K., & Ro, Y., (2004). Enhancement methods of image quality in screen mark attack. Kalker, T., et al., (eds.), *IWDW 2003, LNCS, 2939* (pp. 474–482).
2. Hartung, F., & Kutter, M., (1999). Multimedia watermarking techniques. *Proceedings of the IEEE, 87*(7), 1085–1103.
3. Bender, W., Gruhl, D., Morimoto, N., & Lu, A., (1996). Techniques for data hiding. *IBM Systems Journal, 35*(3, 4), 313–336.
4. Langelaar, G., Setyawan, I., & Lagendijk, R., (2000). Watermarking of digital image and video data – a state of art review. *IEEE Signal Processing Magazine*, 20–46.
5. Cox, I., Kilian, J., Shamoon, T., & Leighton, F., (1997). Secure spread spectrum watermarking for multimedia. *IEEE Transactions on Image Processing, 6*(12), 1673–1687.
6. Wolfgang, R., & Podilchuk, C., (1999). Perceptual watermarks for digital images and video. *Proceedings of the IEEE, 87*(7), 1277–1281.

7. Borra, S., & Swamy, G., (2013). Sensitive digital image watermarking for copyright protection. *International Journal of Network Security, 15*(2), 95–103.
8. Borra, S., & Swamy, G., (2012). A semi-blind image watermarking based on discrete wavelet transform and secret sharing. *Proceedings of IEEE International Conference on Communication, Information, and Computing Technology (ICCICT-2012)* (pp. 1–5).
9. Thanki, R., & Kothari, A., (2016). Digital watermarking – the technical art of hiding a message. *Intelligent Analysis of Multimedia Information*, 426–460.
10. Thanki, R., & Borisagar, K., (2014). Discrete wavelet transform and compressive sensing based multibiometric watermarking–a novel approach to embed watermark into biometric. *Proceedings of 2nd International Conference on Emerging Technology Trends in Electronics, Communication & Networking (ET2ECN-2014)* (pp. 102–107).
11. Swanson, M., Kobayashi, M., & Tewfik, A., (1998). Multimedia data-embedding and watermarking technologies. *Proceedings of the IEEE, 86*(6), 1064–1087.
12. Caldelli, R., Filippini, F., & Barni, M., (2006). Joint near-lossless compression and watermarking of still images for authentication and tamper localization. *Image Communication, 21*(10), 890–903.
13. Fridrich, J., (2002). Security of fragile authentication watermarks with localization. *Security and Watermarking of Multimedia Contents Proc. SPIE, 4675*, 691–700.
14. Li, L., Zhang, C., Li, X., & Li, D., (2008). XOR based LSB watermarking for information protection in land consolidation project. *Proceedings of the 13th WSEAS International Conference on Applied Mathematics (MATH'08)* (pp. 279–284).
15. Yu, C., (2017). Steganography of digital watermark by Arnold scrambling transform with blind source separation morphological component analysis. *Multimedia Tools and Applications, 76*(5), 6821–6842.
16. Elad, M., Starck, J., Querre, P., & Donoho, D., (2005). Simultaneous cartoon and texture image inpainting using morphological component analysis (MCA). *Appl. Comput. Harmon. Anal., 19*(3), 340–358.
17. Nazmudeen, N., & Farsana, F., (2014). Satellite image security improvement by combining DWT-DCT watermarking and AES encryption. *International Journal of Advanced Computer Research, 4*(2), 645–652.
18. Al-Mansoori, S., & Kunhu, A., (2012). Robust watermarking technique based on DCT to protect the ownership of DubaiSat-1 images against attacks. *International Journal of Computer Science and Network Security, 12*(6), 1–9.
19. Hsu, P., & Chen, C., (2012). Feature based digital watermarking for remote sensing images. *International Archives of the Photogrammetry, Remote Sensing and Spatial Information Sciences* (Vol. 39, pp. 473–478).
20. Jiang, L., & Xu, Z., (2012). Commutative encryption and watermarking for remote sensing image. *International Journal of Digital Content Technology and its Applications, 6*(4), 197–205. doi: 10.4156/jdcta.vol6.issue4.24.
21. Miyazaki, A., & Okamoto, A., (2001). *Analysis of Watermarking Systems in the Frequency Domain and its Application to Design of Robust Watermarking Systems* (pp. 506–509). ICASSP'01.
22. Chen, Q., Xie, P., Ma, W., & Wei, W., (2010). A digital watermarking algorithm based on characters of the remote-sensing imagery. *International Conference on MASS 2010*, 1–4.
23. Hemalatha, T., Joevivek, V., Sukumar, K., & Soman, K., (2009). Robust watermarking of remote sensing images without the loss of spatial information. *10th ESRI India User Conference.*

24. Kaarna, A., Krasavin, K., Parkkinen, J., Jaaskelainen, T., & Schyndel, R., (2008). Watermarking spectral images with three-dimensional wavelet transform subject to various illumination conditions. *Journal of Imaging Science and Technology*, *52*(3), 030502-1–030502-18.
25. Kbaier, I., & Belhadj, Z., (2006). A novel content preserving watermarking scheme for multispectral images. *Proceedings of the ICTTA*, 322–327.
26. Ma, L., Li, C., & Song, S., (2005). Digital watermarking of spectral images using SVD in PCA transform domain. *Proceedings of ISCIT 2005* (pp. 1442–1445).
27. Panyavarapom, J., & Rangsanseri, Y., (2005). Digital image-in-image watermarking of remote sensing images. In: *26th Asian Conference on Remote Sensing and 2nd Asian Space Conference* (Vol. 2, pp. 1145–1150).
28. Piva, A., Caldelli, R., Bartolini, F., & Barni, M., (2004). Semi-fragile watermarking for still images authentication and content recovery. *International Workshop on Image Analysis for Multimedia Interactive Services*, 511–515.
29. Kaarna, A., & Toivanen, P., (2003). Digital watermarking of spectral images in PCA/wavelet transform domain. In: *2003 IEEE International Geoscience and Remote Sensing Symposium* (Vol. 6, pp. 3564–3567).
30. Ziegeler, S., Tamhankar, H., Fowler, J., & Bruce, L., (2003). Wavelet-based watermarking of remotely sensed imagery tailored to classification performance. *Proceedings of the IEEE Workshop on Advances in Techniques for Analysis of Remotely Sensed Data* (pp. 564–579).
31. Ho, A., Shen, H., Tan, S., & Kot, A., (2002). Digital image-in-image watermarking for copyright protection of satellite images using the fast Hadamard transform. In: *2002 IEEE International Geoscience and Remote Sensing Symposium* (Vol. 6, pp. 3311–3313).
32. Barni, M., Bartolini, F., Cappellini, V., Maglic, E., & Olmoc, G., (2001). Watermarking based protection of remote sensing images: Requirements and possible solutions. *International Symposium on Optical Science and Technology*, 191–202.
33. Roy, S., & Pal, A., (2017). A robust blind hybrid image watermarking scheme in RDWT-DCT domain using Arnold scrambling. *Multimedia Tools and Application*, *76*(3), 3577–3616. doi: 10.1007/s11042-016-3902-4.
34. Lagzian, S., Soryani, M., & Fathy, M., (2011). A new robust watermarking scheme based on RDWT-SVD. *International Journal of Intelligent Information Processing*, *2*(1), 22–29.
35. Hiena, T., Nakaoa, Z., & Chen, Y., (2006). Robust multi-logo watermarking by RDWT and ICA. *Signal Processing*, *86*, 2981–2993.
36. *For Remotely Sensed Image: The USC-SIPI Image Database,* (2002). http://sipi.usc.edu/database/database.php?volume=aerials (accessed on 28 September 2023).
37. Wang, Z., Bovik, A., & Simoncelli, E., (2004). Image quality assessment: From error visibility to structural similarity. *IEEE Transactions on Image Processing*, *13*, 600–612.

CHAPTER 6

A FUTURE CYBER SECURITY FOR SECURED SMART HOME APPLICATIONS USING THE INTERNET OF THINGS AND BLOCKCHAIN TECHNOLOGY

V. ARULKUMAR,[1] MARLAPALLI KRISHNA,[2] S. JAYACHITRA,[3] and A. PRASANTH[4]

[1]*School of Computer Science and Engineering, VIT University, Vellore, Tamilnadu, India*

[2]*Department of CSE, Sir C.R. Reddy College of Engineering, Eluru, Andhra Pradesh, India*

[3]*Department of ECE, PSNA College of Engineering and Technology, Dindigul, Tamilnadu, India*

[4]*Department of Computer Science and Engineering, Vel Tech Rangarajan Dr. Sagunthala R&D Institute of Science and Technology, Chennai, Tamil Nadu, India*

ABSTRACT

The increasing demand for comfortable and human living has enhanced smart home development. The Internet creates processes and huge amounts of data to handle user's demands effectively, including many highly intelligent homes. This increased demand gives rise to a great deal of concern regarding scalability, efficiency, and safety for a smart home system. These are all tedious questions, and there are no granular studies to overcome them. Detailed data, or the lowest levels that can be in a target set, are Granular Data. It refers to the size of the data fields, in short, to how specific a single

Multimedia Security: Tools, Techniques, and Applications. Bhaskar Mondal and Shyam Singh Rajput (Eds.)

field is. The more subdivided and more specific data, the more granular is also considered. Given that safety and efficiency are necessary as a challenge, this chapter is presented with the combination of integrating blockchain (BC) technology and cloud computing in an effective manner. BC technology is decentralized because it can serve processing services and copy sensitive user data from smart home. Our model uses a multivariate correlation analysis and identification of correlations between traffic functions to guarantee the safety of the smart home network. The performance of our architecture was evaluated with various performance parameters, and BC was found to be an effective security solution for future networks on the Internet.

6.1 INTRODUCTION

The "Internet of Things" (IoT) consists of devices that generate, process, and exchange large numbers of critical information and information about security, as well as confidentiality. Many new IoT networkable devices are light and low in energy. The majority of available energy and computer systems must be used to implement core applications, making it very challenging to provide affordable security and privacy support [1]. In addition to energy consumption and overhead process, traditional security methods tend to be costly to IoT. Furthermore, most of the state-of-the-art safety frameworks are highly central, and the traffic nature and the one point of failure are due to the difficulty of the magnitude [2] and are therefore not necessarily suitable for IoT. Existing methods often reveal noisy information or incomplete information to protect the security of users, which could prevent certain IoT applications from offering personalized services [3]. In IoT demands, the security and data protection provided be lightweight, scalable, and distributed. With Bitcoin's first crypto-currency system, Blockchain (BC) technology [4] could meet these challenges in terms of its distribution, safety, and privacy. Bitcoins have the potential to meet.

Bitcoins users, known through changing cash transactions, are generated and transmitted to the network by Public Key. The users are dividing transactions into this block. If a block is finished, a mining process is added BC. To denote a block, some specific nodes, called miners, try to solve a cryptographic puzzle that consumes resources called the proof of work (POW) [5], which first solves the puzzle by blocking the new BC block. We argued that the introduction of BC into IoT is not easy and poses a number of major challenges in our previous research [6], such as high demands on resources for the solution of POW, long transaction confirmation latency, and low scalability resulting from transactions and blocks broadcast across the network. The POW concert is removed and the coins are needed, we suggested a new instantiation of BC. In keeping with the safety and privacy of BC while being better adapted to the specific needs of the IoT our framework proposed is based on and distributed on a hierarchical structure confidence. In a smart home, we have exemplified our ideas, but we have an agnostic framework and can also be used in other IoT environments [7].

The design includes three main levels: intelligent home and is central to minefield. Service Providers, the cloud storage, smartphones, and personal computers as shown in Figure 6.1 represent an overly network of Smart Homes [8]. The Overlay network is similar to Bitcoins network peer-to-peer and distributes functionality to our architecture. The overlay nodes are based on clusters in order to reduce overtime and overtime for the network (CH). In conjunction with two key lists, the CHs maintain a BC public overlay. These keys are listed as: Require key lists that are the list of PKs overlay users which can access data for intelligent household connected to this cluster; request key lists which are the list of PKs of smart households combined to this cluster which can be accessed. Cloud storage is used to store and share data on intelligent home devices. In previous work, we discussed details of cloud overlay and storage [9]. Automation is a method of functioning in which no human involvement is required.

Automation is the foundation of creating an office or home automation system. Researchers and industrialists are developing simple and cheap automated systems to monitor and regulate various types of equipment, such as lights, fans, and AC, depending on the occurrence of the requirement. Automation not only is efficient, but also helps keep costs down while using water and energy to their maximum potential.

A grant is offered to individuals and devices that are alive and connected, able to connect at any time, anyplace, and utilizing every network and every service.

Using IoT technology for automation is one more major use of these technologies. Everything uses sensors and actuators to regulate lighting, temperature, and humidity in buildings, schools, workplaces, and museums. Innovative houses are a contemporary design advancement that are known as smart homes, intellectual buildings, integrated home systems, or domestics. More and more, home automation and smart homes use generic devices that control many aspects of the house. A short while ago, environmental control systems, such as home heating and lighting, could be managed using smart home technology. Since the recent technological advances, however, almost all electrical devices inside the house may be integrated into the system.

Furthermore, smart home technology is not only concerned with gadgets on and off; it is also concerned with keeping track of the environment, as well as identifying how the occupants of the house are interacting. The outcomes of this technological development are that a smart house may track the movements of the residents of a home, control various equipment according to a specified plan, or conduct various functions as the user desires [10].

One of the main concepts employed in assistive technology is smart home technology, which utilizes the same gadgets that are used in assistive technology to help construct an environment in which various elements in the house are automated, and various home systems may interact with each other. The source of our communication capabilities is the 'business.' A busline is a cable that links all the devices together and makes them all inter-connected.

6.1.1 TECHNICAL APPROACH

A smart house may be thought of as a home with smart items, which connects to a home network, and a gateway that allows the devices in the home to communicate with the external Internet. It is now possible to monitor or communicate with inhabitants through smart objects.

Smart objects may be anything from a light that can regulate or sense its own condition, to a fridge that is capable of feeding itself, to security systems, to movies that are requested by others. Everything in the house will be connected to the home network so that it may register its own status or get instructions. When you use your home networking, you provide your house the ability to be linked and controlled both inside and outside. A home gateway gives access from the outside through Ethernet or the internet. It is feasible for a person's house to connect to the newest services, including those that provide video content. The service provider ensures the provision of new services for residents, as well as the availability of these services. The present chapter is designed to discuss the details of our intelligent home design extensively. First, we describe how to Initialize IoT devices and process transactions. Local and private BC is used to provide secure IoT device and data access control. In addition, the BC provides an immutable history of transactions, which is linked to other third parties for the provision of certain services. The safety of design is based on a range of functions, such as – (i) indirect defaults; and (ii) the smart home and the overlay have different transaction structures. Symmetrical encryption is used to ensure lightweight security in smart home devices. We are arguing qualitatively that the smart home is confidential, integrity, and accessible and that key safety attacks such like connection attack (10) and DDOS can be discussed. Finally, we present quantitative results using simulations and show that our framework generates relatively small overheads. We present the main design sources in the upcoming remainder of the chapter.

6.2 RELATED WORKS

6.2.1 DISTRIBUTED BLOCK CHAIN

Blockchain (BC) is a process that can keep all members and update the BC ledger in new transactions to ensure coherence. The BC is a real thing accountable mechanizm, without a lost transaction. BC it ensures validity of the transaction by keeping the transaction's node register, but also distributing it to the whole network. The development of Internet and its cryptography technology enables all members to check their transaction liabilities so that dependence on authorized third parties has been remedied and there has been a single failure point been achieved. If data transactions take place across multiple organizations own and managed networks, continuous records allow for data tracking, inherently transparent block chained records. Anybody with network connectivity can track and analyze all activities. In addition, some BC networks like Ethereum offer the "intelligent agreement" functionality to produce contracts when terms are fulfilled. It is promising to retain BC and IoT technology. The industry and the general enthusiastic support of the public sector have already been received. The fact that data is not saved in a central, fully decentralized position is one of the most interesting aspects of BC technology. This avoids the need for a central string agency and provides control to the individual user. Moreover, BC technology offers traceability and economic efficiencies for the management of supply chains. It is used for tracking movement of goods, origin, quality, etc. This brings a new level of transparency for the B2B ecosystem.

The network checks the validation and invalidation of the new transaction in a former transaction is also prevented to ensure that the BC includes only legitimate transactions. The BC will be added to the new data block only if the validity of the transaction [11] is agreed upon by the computer. Consensus within the network is based on other voting methods and the work that most often depends on how far the network users are able to access and verify transactions contained in the block, it can be permanently added in the block. Structured list contains BC that stores data as a distributed database in a similar format. It should be easily manipulated and BC verification as a participant in the network store.

In Figure 6.1 of every block of the book, there is a header and footer. High values are stored in the header section, time stamp on the current and preceding blocks. Transactions are part of the corpus. The index method for retrieving block data is used [12].

6.2.2 *GREEN CLOUD COMPUTING*

Green cloud computing is the study and practice of designing, manufacturing, using, and disposing of computers, servers, and associated subsystems.

Key issues are energy efficiency in computing and promoting environmentally friendly computer technologies. Cloud researchers take into consideration the unique benefits of enabling a computer system. Cloud computing offers the pay-per-use service, which eliminates early investment. Cloud computing, virtualization, multiple-lease, cloud storage, hypervisors, and the cloud network are the five main components in creating an efficient smart home.

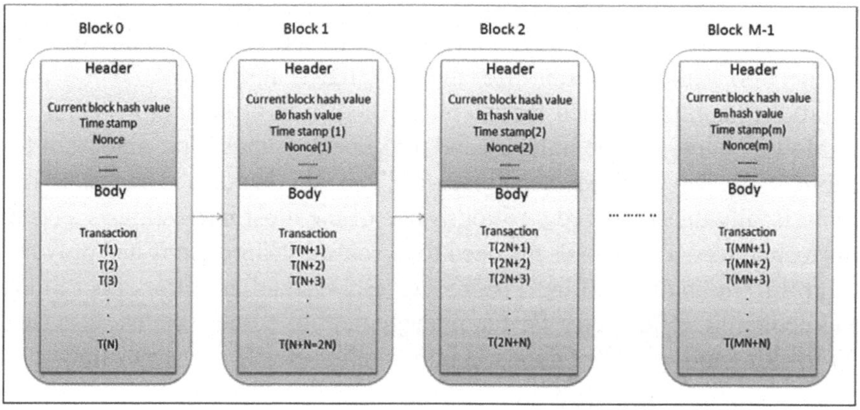

FIGURE 6.1 Structure of blockchain.

It should complete the service level agreement (SLA) with regard to the diverse service of third-party organizations with quality requirements. In terms of a smart home, all the nodes will send the information through an intelligent gateway and is sent to the cloud. Figure 6.1 represents the Structure of BC.

Energy use and cloud infrastructure carbon emissions are key issues. The process is the main driver in a cloud for energy efficiency to present logic groups or computer resources. Flexibly, geographically, cost effectively and energy sensitive cloud computing offers green computing equipment as well. In order to vanish the energy inefficacity disturbance and reduce carbon extractions, there are many approaches to virtualization. According to current study, a migration application in the cloud can reduce carbon emissions to 30% per user.

Various alternate methods also propose a cost effective model in order to calculate service costs and also ideas to minimise energy use. Many factors such as a single unit task, related analysis instruments, the empirical method, and various run-time jobs and workload scheduling will be taken into account when calculating energy consumption. Though architects and system designers have long had a vision of smart, adaptable houses, it has only just been materialized. Early experiments in the 1970s failed because of a notable advancement in science. Current injected framework advancements have emancipated some of the development of skilled homes and their automation. Whether or not that's true, sophisticated mobile phones are widely expected to spread, given how commonplace they already are. It is referred to as 'shrewd house' because it adds sensory devices and actuators to increase a tenant's knowledge of their surroundings, allowing them to intuitively manage their home technology. In general, 'home automation' is considered to be the method to operate many disparate systems through a computer. Besides, you can trust that the drive that empowers brilliant residences is greater, as it also contains something more extensive: beyond executing the instructions, but as well as being comfortable and being flexible.

When seen from three different viewpoints, smart houses have three different traits: utilitarian, instrumental, and socially specialized. Utilitarianism is concerned with implementing and mechanizing day-to-day functions in a house to help people. Competent abilities, for example, vitality administration and security management, are considered instruments in the instrumental perspective. The socio-specialized perspective aims to digitize daily life by stimulating human-building communication and aims to introduce the digital era of human-building communication.

An ideal house provides comfort and consolation to its clients and need to have a heating system that includes both air and water, as well as ventilation, light management, and window ornamentation. However, regardless of that, it is important to keep in mind that having control over your home, as though it were a machine, is only a part of the overall satisfaction with home computing; much of it originates from connecting with your family, so a concealed framework is imperative in order to ensure clients are not displeased or let down, no matter how quickly their home may update. Innovation should be made to liberate people rather than constrain them. At the conclusion of the day, it is important to ask, "how keen ought to be the wonderful home?" generally speaking, a person who puts in place a comprehensive home framework will have to accomplish one of these basic objectives: a real feeling of serenity when managing home affairs, streamlining, and energy effectiveness, entertainment or the awe of others, personalization, and supporting individual needs mechanized at home.

At the outset, the smart house concept was confined to concepts and ideas. The scientific writing community has studied home automation for a long period of time.

Industrious essayists predicted a future when houses would be automatically managed, and allegedly had developed intelligence of their own. This poem has a vision of a robotized house that continues to function even when humans are no longer living there. Everything was all right, and then it occurred to me that there were benefits to home computerization that were better than scary.

While smart houses have been around for quite some time, the really clever ones are just just beginning. This event is focused on equipment; this implies new, real things which may predict future developments. The arrival of household machines in the first decade of the 20th century Even when it comes to household technology, a great achievement was made in the mid-20th century.

All of these achievements may be traced back to the vacuum cleaner, which was first marketed in the year 1901. A down-to-earth power source was born in the year 1907 when an additional down was added to the machine. Over the course of two decades, more modern laundry equipment, including dryers, clothes washers, irons, and toasters, were all developed. Filling in as a servant by an extremely wealthy household was an incomprehensible moment for any person.

ECHO IV and the Kitchen Computer — It was never really available for sale, but the leading smart device was ECHO IV from 1966 to 1967. This cunning device is able to glean purchasing data, regulate the home's temperature, and start and stop various equipment. It came out a year later, and was known as the Kitchen Computer. It stored formula, but its tagline was "In the event that she can only cook and Honeywell can PC," and they didn't sell any units.

Join gerontology and innovation to relieve older people of their burdens. There was a wealth of fresh research and invention in 1990, as shown by a cluster of new projects and endeavors.

Late 1990s to the mid-2000s—The Development of Smart Homes Smart houses, or home robotization, saw a sudden increase in popularity in the early to mid-2000s. The road to countless advancements has been paved by everything. Savvy homebuyers were able to rapidly convert into a more realistic choice, which inspired other homebuyers to consider it as well. Manufacturers begin to include residential technologies and home systems management in retail stock.

Present-day smart houses—that is, smart homes that have a healthy regard for security and environmental sustainability—tend to be more about both security and environmental friendliness. This ingenious building helps to ensure that a house is not draining meaningless energy. This also serves as a warning to those who may have breached our perimeter.

The technology behind home automation has been modernized in recent years to include remote portable control, computerized lighting, mechanized indoor regulator adjustment, and computer-controlled mechanized valves and accessories.

6.3 IoT SMART HOME CHALLENGES

1. **Privacy and Safety:** Real-life communication poses enormous challenges for security, confidence, and confidentiality. IoT already faced numerous security threats and attacks. Due to huge transmissions some opponents can attack the communications of significant network data, e.g., a man-in-the-middle (MitM) attack and DoS/DDoS attack. IoT presents several unique privacy issues, including privacy issues, telephone, and automotive tracking systems. Voice recognition is also integrated to actively transmit the data to cloud storage for processing. The problem with India's IoT consumers is that they are facing many fundamental problems and obstacles, which is why it would be difficult to deploy IoT in India. Additional IoT adoption difficulties include data security, data storage, and ownership issues. You will find many of them in other parts of the world as well. Notwithstanding these difficulties, here are some additional issues facing IoT in India.

   Amongst the important issues in India these days are the existence of the Internet, reliance on necessary association transmission speed, and open availability of the Internet. With respect to being a smart property owner or being an IoT customer, this may remain the biggest challenge.

2. **Cost of IoT-Enabled Smart Homes:** Clients in India with respect to any innovation tend to speculate on the level of convenience and innovation offered by it.

   Global merchants believe Indian customers don't have the skills to push technology. This is particularly noticeable when there is an instance of great home innovation and IoT space implementation, with any variation in seldom bought stock movement. Due to this, we will see minimal familiarity with IoT systems among Indian customers, and a spectacular new device in the homes of these customers.

3. **There are Infrastructure Problems that are Unavoidable:** Regardless of the fact that the supporting web foundation, Smart Power Matrices, Smart Water Supply, and others are same, they cannot be prepared in time.

4. **Compelled by Shortage of Expertise:** Because IoT applications need talented personnel throughout the country to put them into practice, use is further inhibited by a scarcity of such workers. According to the Labor Bureau of 2014, the skilled workforce in India accounts for only 2% of the total population. With these assumptions in mind, we may reasonably anticipate working on some instructional endeavors that will answer the workforce's requirements, meeting the objective of helping the workforce make the most of this innovation.

   While it may be new for a certain audience, hacker attacks have historically been a feature of the more sophisticated society. In any case, the programmers' assaults in the smart house may be what causes enormous losses of security and well-being for the property holder and merchant alike. There are a few information hazards, such as false information, denial of administration, espionage, performance overload, malicious alterations, assaults, and so on. Doing so must be a priority for home-owning proprietors.

5. **Necessity vs. Security:** Does the homeowner really need to allow remote access to his residence? Do webcams need to be installed inside the house, or is just a radio set up at the entrance required? In the event that the owner is anxious about interlopers, a function known as "nearness sensors" works extremely effectively. This feature does not need web cameras.

6. **Limit the Number of Open Ports on Devices:** Making ensuring a house doesn't have any open ports in its switch is a good way to

guarantee privacy at home. Once the owner chooses to limit remote access to his house, the use of a VPN rather than opening ports in the switch may be implemented.

7. **We Use Trustworthy Third-Party Tools:** When using the web interface remotely, it is recommended to always utilize the seller's machinery, software, and appliances. Therefore, in this case, the owner may have skin on his/her outer layer, for example, in this case, a "occasion mode" that is not opened to cameras allows cameras to be removed from the area (LAN) the risk of data ownership.

   Biometrics have traditionally been the most widely accepted security mechanism, according to the security assessment Intel did in 2015. Around 48% of respondents chose fingerprints as the best option, while another 44% chose voice acknowledgement, and still another 42% chose eye examinations. At the same time, however, a key problem related to the ownership of information is whether or not it ought to keep the data on the server at a local location for increased use of neighborhood reference information or store it on a merchant's server for improved security.

   When it comes to hacking information, there's a possibility of information theft. To this day, there is no device that can prevent information hacking entirely. To meet this need, owner has to invest in a noteworthy server security framework in order to keep critical information on the proprietor's nearby server or use virtual private system technology if the merchant chooses to use the distant server.

8. **A Future Cyber Security for Secured Smart Home Applications Using Internet of Things and Blockchain Technology:** Given the concentration of the information store, application preparation, and application stockpiling, it is vital to have a strong security framework in place. While decentralized control systems do have the challenge of the single point of failure, and this is a disadvantage to control, the odds of power loss also increase. Unfortunately, this is not all destiny and sadness, since there are many reasonable ways to circumvent this. Greatest design is when an uninterruptible power supply (UPS) is used so the framework continues running in the event of a power outage. Informal phrasing: Mini-servers may act as a reinforcing framework for more smart house owners. Course of action when dealing with a certain kind of crisis, for example, by wirelessly unsettling power circuits on circuit systems without a memory card, a business may ensure dependable memory card storage.

6.4 PROPOSED METHODOLOGY

In order to increase the performance, scalability, and availability in the smartest home, the proposed SH-Block CC block will use cloud calculation and the technology BC. Figure 6.2 represents the smart home architecture. This wireless and embedded system was designed utilizing integrated wireless and system technologies. In this model, two different approaches are used to communicate between the mobile user and the visitor. A visitor motion is recognized by the user of the house, who may be located anywhere, and the residence is protected against unauthorized and unwanted visitors. Mobile with different choices, such as "active" and "hibernates" mode is also included. In order to observe the movement of visitors as they utilize electrical equipment, such as light switches, over a period of time, different choices are offered to users through the mobile app. This system detects any violations and issues an alert if they occur.

Wireless-based home automation is utilized to reduce costs and effort while enhancing system flexibility. Mobile devices are used in a home automation system to offer a user interface. Home automation is something apparent to our lifestyle, which is both growing and established.

The incidence of theft is higher in an area where security is provided, as a result, there is greater security.

Using GSM, mobile phones, and electrical circuits, this system tends to rely on an automated system designed to function as a thinking device. For security, a strong security system must detect as well as anticipate threats. Low-power consumption motion sensors are located where an intruder must go.

A text message has been sent to the mobile station, which is equipped with a GSM modem, and therefore is used to notify the owner that an unauthorized user is present in the house.

In order to provide users a way to control their home appliances, Arduino-based home automation has been developed. This project demonstrates the design, implementation, and flowcharts of a home automation system that uses Arduino. The primary objective of a home automation system is to make people's lives easier by enabling them to manage home appliances using their Android phones, while also saving money, time, and energy. The burglar alarm and mobile home notification features work together to keep houses protected while owners are away. The alarm will sound if a burglar attempts to enter the property, and the homeowner will get a notification on their mobile phone anytime someone else tries to enter their home.

Remote users may greatly benefit from IoT-based home security and automation. This chapter's prototype may be implemented in any household. The basic block of all internet-based varied applications is this IoT-based system. The proposed system is a cost-effective solution for IoT applications. It was built using low-cost, simple-to-use components. In addition, it makes things easier to operate and quicker to obtain information. It allows the user to access the file from anywhere in the globe with a simple computer connection.

The home automation system has been built using the Arduino microcontroller, and it has worked to perfection. The Android application was used to turn the home appliances on and off. Results were promising on several phones.

6.4.1 ARDUINO MICROCONTROLLER

As an alternative to microprocessors and microcontrollers, Arduino was developed to accommodate different microprocessors and microcontrollers. Digital and analogue input/output pins are available on the boards and may be connected to other boards or circuits. Universal serial bus (USB) is used for loading applications from personal computers on certain models, which utilize serial communications interfaces. The microcontrollers are programmed using the capabilities of C and C++.

6.4.2 ELECTRONIC VOLTAGE REGULATOR

The voltage regulator is used to maintain the consistent voltage level. Additionally, it is also used for stabilizing the AC and DC voltage – feed-forward design or feedback control.

An automated and intelligent home system consists of home gadgets that are automated and managed from anywhere. Security is the most important aspect of every business, including factories. To safeguard the property from different obstacles such as crime and loss, the primary objective is to achieve this goal. This chapter provides information about the microcontroller's security mechanism. It's also very dependable since it is never in need of repair. This chapter explains the cheap and simple setup of a smart home automation system. This section aims to create a home security system that relies on cloud computing while keeping the total cost low.

6.5 COST EFFICIENT COMPUTING

As a matter of fact, cloud computing makes use of the "computing as a service" model and enables clients to access a multitude of resources, such as memory and software, that are shared over the internet via different internet protocols. When you call the service, your customers will be able to activate their local apps or their mobile phones.

Cloud computing delivers web-based applications to customers using software as a service. In addition to its use as a software development platform, the cloud offers additional function that classifies it as a platform as a service for developing apps. It also offers a service architecture for generating virtual machines, which may be useful if one is building a custom environment.

Operating costs are reduced, and maintenance costs are decreased via cloud computing. It has both the operating system and virtual desktops to clone new systems in a fraction of the time. By virtualizing the server and desktop, the administrator, software developer, and end-user may have full control over the operating system or desktop that they use. Another option is to store their work as software field and access it whenever they want. Virtualization is especially useful since the whole system may be housed in a few strong operating boxes. There are many components to cloud services, which do not stand alone. Cloud storage has many significant advantages:

A consumer wants to be able to transfer a file from cloud storage to their local hard drive. Bandwidth: Don't send e-mail attachments to other individuals and instead provide everyone with a direct link to the information.

- It is possible to save and view files anywhere online with an internet connection;
- It provides a second copy of critical files in the event of a disaster;
- Cost Saving: No extra storage is required.

The system's purpose is to provide an automated smart home that's safe against theft and unauthorized access. This system is used to regulate many house features such as fire, intrusion, gas leakage, and various types of smoke. An alert is automatically delivered to the homeowner's e-mail inbox if anything occurs in the house.

This system utilizes the Intel Edison Board as a foundation, and it uses sensors with various specifications to meet the requirements of the end user. Unusual circumstances within the house, such as excessive temperature or gas leaks, are detected by the system. A hazard alert is sent to the user's email inbox with an indicator of the danger level, which falls into one of three risk categories: low, moderate, or high. This motion-detecting technology utilizes image processing to perform its operations. The image processing idea is utilized as the foundation for the algorithm. The goal is to find out whether anything new is happening and set off a buzzer sound if it is.

When an unknown source is found, the buzzer will sound, and those who found it will be aware of it. In order to identify hidden sources, an image processing algorithm is used. The first thing to do is complete the video camera interface. Each frame of video is processed if there is enough input from the camera. Blurring the input feed is how the input picture gets blurred. The foundation picture is already set. Image subtraction is done by considering the incoming input feed and base image. Conversion of the RGB values to grayscale has now been completed. The threshold values are set.

6.6 SMART HOME LAYER

Smart home is a single family of intelligent families consisting of numerous IoT devices.

6.6.1 DISTRIBUTED BLOCKCHAIN (BC) LAYER

In a number of industries, BC has just received stakeholders' attention. The reason is that BC technology permits distributed manipulation of applications that used as an intermediary trustworthy. This is a distributed directory, which data multiple transactions throughout the chain effectively and checkable without a master host. Similar practice may be used without a central authority under the same service contract. BC offers a distributed network in which people can communicate empirically without trusted intermediaries [13].

6.6.2 DISTRIBUTION OF BLOCKCHAIN (BC) IN CLOUD COMPUTING

The Cloud Distribution model allows users to safely and decentralized store data. This is achieved through the use of BC features like books, public/private key encryption, etc., which we discussed previously in this chapter. There was a mistake. The decentralized aspect makes it impossible to compromise the central servers.

1. **Decentralized and Trust:** BC gives together a decentralized structure without one transactional authority. In accepting the transaction in order to secure BC confidence, the nodes in the network should agree [14].
2. **Complete Privacy:** Each user keeps his or her own key in BC. This stores crypto currency systems encrypted user data fragments and various privacy issues can be overcome [15].
3. **High Quality and Facilitate Usage Resource:** With its distributed application, it easily measures resources. For example, with some encrypted functions, an intelligent agreement must fulfil a demanding resource algorithm. This request provides for a clever computer contract with a resource algorithm upon demand that automatically pays for the payment after the performance of the function.

Summarizing, the purpose of BC services qualitative traceability of resource use can be improved [15, 16] so that SLA can be checked by both clients and the reported faults may be determined by service providers and the party responsible.

6.6.2.1 HANDLING TRANSACTION OF BLOCKCHAIN (BC) IN SMART HOME

All devices are managed and stored in the local BC via transactions. Local communication between devices or between overlay knots can do this. For some tasks, like shopping and access, monitoring, and formation mode, each transaction is programmed and removed.

The entire Diffie-Hellman algorithm transaction uses a shared key. In an intelligent home, a genuine transaction is added to create an intelligent house using a shared key. The unit is added to the minor. The weight hacking process is intended to detect changes in the content of transactions. The device is assigned to minors to manage the transaction user with a common key. The children inspect the policy header to give the key and ask the owner for their permission.

Therefore, the device can communicate while the key is valid. The process of genesis employs two approaches to data access in order for the addition of a new device; local access; and secondly cloud access. The equipment sends the miner an application to check and store local data for local permission. A miner can request cloud storage information and return to cloud. The home layer, blockchain network, cloud, and service layer are presented in Figure 6.2, and four elements are included.

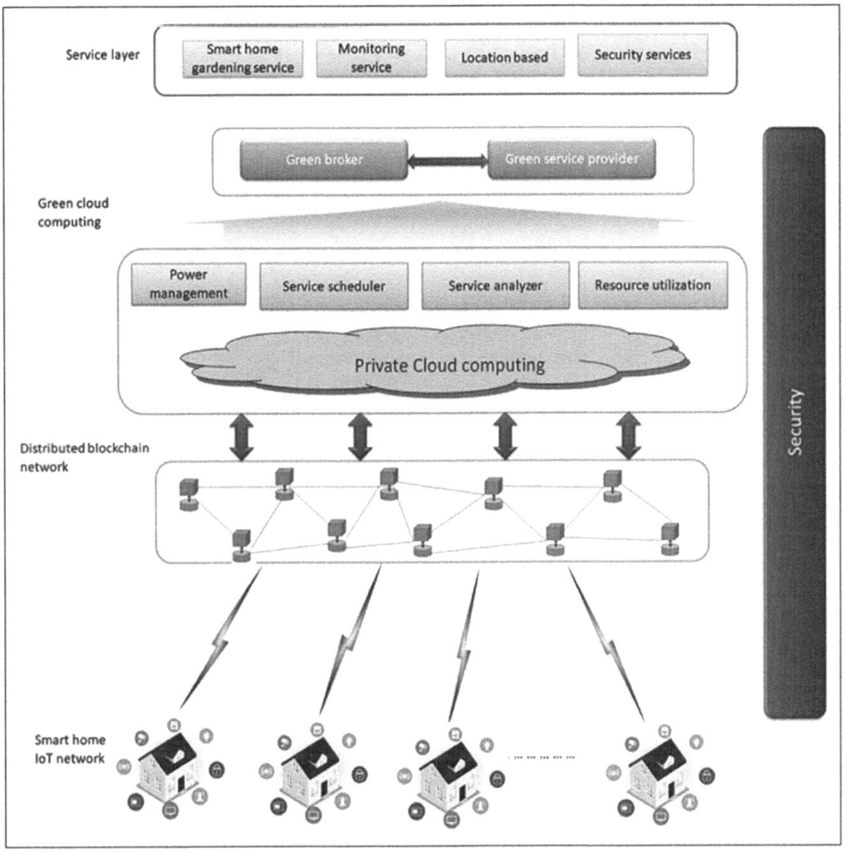

FIGURE 6.2 Smart home architecture.

The process is known as a transaction. The device needs local authentication, and sends miners to save the weather device for permission. The key is shared with the local storage device when permitted, and the data is stored directly in the local storage. To store data in the cloud, you need the same blocks. For saving data, the user will authenticate the block and hash numbers. User data packages are saved in block process after the successful authentication in the first-in-first out order (FIFO) together with hash (Figure 6.3).

6.7 EXPERIMENTAL RESULTS

The proposed architecture has been thoroughly evaluated in a number of scenarios. We have established an accuracy test of a network traffic attack detection model. The security model analyzes the detection of anomalies in the network with intelligence. This section presents the assessment records. Furthermore, we evaluate performance assessment in the intelligent and independent home use of BC technology.

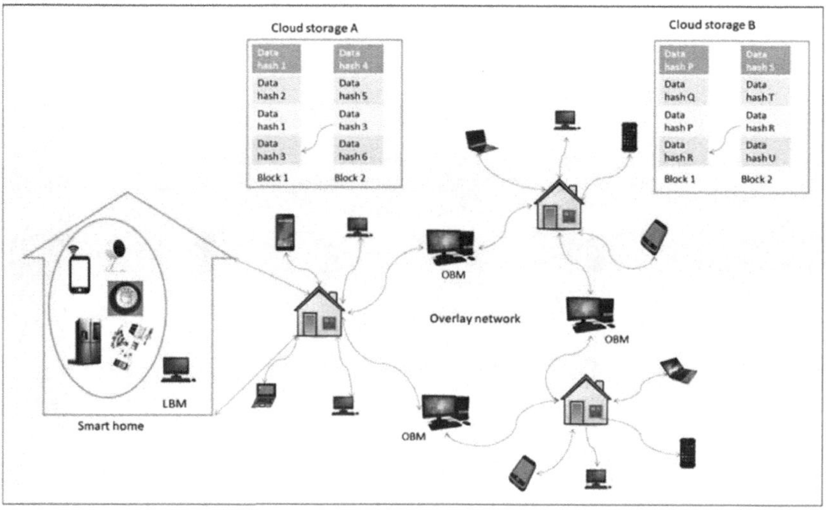

FIGURE 6.3 An overview of the smart home, OBM, and cloud-based overlay network.

6.7.1 PERFORMANCE EVALUATION USING BLOCKCHAIN (BC) TECHNOLOGY

In this model, we assess the different models of the proposed performance. For independent operation, we calculate privately the smart home and the overlay. Two simulators are used here:

1. **Cooja:** It is constructed through Contiki OS, one of the optimized OS for a small amount of memory and minimum processing capacity. Cooja is suitable for IoT devices with resource restrictions. It emulates different IoT device functions, including lighting, heating, and pressure.
2. **Netsim:** The IoT network applications, the simulator and emulator is being tested. We have been evaluating overlay performance here.

The proposed BC model has overhead and time calculations on a smart home device. Cooja and Netsim use the local block manager time to evaluate this, since the local block manager processes all transactions and encrypts them symmetrically and asymmetrically. Figure 6.4 represents the network delay.

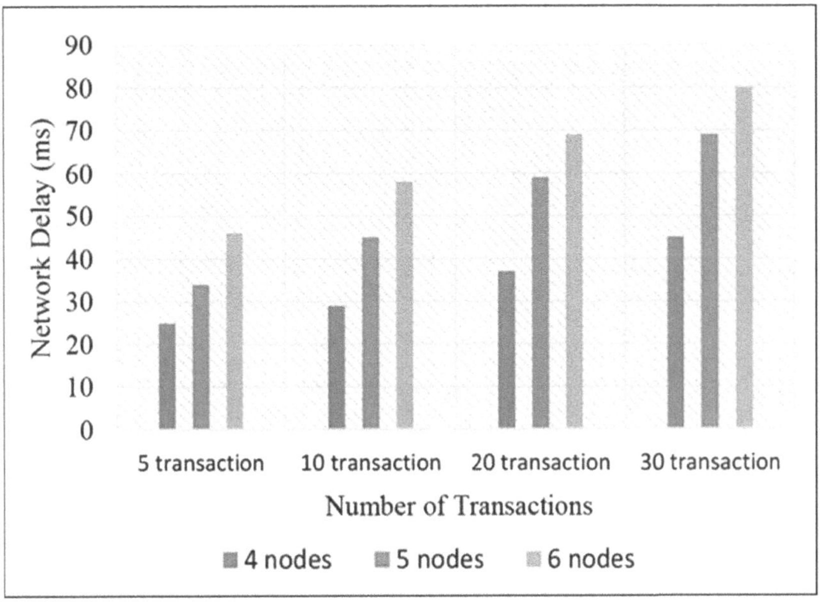

FIGURE 6.4 Network delay.

Figure 6.5 represents bandwidth, and network overhead is represented in Figure 6.6. This figure shows a single, two, and five blocks of memory transaction for various block sizes. Increasing the transaction per block reduces the memory. When block numbers increase, it requires more memory for a fixed transaction. The results show that less is used with higher memory and vice versa. The new block automatically forms if the transaction increases by blocks to limit when this meets the block size. But the consumption is increasing. This reduces storage utilization by more transactions per block.

6.8 SCALING THE BLOCKCHAIN (BC)

The number of resources each sensor node uses within a network, like bandwidth, energy, memory, and time, is defined as Network overhead. We took approximately 100 nodes to scale the BC. The measurements of latency and bandwidth consumption per node in the simulation. In relation to communicated nodes, we increase the network size. The latency of various

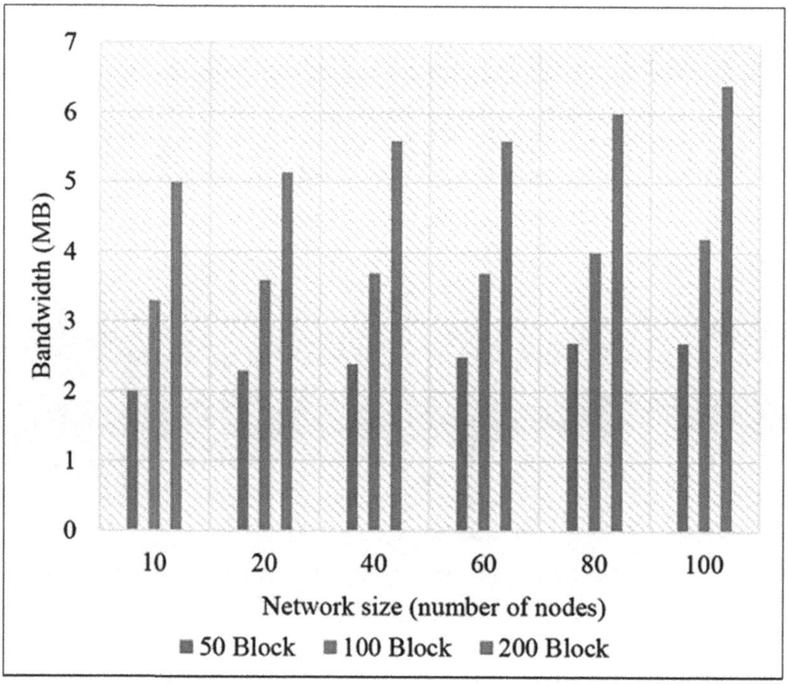

FIGURE 6.5 Bandwidth.

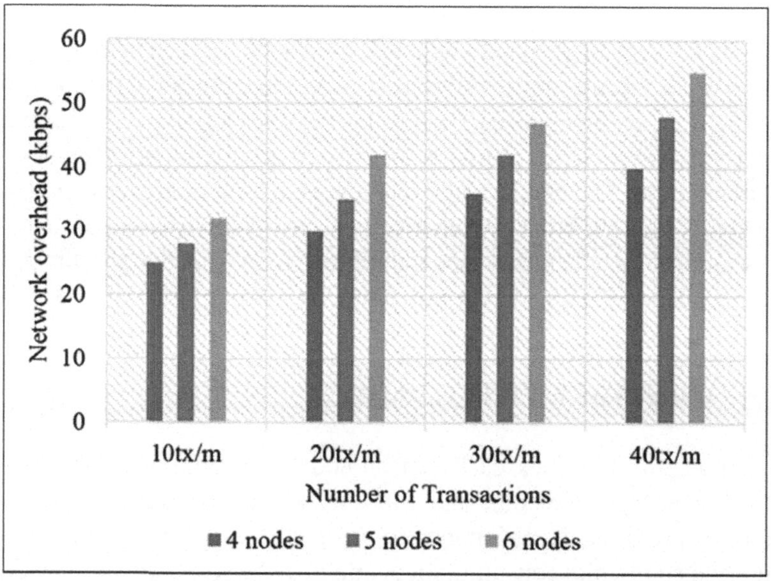

FIGURE 6.6 Network overhead.

micro-block numbers is shown. The latency rate is initially small because in the smaller network there are several knots. The latency rate increases, however with the size of the network increases because the nodes take longer. As shown in figure, with increased network size and block number, network bandwidth consumption is increasing, it is because all blocks are transferred to the networks by radio. The framework proposed contains the BC concept to enhance its performance.

Figures 6.7 and 6.8 are the comparisons between the supportive framework and the core model based on latency and throughput. The framework proposed BC concept includes its perforation expansion.

TRP and FPR are the measurements of performance exactness it is to be determined the proposed model. The FRT is the measurement (false positive ratio). The positive probability and negative probability of the TPR-FPR and FR-to-TNR (true negative relationship) is measured. The SH-BlockCC is found in comparison to TRW-CB, NetAD, and MaxEnt better results.

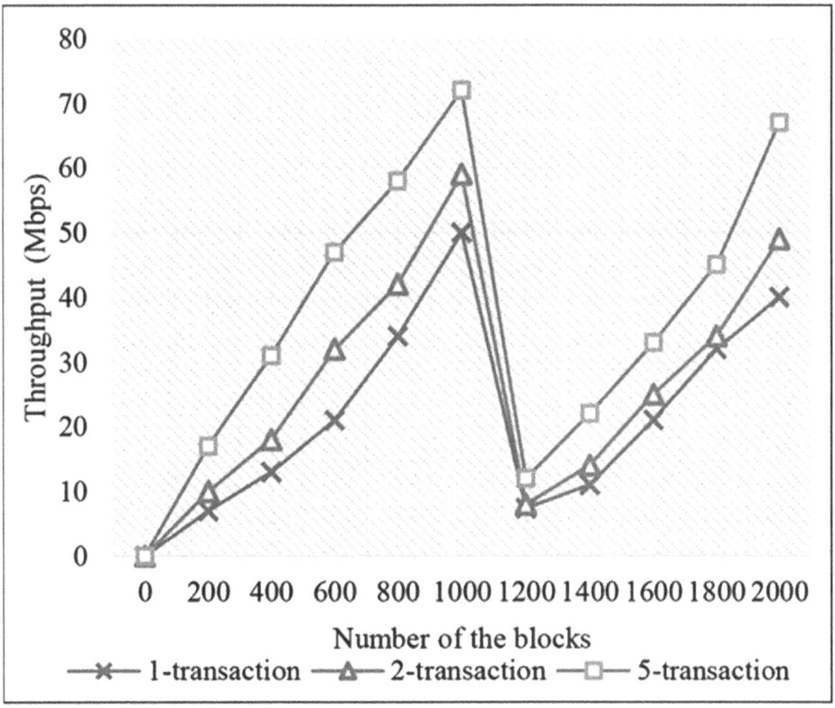

FIGURE 6.7 Overall latency.

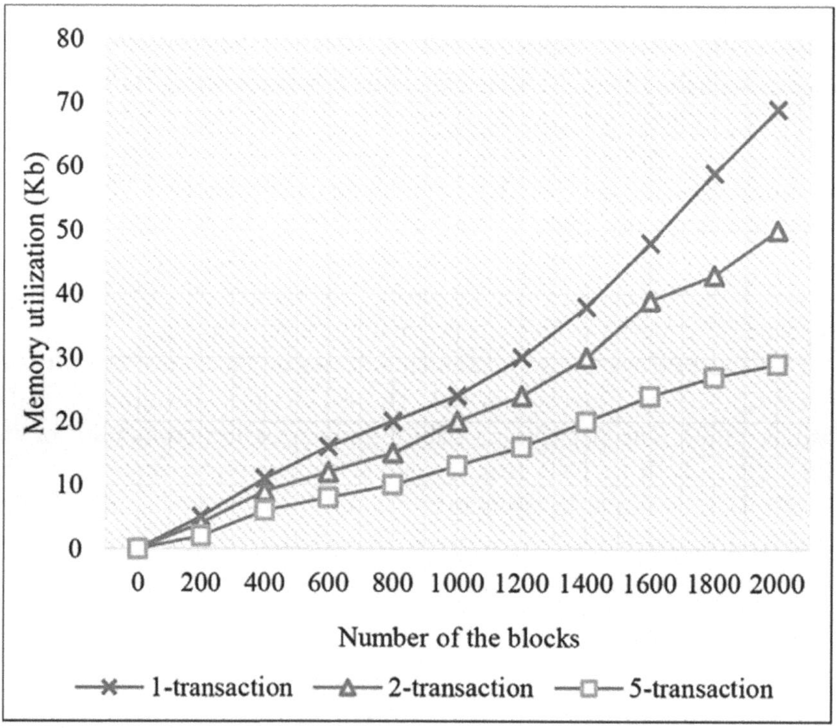

FIGURE 6.8 Throughput (normal case).

The normal case increases throughput, blocks are increasing and the transaction is increased. If the number is 1000, the network will be attacked by a flood, the system's output decreases immediately due to too much network packet management and processing. After some time, to some degree, it becomes uniform and increases and gradually recovers.

The IoT connects all things, both people and things, to the internet so that all items may be interconnected to allow a kind of communication never before seen. Since IoT has taken on a new dimension as a part of the overall community of information and communication technologies, it is essential that we make design a key part of our efforts in this sector. Due to a 35% growth in the number of internet-connected devices per year, it is anticipated that 101.3 million devices were linked to the internet in the year 2011, and it is estimated that in 2021, there will be 1.9 billion devices, which is increasing at a rate of 35%. Machine-to-machine (M2M) connection will grow from 80% in 2011 to 93% in 2021, since the cost of an M2M connection is much lower than the fixed network.

The IoT may be used to design a simple, useful place in the house to provide comfort, as well as improving the quality of life. Due to the appliances being managed remotely, the house is made up of a wide variety of various appliances, all of which are linked to the web. Maintaining the desired temperature in the home may be accomplished by controlling and monitoring the whole environment. Reducing the total cost and conserving energy will be the result of this.

The smart home system detailed in this section can operate gadgets which are inside the house using an Android-based mobile phone at a cheap cost. No personal computer is required as a server with this suggested method. This system utilizes switching technologies to offer a new communication channel for controlling the home environment.

- With your Bluetooth-enabled mobile phone, pair, and open a Bluetooth connection.
- Turn on the mobile-based smart home automation.
- Connect your device by scanning for 'new devices' and choosing 'connect device' from the options menu.

If you want to connect a serial device to your Raspberry Pi, there is a Bluetooth serial module accessible under the "choose device to connect" option. In this case, you would write: Pair 1234, Serial module. Select a device from the list menu to enable/disable on/off.

Incorporating web services that use REST architecture as an interoperable application layer, an innovative, versatile, and inexpensive system is developed and built that controls and monitors all of the home gadgets. This suggested system is made up of micro-sized web servers that are running on Arduino Ethernet, a hardware/hardware-type interface, and Android, which is used by smartphones. This model allows you to customize it in many ways to automatically add new devices to the mini web server that's generated when you get a new smart mobile. The implementation's goal is to limit the usage of costly equipment like personal computers.

All 3G and 4G-enabled mobile devices with Java compatibility are allowed for home usage only. The mobile platform offers a GUI for the homeowner to invoke and control home devices by utilizing the company's server IP.

The home user may connect remotely via Wi-Fi on their mobile app to the system they are using. In your house, there are two essential components: the home gateway and an interface module. The most important duty of the home gateway is to help to facilitate data translation between the Internet and a home network. The home gateway is the core of the network, and it is a small Arduino-based micro-web server. The primary goal of the server is to perform functions such as controlling and monitoring system components, such as hardware modules, that allow hardware modules to execute the job with actuators and report on any event to the server.

Interfacing sensors and actuators with the hardware module involves directly connecting wires. The hardware module is able to handle aspects such as plug load management, lighting systems, and security systems, such as locks and gates. Sensors, including temperature, current, and humidity, may be used to keep an eye on the environment.

6.9 CONCLUSION

The IoT security industry and academia have gained a great deal of attention these days. Existing security solutions are not necessarily appropriate for IoT because of the high energy consumption and overhead processing. In the past, we proposed to use Bitcoins BC, the immutable block ledger, to address these challenges. A representative case study used a smart home idea to talk about. In this chapter, we describe the different core components of the intelligent home industry and discussed various transactions and procedures. We also provided an all-inclusive safety and privacy analysis. Cloud computing extends the intelligent domain and allows users to take advantage of the cloud. The efficient broker manages energy-efficient service provider's selection and BC technology for end users and creates peer-to-peer connections in which untrustworthy nodes can be connected to an economical processing network. In BC technology, we have used encryption, hacking, and confidential smart home network and network overlay. The policy header and shared device/minor key authorization will be obtained and acceptable transactions between device and operator are made available. We talked about MCA algorithms in the intelligent home traffic relations detection network. Most of the issues in Smart Home Automation are going to be solved by the IoT. The suggested technology was capable of remotely controlling and monitoring the house. By supporting a cloud-based monitoring mode, the functionality also gave manufacturers more flexibility when developing smart home automation systems. Automatic setup security elements and scalability are further emphasized by the home's use of Android-based mobile phones.

The proposed architecture generally provides a network attack and response system in intelligent homes. Our proposed architecture can secure an intelligent home with efficiency due to the results of the receiver operation (ROC), use of CPU, overhead transmission time and overhead network. Home automation goods try to enhance the daily lives of people living in their homes, especially those with elderly family members. There will be more sensors available in the future that can be used to better secure and manage the house. These home automation systems, which until now have struggled to "comprehend" how the human life is lived, are now gaining a deeper understanding of those aspects of daily living and automatically changing the environment to suit. It is the goal in the future to develop this system as smart, to aid the inhabitants with thoughts and feelings they express in their own minds, and to assist the homeless in doing the things they want to do.

KEYWORDS

- **blockchain**
- **cloud computing**
- **cyber security**
- **electronic voltage regulator**
- **green cloud computing**
- **Internet of things**
- **security attack**

REFERENCES

1. Moghaddam, E., Ahvar, E., Ortiz, A. M., et al., (2016). To analyze discovery methods for users' locations: A taxonomy and survey in smart homes. *Appl. J. Netw. Comput., 76*, 75–86.
2. Lutolf, R., (1992). Smart home and energy meter integration into a home-based system. In: *Seventh International Conference, GLASGOW, on Metering Devices and Tariffs for Electricity Supplies*. IEEE: IEEE, New York.
3. Wei, W., Ke, X., et al., (2016). Intelligent home defined toward software. *Mag., 54.*, 116–122. IEEE Commun.
4. Sharma, V., Kim, J., Shin, D., et al., (2017). Smart home IoT networks based on PMIPv6 are a secure and efficient route optimization protocol. *Access to IEEE, 5*, 11100–1117.
5. Danda, R. S., Sabina, J., et al., (2016). *The Foundation, Principles, and Applications for the Cyber-Physical Systems.* Academic Press, Boston, MA.
6. Pishva, D., & Takeda, K., (2008). Smart home appliances product-based security model. *IEEE Electron Aerospace Syst. Mag., 23*(10), 32–41.
7. Singh, S., Sharma, P. K., & Park, J. H., (2017). SH-SecNet: An improved, secure, smart-home security threat diagnosis network architecture. Sustainability, *9,* 513–531.
8. Kumar, C. V A., Manoj, K. D., & Prithiviraj, P., (2016). Privacy policy multiparty access control on content sharing sites. *Imperial Journal of Interdisciplinary Research, 2*, 6.

9. Sharma, V. I., & Kul, G., (2017). Socialize drones in ultra-dense wireless networks using blockchain for interservice operation. *Dallas, TX, 30 October 2017: The 2017 International Workshop on the Management of Insider Safety Threats* (pp. 81–84). The ACM in New York.
10. He, Q., Xu, Y., Liu, Z., et al., (2018). Internet of stuff management device system based on block chain is privacy-preserving. *Int. J. Sens. Distrib. N., 14*, 1–12.
11. Arulkumar, V., Charlyn, P. L., & Daniel, Jr. D., (2019). Concept of implementing big data in smart city: Applications, services, data security in accordance with internet of things and AI. *International Journal of Recent Technology and Engineering (IJRTE), 9*.
12. Jacobsson, A., & Davidsson, P., (2015). Drive towards a smart home privacy and security model. In: *IEEE Second World Internet Forum Subjects (WF-IoT)*.
13. Dorri, A., Kanher, S. S., & Jurdak, R., (2012). *Blockchain and in the Interweaving of Challenges and Solutions*. arXiv:1608.05187.
14. Chen, T., Li, X., Jiang, P., et al., (2012). A blockchain system security survey. *Computer System Future Generator.* https:// arXiv.org/abs/1802.06993.
15. Singh, S., Jeong, Y. S., & Park, J. H., (2016). Cloud computer security survey: Threats, problems, and solutions. *Appl. Netw. J. Comput., 75*, 200–222.
16. Hegade, R., & Patil, V., (2015). Green cloud computing. *Int. J. Eng. Adv. Technol., 4*, 1–4.
17. Sharma, P. K., Singh, S., Moon, S. Y., et al., (2017). EH-GC: An efficient and safe energy harvesting infrastructure for green cloud. *Sustainability, 9*, 673–691.

CHAPTER 7

A SECURE AND EFFICIENT QUANTUM ADOPTED CHAOTIC CRYPTOSYSTEM FOR VIDEO SURVEILLANCE APPLICATIONS

T. JANANI, VINOD RAMESH FALMARI, and M. BRINDHA

Department of Computer Science and Engineering, National Institute of Technology, Tiruchirappalli, Tamil Nadu, India

ABSTRACT

The recent advances in digital services such as telehealth, video conferencing, military imaging, and video surveillance require highly reliable, secure storage and transmission of multimedia images and video. Specifically, video surveillance systems are commonly adopted for real-time event detection and video analytics. The video captured by video surveillance cameras often holds personal information, which requires privacy protection. Unlike text, digital images and video frames have many special properties, for instance, pixel correlations, textures, and huge size, which results in conventional cryptosystems like AES, data encryption standard (DES), etc., being appropriate for performing image encryption tasks as they possess very low efficiency when the data size is large. Chaotic cryptosystem has recommended a proficient approach for performing multimedia encryption with special characteristics like randomness and dynamic behavior. Though chaotic cryptosystem provides better security, these frameworks are vulnerable to cryptanalytic attacks with high computation machines. Thus, the recent advancement of quantum methodologies for quantum information processing has gained popularity in many fields such as information security, video processing, etc.

Multimedia Security: Tools, Techniques, and Applications. Bhaskar Mondal and Shyam Singh Rajput (Eds.)

In this work, an attempt was made to investigate the proficiency of quantum cryptography in the field of secure video surveillance applications. To ensure the privacy of surveillance video, the proposed framework presented a novel Quantum based video cryptosystem for encrypting the digital video frames. Initially, the video frames are converted into quantum qubit frames by processing the pixel values. Next, to guarantee the secrecy of digital video information, the qubit planes are encrypted by utilizing series of quantum gates such as swap gate, controlled not gate, Toffoli gate and Hadamard gates under confusion and diffusion phases. The secrecy of digital video frames is reinforced by utilizing plain image-based quantum chaos for secret key generation. The traditional chaotic maps are quantized to obtain the quantum-based random sequences and these random sequences are further utilized to construct the secret keys for performing encryption and decryption operation. The numerical simulations and security analysis prove that the proposed system ensures secure video surveillance against cryptanalytic attacks. Also, the proposed quantum-based secure video surveillance system has ensured the effectiveness of quantum image processing in real-time video surveillance applications.

7.1 INTRODUCTION

With the recent advancements of network and communication technology, the use of surveillance systems has significantly increased over the last two decades. The advancements in surveillance cameras technology make their usage not only in public gatherings such as railway stations, airports, malls but also in personal places such as offices, home, etc. With the help of these vigilance cameras, the authorized users or organization can observe or track the objects that are in a distant area. For instance, the house owner can monitor the live or stored video on their systems or mobile devices to realize what has occurred or occurring in their house and close by territory. A significant driving element for this extensive deployment of surveillance or vigilance cameras is the video reconnaissance hardware, as it becomes progressively modest and easy to utilize. As a result of this progressive development, modern video surveillance systems are no longer closed, and not for single purpose usage. Furthermore, these frameworks often utilize wireless network mediums such as the internet for transmitting live or stored video information. The improvement in network technology makes digital video transmission easier, but also it is quite vulnerable to various security attacks. More specifically, the increasing usage of video processing and the relative openness of the network medium cause a large number of current vigilance camera frameworks as targets for both accidental and intentional attackers. The openness in network medium allows the intruders to copy the digital video content, format, modify it and share it to unauthorized users. As a result, confidentiality, integrity, and authentication concerns for digital video content has been increased and this raises to the requirement of security tool against unauthorized access.

With the advancement of technology in modern surveillance cameras, it is feasible to embed the privacy protection methodology as a feature of the vigilance framework. There are various approaches existing for safeguarding sensitive digital information, such as blanking, hiding, and encryption. Surveillance security is majorly categorized into two approaches: 1. Server-side privacy protection 2. On-camera privacy protection. The on-camera security protection is widely adopted in real-time scenarios as it protects digital information against transmission attacks.

7.1.1 SECURITY REQUIREMENTS FOR VIDEO SURVEILLANCE

The major security requirements of video surveillance frameworks are confidentiality, integrity, authenticity, timestamping, authorization, and availability.

7.1.1.1 CONFIDENTIALITY

The confidentiality of the digital video information is preserved by adopting various security tools or methodology which ensures that no attacker or intruders can eavesdrop on video content that is transmitted between cameras or cameras to the monitoring station. A common approach to ensure digital video confidentiality is data encryption.

7.1.1.2 INTEGRITY

The video or frames recorded by surveillance or vigilance cameras ought to be transmitted to another base station, say a control room or monitoring station, for processing and analysis. During transmission, the frames produced by the camera can be accidentally or intentionally altered by the intruders. These alterations can be identified by utilizing digital signatures, watermarks, etc. The integrity of the video system is not only important for the single frame but also for sequences, as simple re-ordering of frames will lead to a different meaning for captured video.

7.1.1.3 AUTHENTICITY

The authenticity of the frame is important for applications such as traffic monitoring and law enforcement as it ensures the origin of the frames. The authenticity of frames are guaranteed by investigating the camera network medium.

7.1.1.4 TIMESTAMPING

The timestamp of the video surveillance framework ensures the sequence of the video frames. The timestamp prevents the replay attack in wireless network medium. Generally, a replay attack is injected during live video streaming applications.

7.1.1.5 AVAILABILITY

The surveillance or vigilance camera network should guarantee the availability of live or recorded videos. The video surveillance framework should resist denial-of-service attacks reasonably.

7.1.2 CHAOTIC CRYPTOGRAPHY

The network is an open medium, and while transmitting the video through the internet, there is a chance of attacks and leakage of sensitive data. Thus, guaranteeing the digital video information, more specifically sensitive frames from intruders or attackers, is the most significant task. Generally, due to the special properties of digital images, traditional data encryption algorithms are not recommended for an image or video encryption. Thus, to guarantee the confidentiality of digital video information, chaotic systems are adopted for encrypting the images in the digital forum. The chaotic encryption procedure follows Friedrich's architecture for performing pixel shuffling and diffusion. The process involved in chaotic image encryption are shown in Figure 7.1.

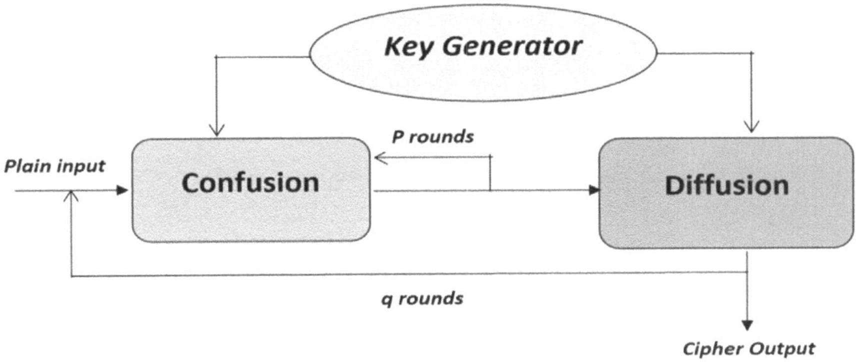

FIGURE 7.1 Chaotic image encryption.

In general, chaotic maps have been popularly utilized in chaotic cryptography for performing image pixel scrambling. The image pixel confusion stage is carried out by generating random numbers with the assistance of chaotic maps. Several chaotic maps are available for performing chaotic cryptography, and some of them are listed in subsections.

7.1.2.1 ARNOLD'S CAT MAP

Arnold's cat map is a two-dimensional chaotic map which can transform the pixel (x, y) in to another value (x', y') by following the linear transformation such as:

$$\begin{pmatrix} x' \\ y' \end{pmatrix} = \begin{pmatrix} 1 & A \\ B & AB+1 \end{pmatrix} \begin{pmatrix} x \\ y \end{pmatrix} \; mod \; n \qquad (1)$$

where x and y are the pixel integers, and n is the size of the image.

7.1.2.2 HENON MAP

A two-dimensional chaotic function known as Henon map is an extended version of the Poincare map for the Lorenz model. The mathematical expressions of the Henon map is mentioned as follows:

$$A_{(x+1)} = 1 - \alpha A_x^2 + B_y \tag{2}$$

$$B_{y+1} = \gamma x_n \tag{3}$$

where α & β represent the control parameters, which are examined for various numeric values, and the chaotic nature is established for $\alpha = 1.4$ & $\beta = 0.3$. Figure 7.2 illustrates the phase portraits of Henon map.

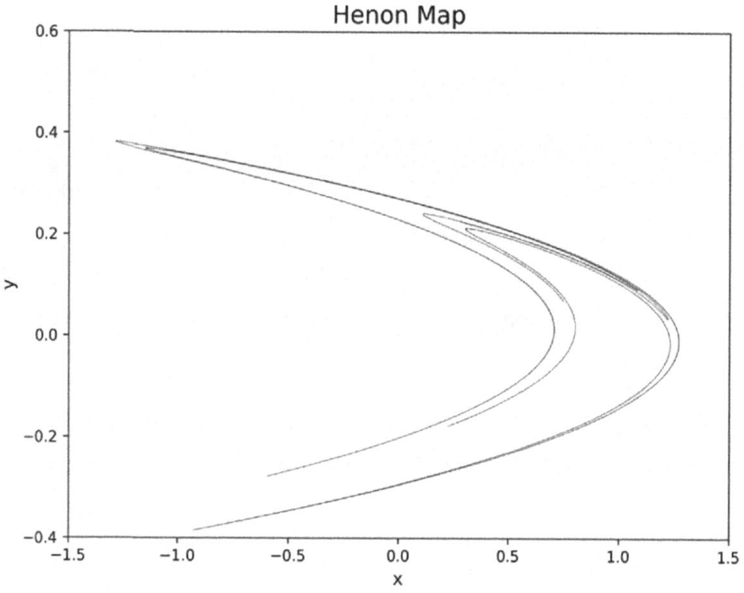

FIGURE 7.2 Henon map.

To attain high randomness, the system parameter will be chosen as not excessively huge or excessively small. The region of contraction will be large if β is nearer to zero. On the other hand, the folding won't be sufficient if it is too large. The proposed system uses Henon for image encryption, and the random values are chosen for permutations, and the diffusion is carried out with the help of the permuted pixel values. Here α & β are chosen as seeds for the chaotic encryption.

7.1.3 QUANTUM CRYPTOGRAPHY

The strength of classical cryptography algorithms is solely based on the fundamental property called factorization of large prime numbers which is indicated as interactable. However, these classical cryptography algorithms are vulnerable to progress in computing power and mathematical evolution, which can rapidly inverse the one-way function such as prime factorization. To overcome this and to achieve security against technological progress in computing power and mathematical proof, it is required to introduce Quantum physics into cryptography. The two fundamental principles utilized in Quantum cryptography are: 1. Heisenberg uncertainty principle and 2. Photon polarization principle.

Heisenberg's uncertainty principle states that it is not feasible to analyze the quantum state of any system without disturbing the framework. Thus, the quantum state of any state can be measured at only the given point. This principle attracts numerous researchers to work on the Quantum based security framework for safeguarding digital information transformation. On the other hand, the photon polarization principle investigates how photons are polarized in a specific direction. Recently, Quantum cryptography techniques are popularly used in a key distribution environment. Consider a scenario if Party A sends a message to Party B using a stream of photons through a communication channel. The stream of photons can be created by evaluating the polarizations related to different directions such as horizontal, vertical, diagonal, or opposite directions. On the receiving end, Party B has mutually chosen a photon filter concerning party A and measures the photon polarization and makes an account of it. If any intruders or attackers eavesdrop on the communication channel, they can read the stream of photons only if they choose the same photon filter type as mutually agreed as Party A and Party B. Thus, attackers will not accurately interpret the photons which result in the key for the encryption algorithm (Figure 7.3).

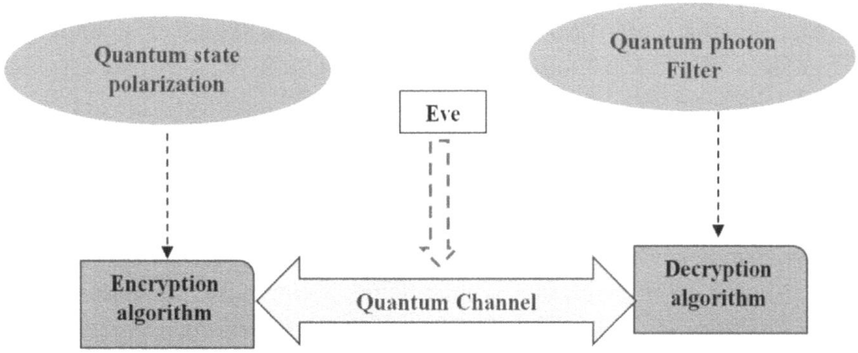

FIGURE 7.3 Quantum cryptography.

7.1.3.1 QUANTUM CRYPTOGRAPHY KEY DISTRIBUTION PROTOCOL

The basic Woking principle of quantum cryptography is listed below:

- Party A transmits a message as a random sequence of photons denoted by four directions horizontal H, Vertical V, shift S.
- On the receiving side, Party B measures the polarization by using a photon filter, say rectilinear and circular.
- Party B informs the obtained value and its filter with the sender.
- Party A verify the measurements with the same polarization filter and informs the result.
- Both the party records the correct result and discard the other photon values.
- The final message is obtained by converting the values to the binary sequence with respect to filter and polarization.

The major advantages of Quantum cryptography are the no cloning principle, i.e., information regarding photons cannot be duplicated. This is due to the property that, the photons are destroyed once they are measured or tampered in the communication channel. Secondly, both sender and receiver mutually agreed upon choosing the filter, which in turn increases the difficulty level for an attacker to intercept the information.

7.1.3.2 ADVANTAGES OF QUANTUM CRYPTOGRAPHY OVER CLASSICAL CRYPTOGRAPHY TECHNIQUES

With the agile improvement of technology, Quantum cryptography has become a major tool in secure transmission mediums. There exist numerous advantages of the Quantum encryption technique over classical cryptography, which are listed below:

1. **Key Confidentiality:** The uncertainty principle of Quantum mechanics provides the gateway for utilizing quantum technology for distributing the secret keys. The existing public-key cryptosystem suffers from the uncertainty that mathematical factorization involves in decryption is intractable. Meanwhile, the classical key distribution protocol suffers from internal threats such as eavesdropping on the communication channel, disturbing the key generation process, etc.

2. **Authentication:** Quantum cryptography ensures user authentication by verifying the photon polarization filter. Secure communication can be guaranteed by mutually agreeing on the polarization filters.
3. **Robustness:** Quantum methodology ensures the robustness in the key generation process, i.e., if any attacker or intruder is eavesdropping on the communication medium, the key generation process will get disrupted.

With these benefits, recently, Quantum based image encryption schemes are utilized for encrypting sensitive images, say IoT images, medical images, etc. The quantum image encryption is carried out by utilizing a sequence of quantum gates with quantum keys. The well-known and basic quantum gates used for encryptions are Hadamard gate, Toffoli, CNOT, Swap, and Identity gate.

7.2 NOVEL APPROACH FOR SECURE VIDEO SURVEILLANCE SYSTEM

In recent years, because of the colossal movement in computation and communication technologies, the world has been experiencing enormous improvement in multimedia information such as images, video, audio signals, etc. More specifically, the usage of surveillance or vigilance cameras is increasing dramatically, which leads to performing more analysis on captured video frames for future use. This advancement of computerized data needs more storage space than the personal system. However, while transmitting the video frames, the prevalence of multimedia technology requires serious protection of privacy. The common approach used to ensure privacy in video surveillance systems is video/image encryption, which would protect the sensitive content from unauthorized users. In general, there are two sorts of encrypting digital images such as spatial domain and transform domain. As video is compressed discrete cosine transform (DCT) domain, thus almost all conventional video encryption schemes utilize transform domain for encrypting the video frames. Surveillance video encryption is broadly classified into two types: encrypting whole video information or encrypting the important region, say ROI. The conventional secure video surveillance system mostly uses chaotic cryptography for encrypting the video frames. Though the chaotic cryptosystems provide better privacy protection methods, it is vulnerable to high computation machine attacks. Thus, to withstand the high computational machine attack such as quantum cryptanalytic attack, it is necessary to introduce quantum-based encryption strategy for encrypting the video information. To achieve this, the proposed work utilized sequence of quantum gates for performing encryption operation. The video frames are encrypted with the help of quantized chaotic map for frame permutation and diffusion operation.

7.2.1 RELATED WORKS

To preserve the confidentiality of video information, numerous encryption schemes were suggested in the past decades. Due to the inherent properties like uncertainty, sensitivity to the initial conditions, etc., chaotic cryptosystems are widely adopted techniques for encrypting the multimedia information with substitution and diffusion stages [6–8]. Since the past two decades, selective encryption has been extensively utilized for encrypting multimedia images, video information. Li et al. [9] presented a review article about chaos-based image or video encryption mechanisms. The author listed out available video encryption schemes and their categories such as JPEG/MPEG video compression and encryption, selective encryption and so on. Tosun & Feng [10] suggested that selective encryption is useful for developing error-control, video frame encryption methods. Dawen Xu et al. [11] suggested commutative encryption and data hiding for video compression. The method comprises of both data hiding and video compression-based encryption. The author utilized a quantized transform coefficient for encrypting the video information using high-efficiency video coding. Additionally, Mohamed Abomhara et al. [12] presented a literature article about existing video encryption techniques. The chapter provides insights into a different category of video surveillance systems and their security issues also highlights the possible solution for providing secure video transmission. To ensure the privacy of video-sensitive areas, various cryptographic and ROI extraction techniques are presented in the literature. In general, Chaos cryptographic techniques are applied to encrypt ROIs in Refs. [13, 14]. Qiang Zhang et al. [15] suggested DNA sequence addition and a chaos-based image encryption scheme. However, the scheme possesses a high computation burden for the receiver end.

In general, video monitoring and stealing recorded video streams are two common attacks in video surveillance applications [16]. To ensure the safe installation of video surveillance, it is necessary to install certified devices [17]. In Ref. [18], author presented an article about the dual-channel authentication approach for video camera systems with the enable of RFID mechanizm. Yet, the system is vulnerable to encrypted statistical attacks.

Xing Zhang et al. [19] suggested a lightweight encryption methodology for preserving the privacy of the video surveillance applications. The author utilized layered cellular automata for encrypting the video frames. However, the scheme exhibits minimal NPCR and UACI values which leads to differential attacks. Benoit Boyadjis et al. [20] suggested an extended selective encryption technique for preserving the confidentiality of digital video. The author uses H.264/AVC for scrambling video frames.

Meanwhile, Hao Li et al. [21] proposed a secure video surveillance framework for smart city application scenarios. The author utilized the Chinese remainder theorem for secret key generation and distribution. Additionally, the author introduced a video security decryption suite for large-scale applications which ensures authenticity and confidentiality of the digital video frames. These systems provide enough security for video applications, but it might be vulnerable to quantum cryptanalytic attacks. Thus, to withstand the quantum cryptanalytic attack in video surveillance applications, it is necessary to introduce quantum-based encryption methodology for multimedia information. In general, Quantum based image or video encryption is the combination of parallel quantum mechanics and cryptography techniques that ensures the secrecy of multimedia information. In recent years, researchers have started investigating the quantum cryptographic techniques for processing sensitive information in a secure and efficient way [22–25].

Yan et al. [25] presented an article about quantum image processing which states the benefits of utilizing quantum mechanics in the image processing environment. The author investigated the application of quantum principles over different image processing applications such as watermarking, encryption, data hiding, etc. With the popularity of quantum mechanics in digital image processing, different quantum image representation model has been proposed such as Qubit lattice [1], FRQI [4], NEQR [5], entangled image [2] and so on. Xian-Hua Song [26] suggested a quantum-based video encryption technique based on qubit planes and controlled XOR operation. Author utilized inter-frame permutation, intraframe permutation using geometric transformations. Then improved logistic map is used for encrypting the inter and intra frame pixels. Though the experiments ensure the secrecy of video information through quantum technology, it exhibits very low key space for encryption and decryption operation.

Similarly, Ahmed [27] proposed secure and robust video encryption-based quantum walks. The author used quantum walks for constructing the quantum substitution box and block cipher techniques for encrypting the video information. Luo et al. [28] suggested image encryption based on quantum codes. The scheme comprises of bit-level pixel scrambling and diffusion function for encrypting the images. The pixel confusion stage is solely relied upon quantum codes. Yet the scheme exhibit very low randomness in encrypted image.

7.2.2 QUANTUM IMAGE PREPARATION

As a new computing technology, the quantum computation can read the information, process, store, and transmit it to the other end by utilizing the phenomenon properties of quantum mechanics such as entanglement and superposition state. By witnessing the recent development in an imaging system such as IoT devices, surveillance or vigilance camera, the number of images or video frames have increased dramatically which in turn require high-speed computation image processing techniques. To overcome this, researchers have begun to explore the combination of quantum computation with digital image processing techniques. Thus, quantum computation has become an effective methodology for the digital image processing environment. Numerous quantum image representation models have been suggested by utilizing quantum mechanics for processing and transmitting digital images. Some of the methods are Qubit lattice [1]; Entangled image [2]; Real Ket [3]; flexible representation of quantum image (FRQI) [4], novel enhanced quantum image, color image representation and so on. This section provides an overview of popular quantum image representation techniques.

7.2.2.1 FLEXIBLE REPRESENTATION OF QUANTUM IMAGE (FRQI)

Flexible representation of quantum image (FRQI) is introduced as a digital image representation for quantum computers. The FRQI converts the input image into a quantum image in the form of a normalized state which uses color pixel values and its respective pixel position. The mathematical expression for FRQI quantum image is given as:

$$|M(\theta)\rangle = \frac{1}{2^n}\sum_{i=0}^{2^n-1}(cos\theta_i\,|0\rangle + sin\theta_i\,|1\rangle)\otimes|m\rangle \quad (4)$$

$$\theta_i \in \left[0,\frac{\pi}{2}\right], i = 1,2,..,2^n - 1 \quad (5)$$

Where; $|0\rangle$, $|1\rangle$ are the 2D computational quantum phases and $|m\rangle$ is the 2 dimensional computation basic quantum phases. Similarly, $\theta = 1,2,.., 2^n - 1$ represents the vector angle for image pixel values. The computation angle

$(cos\theta_i\,|0\rangle + \sin\theta_i\,|1\rangle)$represents the pixel color value and its position $|m\rangle$. For clear understanding, FRQI representation of 2×2 is given as follows.

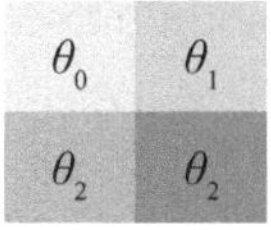

$$|M(\theta)\rangle = \frac{1}{2^n}\sqrt{\sum_{i=0}^{2^n-1} cos^2\theta_i + sin^2\theta_i}$$

$$\begin{aligned}|M\rangle = \frac{1}{2}\big[&cos\theta_0\,|0\rangle + sin\theta_0\,|1\rangle)\otimes|00\rangle + cos\theta_1\,|0\rangle + sin\theta_1\,|1\rangle)\otimes|01\rangle\big] \\ &+cos\theta_2\,|0\rangle + sin\theta_2\,|1\rangle)\otimes|10\rangle + cos\theta_3\,|0\rangle + sin\theta_3\,|1\rangle)\otimes|11\rangle\end{aligned} \tag{6}$$

The FRQI representation uses basic quantum states for encoding the pixel color and position values. Additionally, it considers the geometric angle θ_i for representing color images. The simple two-position encoding in FRQI is given in Figure 7.4.

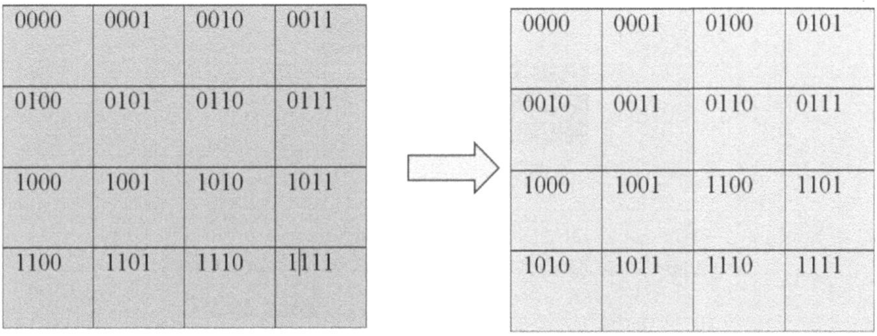

FIGURE 7.4 FRQI representation.

All transforms utilized in the above two position quantum state is a unitary transform which can be decoded by unitary matrices. Unary transforms are applied with respect to three operators, such as operators considering only color values, color at any given pixel position, and color along with pixel positions. Due to its color encoding strategy, FRQI is known as a fundamental quantum representation methodology.

7.2.2.2 NOVEL ENHANCED QUANTUM REPRESENTATION (NEQR)

The quantum representation techniques serve as a fundamental principle for quantum image processing algorithms. The quantum image processing methodology considers the digital image as an input and produces output by performing operations over quantum converted input image. In novel enhanced quantum representation (NEQR), the input image M is converted into grayscale and these grayscale values are converted into qubit sequence. The NEQR model uses quantum basic states for storing grayscale qubit sequences. Thus, to convert the digital image into the quantum image, NEQR requires two entanglement qubits related to grayscale value and its pixel position. The mathematical relation for converting $2^m \times 2^m$ the input image is expressed as below:

$$|M\rangle = \frac{1}{2^n} \sum_{p=0}^{2^m-1} \sum_{q=0}^{2^m-1} |f(p,q)\rangle |qp\rangle \tag{7}$$

Let us consider a sample input image which undergoes NEQR representation and the respective quantum state $|M\rangle$ are illustrated in Figure 7.5.

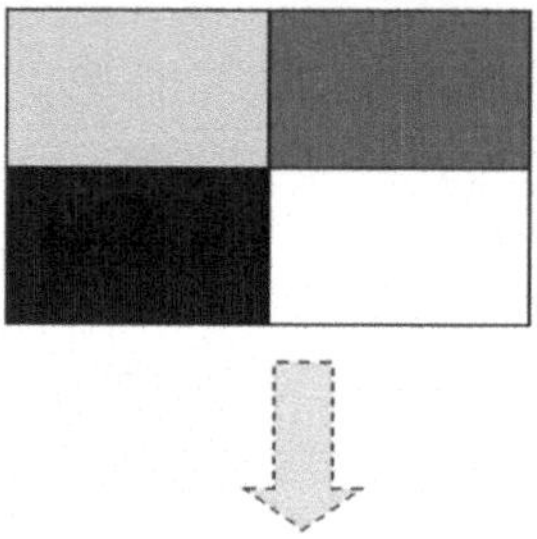

$$|M\rangle = \frac{1}{2^m}(|153\rangle \otimes |00\rangle + |205\rangle \otimes |01\rangle + |255\rangle \otimes |10\rangle + |0\rangle \otimes |11\rangle)$$

FIGURE 7.5 NEQR representation.

Here, the input images are represented as a quantum image by utilizing tensor product over the pixel values. The tensor product yields the qubit sequence, which in turn produces a quantum state image for image processing algorithms. The NEQR utilizes series quantum gates such as Identity gate, Hadamard gate, CNOT gate and so on. However, NEQR records two-qubit entanglement states, it requires $q - 1$ qubits than FRQI.

7.2.3 PROPOSED QUANTUM-BASED SECURE VIDEO SURVEILLANCE APPROACH

To ensure the secrecy of digital video frames, novel quantum-based secure video surveillance approach is introduced and utilized for video applications.

The architecture diagram of the proposed quantum-based secure video encryption scheme is illustrated in Figure 7.6. The suggested secure video encryption scheme takes video frames as input from the digital video recorder. In order to utilize quantum state properties, the obtained frames are converted into quantum representations. The proposed approach uses Flexible Quantum Representation Image methodology (FRQI) for quantum frame preparation.

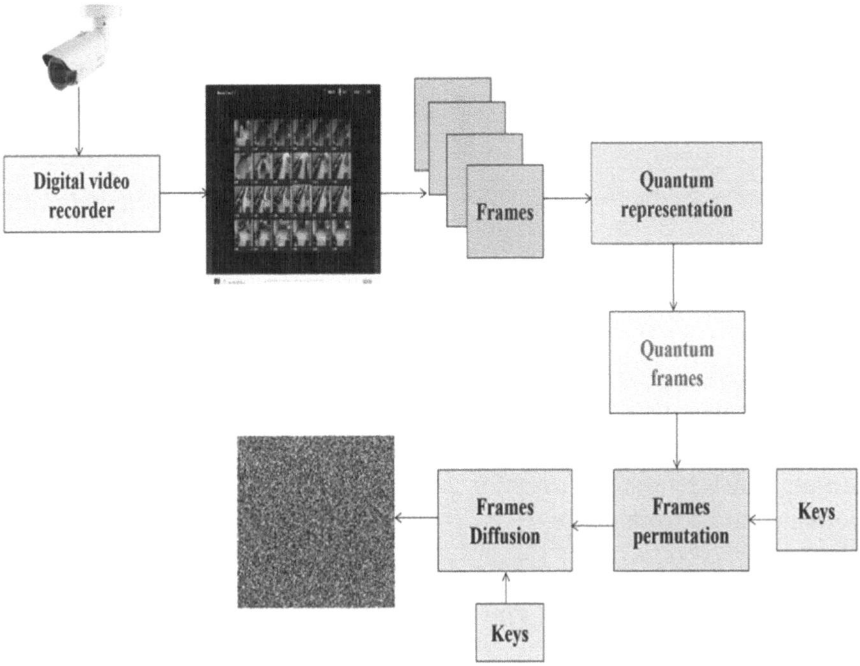

FIGURE 7.6 Block diagram of the proposed quantum-based secure video encryption.

7.2.3.1 QUANTUM FRAME PREPARATION

The proposed framework uses a quantum methodology for encrypting the recorded video frames. To achieve this, the recorded digital video frames are converted into quantum frames by utilizing a quantum image representation model. Initially, the proposed work processes the video from a digital video recorder and obtains individual video frames. Later, these frames are converted into the quantum state by utilizing FRQI representation as given in Eqn. (4). Figure 7.7 shows the quantum representation working principles.

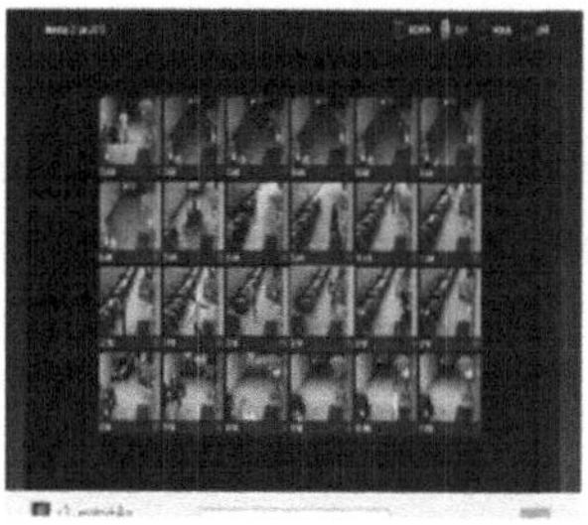

$$|M(\theta)\rangle = \frac{1}{2^n}\sum_{i=0}^{2^n-1}(cos\theta_i|0\rangle + sin\theta_i|1\rangle)\otimes|m\rangle$$

0000	0001	0100	0101
0010	0011	0110	0111
1000	1001	1100	1101
1010	1011	1110	1111

FIGURE 7.7 Quantum frame preparation.

The above diagrammatic representation illustrates the quantum states of the video frames where the computation angle $(cos\theta_i|0\rangle + sin\theta_i|1\rangle)$ represents the pixel color value and its position $|m\rangle$ and these resultant quantum frames are utilized for further encryption processes in the Qubit substitution and permutation network phase.

7.2.3.2 SECRET KEY GENERATION

In general, the effectiveness of the encryption scheme depends upon the time taken to identify the secret keys or encryption function. The conventional encryption schemes are solely based on secret keys, which are developed by utilizing chaos theories. Due to the progressive in high computation machine and mathematical model, the conventional chaos-based secret keys are vulnerable to security attacks like Shor factorizing algorithm. This results in the possibility of breaking traditional cryptographic algorithms using quantum machines. To overcome these challenges, it is important to introduce a quantum resist cryptography technique for encryption methodology. Quantum cryptography is the study of exploiting quantum mechanical properties to perform cryptographic tasks. After the emergence of Quantum cryptography, the traditional chaotic system becomes quantized and termed Quantum chaos.

The primary motivation for quantum chaos is to study the quantum versions of classical chaotic systems. The proposed secure video surveillance framework uses a traditional chaotic Henon map for performing pixel permutation in video frames. The mathematical expression of the Henon map is given as follows:

$$A_{(x+1)} = 1 - \alpha A_x^2 + B_y$$

$$B_{y+1} = \gamma x_n$$

This results in the chaotic sequence A and B. To attain quantum-based video frame encryption, the obtained chaotic sequences are converted into a qubit matrix. The qubit matrix is generated by the following steps. First, the obtained chaotic sequences are converted into binary formats with an upper bound of image size. Secondly, the tensor product is applied over the binary sequence, which results in a qubit secret key matrix. The steps involved in qubit generation is given in Algorithm 7.1. The Quantum secret key generation algorithm takes the input as Initial value and control parameter α & γ and executes Henon map for n rounds for obtaining chaotic sequences A' and B'. The obtained random sequences are converted into binary format, and tensor products are applied over it. The resultant key matrix is called quantum secret key Matrix $|K\rangle$.

ALGORITHM 7.1 Quantum Secret Key Generation

Input: Initial values and control parameters.

Output: Key matrix $|K\rangle$.

Call Henon function and generate keystream.

$$A_{(x+1)} = 1 - \alpha A_x^2 + B_y$$

$$B_{y+1} = \gamma x_n$$

For i in chaotic sequence A and B:
Convert A and B binary vector:

$$|K\rangle = A' \otimes B'$$

$$Return\ |K\rangle$$

7.2.3.3 QUANTUM FRAME PERMUTATION

The proposed secure Quantum based video surveillance encryption framework is composed of two stages: Quantum frame permutation and diffusion.

To improve the effectiveness of the encryption strategy, the suggested system uses separate quantum secret keys for performing permutation and diffusion operations. The permutation operation takes the input as quantum represented frames and performs pixel confusion operation using series of quantum gates such as swap gate, Hadamard, Toffoli gates.

The first round of permutation is performed by swapping the frame qubits concerning secret keys. To ensure adequate randomness, the swapped qubits undergo spatial transformations by projecting the qubits in the spatial domain. The steps involved in Quantum frame permutation are illustrated in Algorithm 7.2.

ALGORITHM 7.2 Quantum Frame Permutation.

Input: Quantum frames $|M\rangle$, $|K\rangle$

Output: Permuted frame $|M\rangle'$.

For (i,j) in $|M\rangle$, $|K\rangle$:

Read $|K\rangle$

Perform frame permutation operation $|M_n\rangle$

Specify the angle θ

For (i,j) in $|M_n\rangle$:

Spatial transformation ST:

$|M\rangle' = ST\,(|M\rangle)$

Return $|M\rangle'$

7.2.3.4 QUANTUM FRAME DIFFUSION

The suggested permutation phase in secure video encryption scrambles the quantum pixel positions in the quantum frame by using the quantum secret key and sequence of quantum basic gates. The suggested encryption methodology comprises two stages co:nfusion and diffusion. To improve the effectiveness of the suggested video encryption scheme, the proposed system uses a separate secret key matrix for permutation and diffusion. Thus, to attain better secrecy and to preserve the confidentiality of surveillance video frames, the permuted pixels undergo the frame diffusion stage. The suggested Quantum frame diffusion stage utilizes the Arnold cat map for generating a secret key matrix. The proposed system executes Arnold cat map for n, which results in a chaotic sequence of x' and y'. Algorithm 7.3 illustrates the procedure involved in frame diffusion.

ALGORITHM 7.3 Quantum Frame Diffusion

Input: Permuted frame $|M\rangle'$

Output: Encrypted frame EM

Compute Secret Key $|K_2\rangle$

Iterate Arnold map n rounds

$$\begin{pmatrix} x' \\ y' \end{pmatrix} = \begin{pmatrix} 1 & A \\ B & AB+1 \end{pmatrix} \begin{pmatrix} x \\ y \end{pmatrix} \; mod \; n$$

$$|K_2\rangle = x', y'$$

Perform Diffusion, EM

$$|M\rangle' \; CNOT \; |K_2\rangle$$

Return EM

As depicted in Algorithm 7.3, the diffusion stage utilizes the Arnold cat map and CNOT for diffusion of the permuted pixels obtained from the confusion stage. The suggested secure video encryption uses two sort of secret keys, i.e., quantum chaotic Henon and Arnold map.

7.2.4 RESULTS AND DISCUSSION

The suggested quantum-based secure video surveillance system is experimented using Windows 10 64-bit, Intel Core i7, with 16 GB RAM and equipped with MATLAB R2019B environment. The performance can be analyzed in terms of image encryption. Some commonly used metrics for analyzing the capability of image encryption against cryptographic attacks are key space analysis, key sensitivity analysis, correlation, histogram analysis, differential attack analysis and so on.

7.2.4.1 KEY SPACE ANALYSIS

To break the encryption strategy, intruders usually perform a brute-force attack to identify the secret keys that are used for the image cryptosystem.

Generally, total number of keys used for encrypting and decrypting the plain and cipher images are termed as keyspace. An effective encryption mechanism should possess a large keyspace to make the brute-force search infeasible. Strong image encryption should possess a keyspace of at least 2^{100} keyspace to withstand the brute-force attack. The total number of keys utilized for image cryptosystem always relies on both keys used in image pixel confusion and diffusion methods. The suggested secure quantum-based video encryption framework utilizes two chaotic maps for generating random sequences such as Arnold cat map and Henon map, where quantized Henon map is used to generate $|K\rangle$ and Arnold map generates $|K_2\rangle$. Thus, the proposed system utilizes A,B,X,Y initial seed values and control parameters α, γ, a, b as a secret key value. Each key value is represented in floating-point standard 10^{-15}. Thus, the key space for proposed quantum-based secure video encryption is:

$$key\,space = 10^{-15\,(8)} = 10^{120} \tag{8}$$

7.2.4.2 KEY SENSITIVITY ANALYSIS

A strong encryption methodology should guarantee high sensitivity towards the secret key, i.e., a minimal change in the secret key should produce significant changes in cipher image. In general, the sensitivity of the secret keys can be evaluated by calculating the difference between two enciphered images. The mathematical expression is given as follows:

$$Key_{sen} = \frac{D(E,E_1)+D(E,E_2)}{2(M \times N)} \times 100\% \tag{9}$$

$$D(E,E_1) = \sum_{p=0}^{M} \sum_{q=1}^{N} Dif\left(E(p,q),E_1(p,q)\right) \tag{10}$$

$$Dif\left(E(p,q),E_1(p,q)\right) = \begin{cases} 0;\, if\ E(p,q) = E(p,q) \\ 1;\, if\ E(p,q) \neq E(p,q) \end{cases} \tag{11}$$

In the above mathematical formula, E, E_1, E_2 are the encrypted images in which E is encrypted using the correct secret key whereas E_1 and E_2 are the encrypted images obtained by using modified secret keys. Thus, key sensitivity is examined by computing the differences among different cipher images encrypted using modified secret keys. Theoretically, strong encryption should possess a difference value for cipher images as 99% for any cryptosystem.

7.2.4.3 CORRELATION ANALYSIS

The pixel correlation is the fundamental property of images through which it can convey meaningful information or pictures. However, to withstand the statistical attack, the encrypted video frame should possess zero correlation among adjacency pixels. The zero correlation ensures high randomness in encrypted frames. Meanwhile, if frame pixel correlation are high, the intruder can easily eavesdrop or break the encryption methodology and perceive the pain video information. Generally, correlations are evaluated in three directions, namely horizontal, vertical, and diagonal, as given below:

$$corr_{ij} = \frac{E((i-E(j)))-E((i-E(j)))}{\sqrt{V(m)V(n)}} \tag{12}$$

where;

$$E(i) = \frac{1}{x}\sum_{x=1}^{n} i_x \tag{13}$$

Similarly,

$$V(i) = \sum_{x=1}^{n} (E(i)-E(j)2 \tag{14}$$

Here, i, j represents the neighboring pixels, and x is the number of pixels in the images. The analysis is carried out for sample video frames, and the results are tabulated in Table 7.1.

TABLE 7.1 Correlation Analysis

SL. No.	Encrypted Frame	Correlation		
		Horizontal	Vertical	Diagonal
1.	Motion1	0.0313	0.0461	0.0562
2.	Motion2	0.0038	0.0051	–0.073
3.	Car slide1	0.0450	–0.0190	–0.0012
4.	Motor vehicle1	0.0382	0.0542	0.0665
5.	Motor vehicle1	0.0410	0.0043	0.0046

7.2.4.4 HISTOGRAM ANALYSIS

Histogram plots of an image give information about the pixel patterns exhibiting in the image. Thus, an efficient encryption strategy should possess uniform histogram plots to withstand statistical attacks. The qualitative analysis of histograms can be carried out by computing histogram variances. Histogram variance is another important metric to evaluate the image pixel uniformity. The ideal image cryptosystem should possess minimal variances with higher uniformity. The mathematical expression for calculating histogram variance is given below.

$$Var(M) = \frac{1}{x^2}\sum_{i=1}^{x}\sum_{j=1}^{x}\frac{1}{2}(M - M_j)^2 \quad (15)$$

where M is the array of histogram values, and M_i, M_j is the number of pixels at particular cipher images. The sample histogram plots of the plain and cipher images are given in Figure 7.8. The figure depicts that the plain image reveals the pixel pattern, whereas the cipher image possesses uniformity in the image pixel histogram.

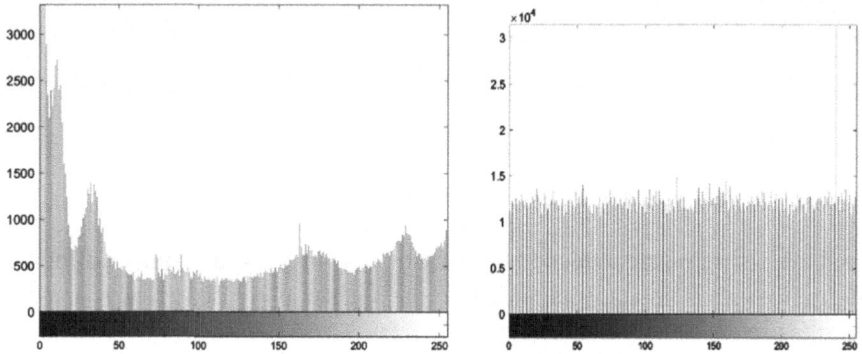

FIGURE 7.8 Histogram plots of plain and cipher image.

7.2.4.5 ENTROPY ANALYSIS

Shannon entropy analysis is the commonly used metric for investigating the randomness in the image encryption scheme. The Shannon entropy or information entropy is evaluated as:

$$S(M) = \sum P(M_i) log \frac{1}{P(M)} \quad (16)$$

where; $P(M_i)$ is the probability of pixel-level values M_i. The maximum entropy value for an effective encryption is 8. Subsequently, it is necessary to measure the randomness among local non-overlapping block pixels of an encrypted image. The local Shannon entropy is determined as:

$$HT_{L,B}^{-} = \sum_{i=1}^{L}\frac{P(M_i)}{k} \quad (17)$$

where H is the number of blocks and B is the size of local blocks. Table 7.2 shows the outcome of Shannon's entropy measure.

TABLE 7.2 Entropy Analysis

SL. No.	Encrypted Frame	Entropy
1.	Motion1	7.8799
2.	Motion2	7.8654
3.	Car slide1	7.8012
4.	Motor vehicle1	7.726
5.	Motor vehicle1	7.775

7.2.4.6 DIFFERENTIAL ATTACK ANALYSIS

Effective image encryption should possess maximum plain image sensitivity, which makes it infeasible for an attacker to perceive original images. The common evaluation metric used for evaluating the plain image sensitivity is the number of pixel change rate (NPCR) and unified average change intensity (UACI). A strong encryption should possess 99% of NPCR and 34% of UACI. The mathematical expression for NPCR and UACI is given below:

$$NPCR = \frac{\sum_{m=1}^{J}\sum_{n=1}^{J} S(i,j)}{I\,X\,J} \times 100\% \tag{18}$$

where; $S(i,j) = \begin{cases} 0; if\ M_1(i,j) = M_2(i,j) \\ 1; if\ M_1(i,j) \neq M(i,j) \end{cases}$

here; $M_i\ (i,j)$ are the encrypted images whose plain image is differed by one bit. Similarly, UACI is expressed as (Table 7.3):

$$UACI = \frac{1}{I \times J}\sum_{m,n=1}^{I,J} mn\,\frac{|E_1(m,n) - E(m,n)|}{256} \times 100\% \tag{19}$$

TABLE 7.3 Differential Attack Analysis

Sl. No.	Encrypted Frame	NPCR	UACI
1.	Motion1	98.9743	33.03
2.	Motion2	99.092	33.0128
3.	Car slide1	99.163	33.3420
4.	Motor vehicle1	99.4965	32.89723
5.	Motor vehicle1	98.1754	32.9524

7.3 CONCLUSION

In the video surveillance application, the suggested Quantum-based secure video encryption scheme can act as an efficient secure mechanism for processing and analyzing digital video information. To guarantee the privacy of video information, the suggested system utilized quantum-based video frame encryption which comprises of initial secret key generation, quantum frames permutation followed by quantum frame diffusion. The proposed work investigates various performance metrics for the efficacy of the framework with respect to video and image security. This prototype guarantees the secrecy, availability, and access controls over sensitive video information.

KEYWORDS

- **chaos**
- **differential attack analysis**
- **encryption**
- **quantum**
- **secret key generation**
- **security**
- **video surveillance**

REFERENCES

1. Venegas-Andraca, S. E., & Bose, S., (2003). Storing, processing and retrieving an image using quantum mechanics. *Quantum Information and Computation, International Society for Optics and Photonics, 5105,* 137–147.
2. Venegas-Andraca, S. E., & Ball, J. L., (2010). Processing images in entangled quantum systems. *Quantum Information Processing*, *9*(1), 1–11.
3. Latorre, J. I., (2005). *Image Compression and Entanglement.* arXiv preprint quant-ph, 0510031.
4. Le, P. Q., Dong, F., & Hirota, K., (2011). A flexible representation of quantum images for polynomial preparation, image compression, and processing operations. *Quantum Information Processing*, *10*(1), 63–84.
5. Zhang, Y., Lu, K., Gao, Y., & Wang, M., (2013). NEQR: A novel enhanced quantum representation of digital images. *Quantum Information Processing*, *12*(8), 2833–2860.

6. Guan, Z. H., Huang, F., & Guan, W., (2005). Chaos-based image encryption algorithm. *Physics Letters A*, *346*(1–3), 153–157.
7. Liu, L., Lei, Y., & Wang, D., (2020). A fast, chaotic image encryption scheme with simultaneous permutation-diffusion operation. *IEEE Access*, *8*, 27361–27374.
8. Pareek, N. K., Patidar, V., & Sud, K. K., (2006). Image encryption using chaotic logistic map. *Image and Vision Computing*, *24*(9), 926–934.
9. Li, S., Chen, G., & Zheng, X., (2006). Chaos-based encryption for digital image and video. In: *Multimedia Encryption and Authentication Techniques and Applications* (pp. 129–163). Auerbach Publications.
10. Tosun, A. S., & Feng, W. C., (2001). Lightweight security mechanisms for wireless video transmission. In: *Proceedings International Conference on Information Technology: Coding and Computing* (pp. 157–161). IEEE.
11. Xu, D., (2019). Commutative encryption and data hiding in HEVC video compression. *IEEE Access*, *7*, 66028–66041.
12. Abomhara, M., Zakaria, O., & Khalifa, O. O., (2010). An overview of video encryption techniques. *International Journal of Computer Theory and Engineering*, *2*(1), 103.
13. Rahman, S. M. M., Hossain, M. A., Mouftah, H., El Saddik, A., & Okamoto, E., (2010). A real-time privacy-sensitive data hiding approach based on chaos cryptography. In: *2010 IEEE International Conference on Multimedia and Expo*, *IEEE, 2010*, 72–77.
14. Rahman, S. M. M., Hossain, M. A., Mouftah, H., El Saddik, A., & Okamoto, E., (2012). Chaos-cryptography-based privacy preservation technique for video surveillance. *Multimedia Systems*, *18*(2), 145–155.
15. Zhang, Q., Guo, L., & Wei, X., (2010). Image encryption using DNA addition combining with chaotic maps. *Mathematical and Computer Modeling*, *52*(11, 12), 2028–2035.
16. Kalbo, N., Mirsky, Y., Shabtai, A., & Elovici, Y., (2020). The security of IP-based video surveillance systems. *Sensors*, *20*(17), 4806.
17. Rashwan, H. A., Solanas, A., Puig, D., & Martínez-Ballesté, A., (2016). Understanding trust in privacy-aware video surveillance systems. *International Journal of Information Security*, *15*(3), 225–234.
18. Kim, J., Lee, D., & Park, N., (2020). CCTV-RFID enabled multifactor authentication model for secure differential level video access control. *Multimedia Tools and Applications*, *79*, 23461–23481.
19. Zhang, X., Seo, S. H., & Wang, C., (2018). A lightweight encryption method for privacy protection in surveillance videos. *IEEE Access*, *6*, 18074–18087.
20. Boyadjis, B., Bergeron, C., Pesquet-Popescu, B., & Dufaux, F., (2016). Extended selective encryption of H. 264/AVC (CABAC)-and HEVC-encoded video streams. *IEEE Transactions on Circuits and Systems for Video Technology*, *27*(4), 892–906.
21. Li, H., Xiezhang, T., Yang, C., Deng, L., & Yi, P., (2021). Secure video surveillance framework in smart city. *Sensors*, *21*(13), 4419.
22. Cho, K., & Miyano, T., (2014). Chaotic cryptography using augmented Lorenz equations aided by quantum key distribution. *IEEE Transactions on Circuits and Systems I: Regular Papers*, *62*(2), 478–487.
23. Akhshani, A., Akhavan, A., Lim, S. C., & Hassan, Z., (2012). An image encryption scheme based on quantum logistic map. *Communications in Nonlinear Science and Numerical Simulation*, *17*(12), 46534661.

24. Xu, J., Li, P., Yang, F., & Yan, H., (2019). High intensity image encryption scheme based on quantum logistic chaotic map and complex hyperchaotic system. *IEEE Access*, *7*, 167904–167918.
25. Yan, F., Iliyasu, A. M., & Le, P. Q., (2017). Quantum image processing: A review of advances in its security technologies. *International Journal of Quantum Information*, *15*(3), 1730001.
26. Song, X. H., Wang, H. Q., Venegas-Andraca, S. E., & Abd El-Latif, A. A., (2020). Quantum video encryption based on qubit-planes controlled-XOR operations and improved logistic map. *Physica A: Statistical Mechanics and its Applications*, *537*, 122660.
27. Abd El-Latif, A. A., Abd-El-Atty, B., Mazurczyk, W., Fung, C., & Venegas-Andraca, S. E., (2020). Secure data encryption based on quantum walks for 5G Internet of Things scenario. *IEEE Transactions on Network and Service Management*, *17*(1), 118–131.
28. Luo, Y., Tang, S., Liu, J., Cao, L., & Qiu, S., (2020). Image encryption scheme by combining the hyper-chaotic system with quantum coding. *Optics and Lasers in Engineering*, *124*, 105836.

CHAPTER 8

EVALUATION OF PANSHARPENING TECHNIQUES FOR MULTIMEDIA IMAGE ENHANCEMENT

KULDEEP CHAURASIA and SHAKTI SHARMA

Department of Computer Science Engineering, Bennett University, Greater Noida, Uttar Pradesh, India

ABSTRACT

Pansharpening is the process of sharpening a low spatial resolution multi-spectral (MS) image and high spatial resolution panchromatic (PAN) image to yield a color image with high resolution in both spatial and spectral domains. Such color images are significant to several applications including pattern recognition, target identification, landuse/landcover classification and change detection. It is one of the fundamental and dynamic research areas in the geospatial domain. Methods of pansharpening have been developed for more than two decades. These Pansharpening methods, on the other hand, add spatial and spectral distortions to the resulting images, necessitating careful selection of the Pansharpening method for any application that uses the pansharpened image as an input. Furthermore, how well the pansharpening technique copes with the large amount of image data is critical. The current study mainly focuses on the evaluation of the six pansharpening approaches for satellite image enhancement and attempts to determine the best method for landuse classification using Indian Remote Sensing dataset. In this research, Resourcesat-1 satellite images acquired by LISS-IV sensor and IRS-P5 (Cartosat-1) images acquired by PAN sensor are used to obtain the high-resolution color imagery. The study also identifies which landuse characteristics are improved as a result of using

Multimedia Security: Tools, Techniques, and Applications. Bhaskar Mondal and Shyam Singh Rajput (Eds.)

pansharpening techniques. The findings of this study can be very useful in determining the pansharpening method's suitability for various geospatial applications.

8.1 INTRODUCTION

The advancement in satellite technology has led to the increased availability of satellite images acquired by different earth observation (EO) satellite systems [1]. In the current scenariao, space agencies of various countries, including India, have developed several advanced technologies which enable them to launch multiple satellites in different orbits in a single mission [2]. After the successful launch of the satellite and ejecting the same in one of the orbital plane around the Earth, it starts capturing the images of the Earth features and sending these images to the ground stations [3]. The satellite images captured by various sensors mounted on the satellite are the raw images that cant be used as it is without pre-processing. Preprocessing of the satellite images involves the elimination of the radiometric and geometric distortion introduced in the image during the acquisition [4–6]. The geometric distortions are introduced in the image due to several factors including random variations in platform altitude and attitude along the direction of flight, relief displacement and effect of Earth's rotation [7] whereas radiometric distortions are introduced in the imagery due to sun's azimuth and elevation and atmospheric conditions that can influence the energy observed by the sensor [8].

The raw image, after geometric and radiometric distortions have been removed, is used for analysis with any application. These procedures are referred to as pre-processing methods, and they are required for any satellite image analysis task [6].

Satellite images contain an abundant amount of spectral and spatial information. Now a days, the satellites are launched with a payload consisting of multiple sensors, including high resolution Panchromatic (PAN) that is 0.5–0.85 μm and MS sensor, which contains multiple camaras operating in the visible range (0.4–0.7 μm) and near infrared (NIR) wavelength. EO sensors that capture data in multiple bands produce the MS images. The MS images contain high spectral details whereas pa,nchromatic images are rich in spatial information. There are several applications which requires both high spectral and spatial information for better visualization of the different landcover features present on the satellite images [10].

For remote sensing systems, four types of resolutions are known: spatial, spectral, temporal, and radiometric. Spatial resolution is defined as the lowest dimension on the Earth's surface across which the sensor can make an independent measurement. It is represented in meters by the size of the pixel on the ground. The instantaneous field of view (IFOV) governs it. It is represented in meters by the size of the pixel on the ground. It is governed by the IFOV, which is defined as the region of the Earth's surface visible at any one instant [11–13].

The capacity of a sensor to define narrow wavelength intervals is referred to as spectral resolution. It may also be defined as a sensor's capacity to resolve the energy received in a spectral bandwidth in order to describe distinct constituents of the earth's surface. It is determined by the spectral band width of the filter and the sensitivity of the detector. The narrower the wavelength range for a given channel or band, the finer the spectral resolution [14, 15]. Radiometric resolution is defined as the sensor's sensitivity to the magnitude of electromagnetic radiation. The greater a sensor's radiometric resolution, the more sensitive it is to detecting tiny changes in energy [16]. Temporal resolution is defined as the number of times an item is sampled or the frequency with which data for the same region is collected [17, 18].

Pansharpening is the process of generating high resolution color imagery (HRMX) from a low resolution multispectral (LRMX) image and an HSRP image. Image Pansharpening technology is required to achieve high-resolution multispectral (HRMX) images. The complementary information of PAN and LRMX images can be integrated by merging them. This is known as Pansharpening, and it is one branch of image fusion that aims to obtain HRMX images by injecting spatial details from PAN into LRMX. Furthermore, more precise scene descriptions and more trustworthy interpretations can be provided [19–22].

Six Pansharpening techniques are explained in this work, and their performance is also evaluated. The remainder of the manuscript is divided into five sections. The first section provides a detailed review of the literature on traditional and advanced image Pansharpening techniques. The second section describes the study area and datasets used in the analysis. The third section describes the methodology used to assess the performance of the pansahrpening techniques, as well as the mathematical details of each technique. The results and their explanation are explained in the fourth section of the manuscript. The final section of the chapter deduces the research work's conclusion.

8.2 RELATED WORK

High resolution satellite images in the remote sensing domain have the potential to help decision making in several socital benefit areas such as transport infrastructure development [23], urban studies [24], sustainable and smart agriculture [25], and water resources development [26]. The growing demand for products and services requiring exquisite spectral and spatial accuracy for feature identification/extraction has prompted EO service providers to build and develop such spatial products. These products make use of pansharpened or fused satellite images derived from low spatial resolution MS and HSRP bands [27].

Wang et al. [28] have carried out a comparative study of image pansharpening techniques and assessed the adopted methods using image quality metrics such as spectral angular mapper, relative dimensionless global error, structural similarity index measure (SSIM). They compared the most commonly used methods such as generalized IHS image fusion, principal components (PCs), and Wavelet. They found the wavelet method as the best method for obtaining the pansharpened image with the minimal spectral distortions. They found comparable image quality for PCs and generalized IHS image fusion methods, whereas spatial qualities of the pansharpened imagery employing the wavelet technique were found to be lower than those of the others. Ruta & Bogdan [29] presented classifier fusion techniques and attempted to identify new trends that might influence this area of research in coming years. They also provided the taxonomy of fusion methods and compared with the existing methods.

Tu et al. [30] carried out a research work on IHS-like image fusion method. They utilized MS and panchromatic images acquired from SPOT satellite sensor for obtaining the color imagery. The MS and panchromatic dataset have spatial resolutions of 20 m and 10 m, respectively. The study was conducted in some rural areas in Taichung, Taiwan, on the satellite images procured on 04 march 1994. It was reported that the proposed approach was highly reliable and could be used as a reference for developing the new image pansharpening algorthms. Gan & Chris [31] compared two measurement pansharpening techniques for Kalman-Filter-based fusion of multisensor data. They determined that if the sensors had similar measurement matrices, the two measurement pansharpening approaches were functionally equal. Method II was shown to be more efficient than Method I in this situation due to its reduced computing cost. The performance of the Method-I was found better than the Method-II in case of different measurement matrices of the sensors.

A detailed review of the different image Pansharpening techniques can be found in the literature [32–34].

Zhang [35] presented a review on current methods of multi-source satellite dataset fusion and explained concept of hierarchical classification to present the future trends and challenges of the multi-source satellite dataset fusion methods. They classified the fusion methods into three parts, viz. pixel/data level, feature level, and decision level. They also discussed the approaches which were frequently used with optical panchromatic and multi-spectral datasets including synthetic aperture radar and LiDAR. Belgiu & Alfred [36] carried out a study on spatiotemporal data fusion methods used with remote sensing datasets. MODIS and Landsat satellite images were comobined to obtain the fine resolution fused image. They emphasized the use of more advanced machine learning (ML) algorithms including convolutional neural network (CNN) capable of fusing the images with varying spectral values for spatiotemporal pansharpening.

Crowson et al. [37] showed the comparison of fusion methods tested over satellite images to measure peat swamp forest loss in Sumatra located in Indonesia. They compare three different classification techniques to measure the residual spread of peat swamp forest. The satellite data which adopted for analysis was acquired from Sentinel-1 and Sentinel-2 satellites covering the surrounding region of Sungai Buluh Protection Forest located in Sumatra, Indonesia. By employing PCA for object-based classification/pixel-based classification, data fusion was not found beneficial.

Zhu et al. [38] presented a study on Spatiotemporal Fusion of Multisource Remote Sensing Data. They investigated literature on current spatiotemporal data fusion methods and divided the existing techniques into different categories. Principal laws underlying these techniques were also discussed in this chapter, along with the summary of the potential applications of the fusion methods.

Bretschneider & Odej [39] carried out a study on image fusion method used in remote sensing. They talked about a method that was developed based on a sub-pixel misalignment of concurrently acquired spectral bands and used the displacement as additional information to effectively increase the sampling frequency.

To map land cover, several techniques, such as indices generated from remote sensing datasets, have been utilized. One of the very popular index called normalized difference vegetation index (NDVI) is computed as the ratio of the difference and summation of the red and NIR bands. Achard & Estreguil [40] used multitemporal dataset prepared by mosaicing of AVHRR images for NDVI mapping in tropical forests. Fernandez et al. [41] used NDVI for damaged surface mapping caused by significant forest fires. Using Landsat thematic mapper (TM) bands-2 and band-5, the normalized difference snow index (NDSI) was effectively used to extract glaciers [42].

NDSI was calculated using the differential between the strong reflection of visible light and the near total absorption of middle infrared wavelengths by snow [43]. The bright surfaces, such as plants, rock, and dirt, were clearly distinguishable from snow using NDSI [44]. Normalized difference water index (NDWI) was used as the SIR for identifying water classes, extracting water bodies, and distinguishing natural and artificial characteristics [45]. The NDWI, defined as the ration of difference and summation of green and red bands, was created to distinguish water features and improve their visibility in satellite dataset using reflected NIR and visible green wavelengths [46]. Because NDWI is sensitive to variations in plant canopy liquid water content., it should be used in conjunction with, rather than as a substitute for, NDVI [47].

The current study is distinct from prior investigations based on consideration of IRS images for 3-band LISS-IV sensor image and single band panchromatic image of Cartosat-1 satellite for image fusion which are suitable for landcover mapping and other other societal applications.

In addition, Image fusion is being frequently utilized in medical domain also. Medical image fusion includes a wide range of approaches ranging from image fusion to generic information fusion used to solve medical concerns represented in images of the organs and cells of human body. Imaging technologies are gaining popularity and use in medical diagnosis, analysis, and historical documentation [48].

There are a variety of medical imaging modalities that can be utilized as major contributions to medical image fusion studies. The selection of an imaging modality for a focused clinical research necessitates medical knowledge relevant to the organs under investigation [49]. It is very hard to acquire all of the information from a single imaging modality in order to assure clinical accuracy and robustness of the analysis and resultant diagnosis. To create a more trustworthy and accurate evaluation, it is apparent to look at photos from several modalities. The following are some of the major modalities utilized in clinical practice: (i) quantitative coronary angiography, a kind of angiography [50] and quantitative vascular angiography [51]; (ii) computer tomography, a type of angiography [52]; (iii) dual-energy X-ray absorptiometry [53]; and (iv) magnetic resonance imaging, a kind of angiography [54]. Some of the other important applications of image fusion methods include in logo watermarking, asymmetric cryptosystems, and hiding.

Kundur & Dimitrios [55] investigated the topic of logo watermarking in still pictures and used multiresolution data fusion techniques for watermark embedding and extraction. To conceal a higher energy concealed logo in prominent picture components, a human visual system model based on contrast sensitivity was used. Watermark extraction required both attack characterisation and logo estimate utilizing a rake-like receiver. The statistical study revealed how the suggested technique may be utilized for watermark detection applications to reduce the problem of false negative detection while boosting the rate of false positive detection. They validated theoretical discoveries using simulation data and proved FuseMark's actual performance.

Mehra & Naveen [56] used wavelet transform color picture fusion to secure data using an asymmetric encryption method and image concealing. To get a higher quality recovered color image, the color image bands having distinct wavelengths (red, green and blue) were merged together using DWT.

In conclusion, Pansharpening methods on spatiotemporal and multi-source satellite images are widely used and adopted by various researchers. The satellite images acquired by the optical, microwave, and hyperspectral sensors are used in the various studies. Image Pansharpening technology is also used to combine images with elevation information, such as those obtained by LiDAR. Fusion techniques are also very popular and are being utilized in medical images to solve medical problems, including logo watermarking and asymmetric cryptosystems.

8.3 STUDY AREA AND DATA RESOURCES

The study area belogs to the district of Yamuna nagar of Haryana state, India. The scene central co-ordinates are 30° 19′ 6.7″ N and 77° 35′ 54.75″ E. The study area contains several landcover features, including agricultural fields, a Built-up area, a small part of the river, and roads. The input satellite images (panchromatic and multispectral) are shown in Figures 8.1 and 8.2. In this study, satellite data un,like spatial and spectral resolutions acquired by different sensors, has been used for carrying out analyses. The details of the satellite data used for analyzes have been shown in Table 8.1.

The visualize the range of the DN values of the input image (panchromatic), histogram plot is shown in Figure 8.3. The histogram plot shows the graph between DN values on the X-axis and the frequency of that DN value, i.e., count on the Y-axis.

FIGURE 8.1 High resolution panchromatic (Cartosat-1) image.

FIGURE 8.2 Multispectral (LISS-IV sensor) image.

TABLE 8.1 Details of Satellite Data Used

SL. No.	Satellite	Image Sensor	Spatial Resolution (m)	Spectral Resolution (μm)	Date of Acquisition
1.	IRS P5 (Cartosat-1)	PAN	2.5	0.50–0.85	25 Oct, 2009
2.	Resourcesat-1	LISS-IV	5.8	0.52–0.59 (green) 0.62–0.68 (red) 0.77–0.86 (NIR)	25 Oct, 2009

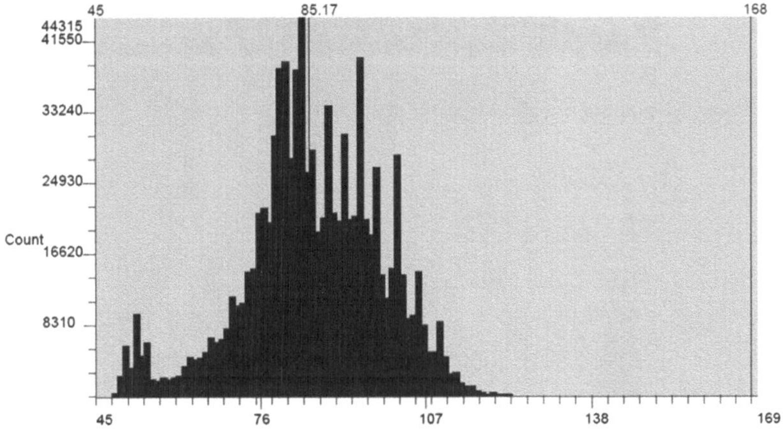

FIGURE 8.3 Histogram of panchromatic (Cartosat-1) image.

Similarly, for MS image, which is composed of three bands viz. Green, Red, and Near-Infrared (NIR), histogram has been plotted for all bands. The first layer (Layer-1) in the MS image is green band and the histogram of the same has been plotted in Figure 8.4.

The second layer (Layer-2) is a red band, and the histogram of the same is shown in Figure 8.5. The third layer (Layer-3) is NIR band and the histogram of the NIR band has been shown in Figure 8.6.

8.4 MATERIAL AND METHODS

The overall methodology adopted for this study is summarized in Figure 8.7. The current study discusses the six popular image pansharpening techniques (Figure 8.8) viz. resolution merge, hyperspherical color space (HCS) resolution merge, modified intensity hue saturation (MIHS), Ehlers Fusion, high pass filter (HPF), and Wavelet Resolution Merge.

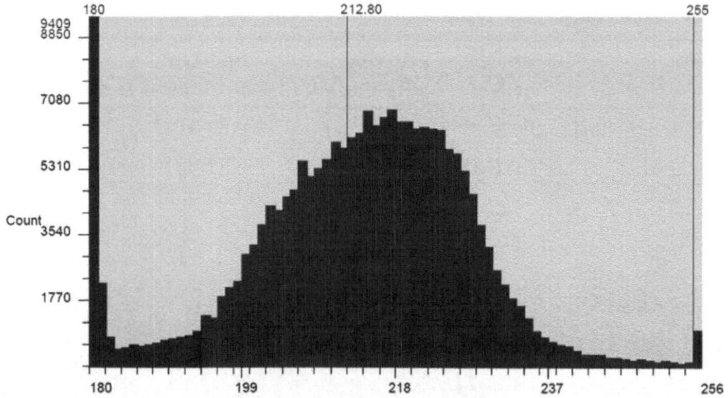

FIGURE 8.4 Histogram of LISS-IV image (Layer-1).

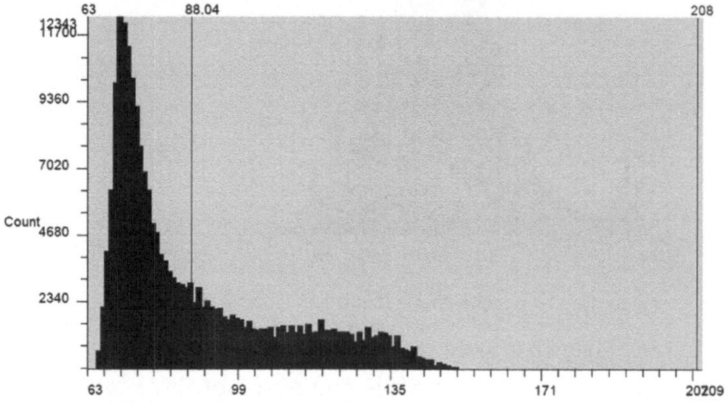

FIGURE 8.5 Histogram of LISS-IV image (Layer-2).

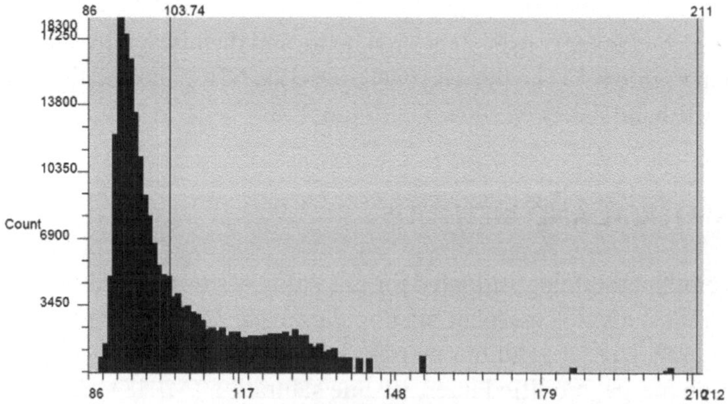

FIGURE 8.6 Histogram of LISS-IV image (Layer-3).

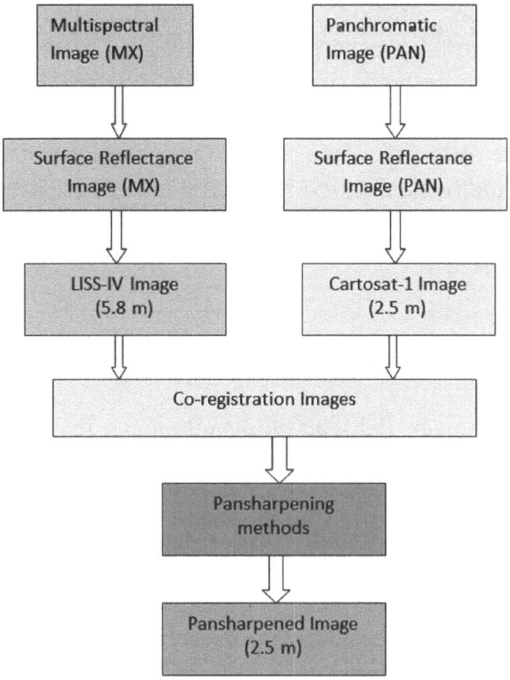

FIGURE 8.7 Methodology.

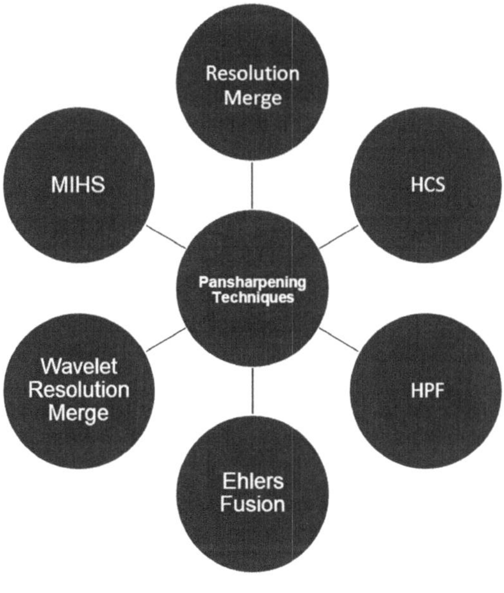

FIGURE 8.8 Pansharpening techniques.

The current study's methodology entails multiple processes to obtain the high resolution colored pansharpened image. (i) The raw satellite pictures (MX and PAN) are acquired; (ii) then raw image digital numbers (DN) are atmospherically corrected, and surface reflectance images are produced; (iii) Both surface reflectance images are then geometrically adjusted using co-registration; and (iv) finally, co-registered images are used in conjunction with the aforementioned pansharpening procedures to obtain the final high resolution color imagery.

8.4.1 RESOLUTION MERGE

This approach of image Pansharpening integrates the images of different spatial resolution. It is used to obtain the fine resolution MS imagery using the fine resolution panchromatic and LRMX satellite images. The interpretability of the data is improved in the resultant image. This strategy includes three methods viz. PCs, multiplicative, and Brovey transform. In PCs technique, the first principal component (PC-1) of the input image is computed, and before executing the inverse PC transformation remapping of the high resolution imagery into the data range of PC-1 is carried out and replaced the same for PC-1. The PC technique is ideal for situations where the original scene radiometry (color balance) of the input MS picture must be kept as nearly as feasible in the output. Unfortunately, this radiometric accuracy comes at a high computational cost. As a result, the PC approach is the slowest and uses the most system resources of the three ways offered [57].

The multiplicative approach is based on the inputs' basic arithmetic integration. The inputs for this are MS and panchromatic images. The most fundamental of the three is the Multiplicative technique. It is generally the quickest technique and requires the fewest system resources because it is computationally simple. The radiometry of the input MS image, however, is not maintained in the merged image. Instead, the intensity component is amplified, making this approach appropriate for highlighting urban characteristics.

The Brovey Transform was created to visually enhancing the contrast in the image's histogram's both low and high points (i.e., water, shadows, and high reflectance regions including urban features appeared with high contrast). As a result, if original scene radiometry need to be preserved, the use of Brovey Transform for image fusion should be avoided. It is, however, useful for generating RGB images high contrast. This technique employs a ratio algorithm to produce the pansharped image. The computation formula for Brovey transform is shown in Eqns. (1) to (3).

$$\left[\frac{DN_{B1}}{DN_{B1} + DN_{B2} + DN_{B3}} \right] * DN_{HRS} = DN_{B1\_n} \quad (1)$$

$$\left[\frac{DN_{B2}}{DN_{B1} + DN_{B2} + DN_{B3}} \right] * DN_{HRS} = DN_{B2\_n} \quad (2)$$

$$\left[\frac{DN_{B3}}{DN_{B1} + DN_{B2} + DN_{B3}} \right] * DN_{HRS} = DN_{B3\_n} \quad (3)$$

where; DN_{B1} is the digital number of the green band; DN_{B2} is the digital number of the red band; DN_{B3} is the digital number of NIR bands of the LISS-IV image; and DN_{HRS} a digital number of the high resolution Cartosat-1 image.

8.4.2 HPF RESOLUTION MERGE

High pass filter (HPF) resolution merge method of image pansharpening is used to combine the fine-resolution panchromatic data with lower resolution MS data to produce an image with both excellent spectral and spatial details. The HPF resolution merge method begins with reading the pixel sizes from picture files and calculating R, which is the ration of cell size of MS and high-resolution images. Then, on the image with fine spatial resolution, use the High-pass filter. Then resampling (high-pass image's pixel size) is performed in MS image. To each MS band, add the HPF image. The HPF image is weighted with reference to the MS band's global standard deviation. At the end, stretching of the new MS image is performed to match the mean and standard deviation of the original (input) MS image [58].

8.4.3 MODIFIED IHS RESOLUTION MERGE

The Modified IHS resolution merge algorithm merges high-resolution panchromatic images with lower-resolution MS images. Here, HIS stands for intensity, hue, and saturation. The output of the modified IHS method contains great details and depict an accurate original MS scene colors [59].

8.4.4 WAVELET RESOLUTION MERGE

Wavelets Resolution Merge sharpens MS data with poor spatial resolution by combining them with a coregistered panchromatic image with high resolution. Fourier transform analysis is analogous to wavelet-based processing.

Long continuous (sine and cosine) waves are utilized as the basis in the Fourier transform. Instead of a lengthy, continuous "wavelet," the wavelet transform employs short, discontinuous wavelets [60]. The details of the wavelet based Pansharpening method can be obtained in the literature [61, 62]. Figure 8.9 shows the workflow of obtaining the panshapened image using 2D discrete wavelet transform (DWT).

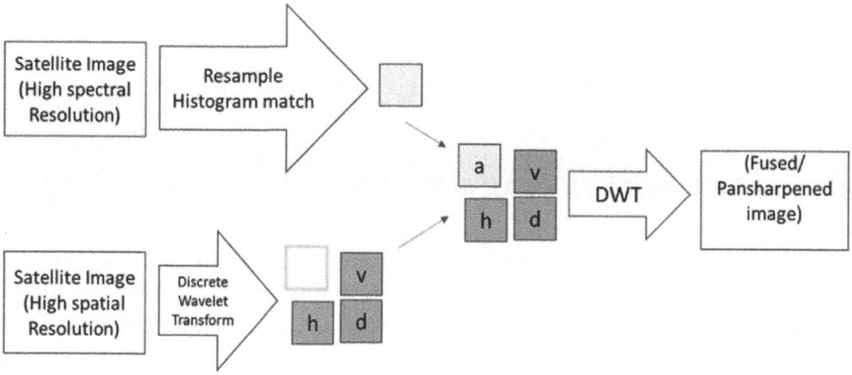

FIGURE 8.9 Wavelet resolution merge.

The histogram match approach is used to resample a high spectral resolution image (MS). DWT is used to breakdown the high spatial resolution (panchromatic) image, resulting in the four components. One low frequency component (a) and three high frequency components (h, v, d) make up the four components. The output image obtained by resampling the MS image replaces the low frequency component that appears similar to the original input image. The pansharpened image is created by combining the resampled image with the three high frequency components [63, 64].

8.4.5 *EHLERS FUSION*

This method allows you to combine images with varying spatial resolutions. Because higher resolution photography is typically single band (e.g., SPOT satellite Panchromatic data with 10 m), and MS data is typically lower resolution (e.g., Landsat satellite Thematic Mapper (TM) sensor 30 m data), these approaches are frequently employed to provide high resolution, MS images. Prof. Manfred Ehlers of the University of Osnabrück devised the Ehlers Fusion algorithm [65].

8.4.6 HYPERSPHERICAL COLOR SPACE (HCS) RESOLUTION MERGE

The Hyperspherical Color Sharpening technique is used to combine fine-resolution panchromatic data with lower-resolution MS data. This option is intended for WorldView-2 sensor 8 band data, but it can be used with any MS data consisting of three or more bands. The data is converted from the native color system to the HCS. The first mode substitutes an intensity-matched version of the pan band for the MS intensity component. The second mode replicates the original MS colors [66].

8.5 RESULTS AND DISCUSSION

Multispectral (MS) and panchromatic images have been fused to obtain the high resolution color imagery using Pansharpening methods. Resolution merge technique of image Pansharpening offers three methods viz. PCs, multiplicative, and brobey transform.

The output of the PCs method is shown in Figure 8.10. In Figure 8.10, several landcover features such as agricultural fields (shown in red color), water body (shown in black color), built-up (shown in cyan color), etc., are visible.

FIGURE 8.10 Resolution merge–principal components.

The landcover features inc,luding built-up areas, are highlighted and clearly visualized in the output image generated using wavelet resolution merge. The output of the multiplicative method is shown in Figures 8.11 and 8.12 represents the output image generated after applying Brovey transform. The output obtained after applying HCS resolution merge technique is shown in Figure 8.13.

FIGURE 8.11 Resolution merge – multiplicative.

Figure 8.14 represents the output image generated after applying HPF resolution merge technique.

The output obtained after applying MIHS technique is shown in Figure 8.15.

Figure 8.16 represents the output image generated after applying Ehlers fusion technique.

The output obtained after applying Wavelet resolution merge technique is shown in Figure 8.17.

The evaluation processes are based on the verification of spectral features preservation and the increase of spatial resolution. First, the pansharpened images are compared visually. Despite the fact that visual appearance is subjective and relies on the human interpreter, the strength of visual cognition as a final background should not be overlooked. Second, we have also evaluated the fusion methods based on some popular statistical measurements including Mean Squared Error (MSE) and Structure Similarity Index (SSIM).

FIGURE 8.12 Resolution merge–Brovey transform.

FIGURE 8.13 Output image of HCS method.

FIGURE 8.14 Output image of HPF method.

FIGURE 8.15 Output image of MIHS method.

FIGURE 8.16 Output image of Ehlers fusion method.

FIGURE 8.17 Output image of wavelet resolution merge method.

According to Lucien [67], the RMS error is calculated as "the difference between the standard deviation and mean of the fused and original images." Again, the best possible value is zero.

The concept of the Structure Similarity Index (SSIM) was introduced by Wang et al. [68]. The SSIM method examines brightness, contrast, and structure in an 8 by 8 square window and is applied locally. This window travels across the entire image pixel by pixel. This window is moved pixel by pixel over the whole imagery. At each step, the local statistics and the SSIM index are computed within the window. The values range from 0 to 1. Values near 1 indicate the greatest correspondence with the source images.

The details of the performance evaluation of the all six techniques using two methods (SSIM and MSE) is summarized in Table 8.2.

TABLE 8.2 Details of the Accuracy Assessment

SL. No.	Method		Structure Similarity Index (SSIM)	Mean Squared Error
1.	Resolution merge	PCA	0.4096	0.0498
		Multiplicative	0.6423	0.0225
		Brovey	0.5252	0.0384
2.	High pass filter resolution (HPF) merge		0.9012	0.0032
3.	Modified intensity hue saturation (MIHS)		0.6060	0.0303
4.	Hyperspherical color space (HCS) resolution merge		0.8586	0.0044
5.	Wavelet resolution merge		0.9302	0.0022
6.	Ehlers fusion		0.7508	0.0092

Table 8.2 shows that the wavelet resolution merge technique beat the other image Pansharpening strategies in terms of structure similarity index and mean squared error. This method yielded the highest similarity index and the lowest mean squared error. The PCs method of resolution merging technique performed the worst of all techniques, producing the lowest similarity index and the biggest mean squared error.

8.6 CONCLUSION

This study demonstrates not just the relevance of consistent evaluation methodologies and the need for a combined strategy for a quantitative assessment of spatial improvement and spectral preservation. Six popular pansharpening methods are used in this study on IRS satellite images acquired from Cartosat-1 and LISS-IV sensors to produce color imagery with a spatial resolution of 2.5 m. The accuracy of the pansharpened images obtained from the various image fusion techniques has also been evaluated through visual analysis and through quantitative evaluation. It is also clear that quantitative evaluation approaches can give outcomes that are not supported by visual inspection. The reason behind this needs to be looked into more. At this point, we can only conclude that visual analysis must be an important part of any quality evaluation procedures, and that diverse evaluation methods should be employed to avoid outliers from skewing the final results.

KEYWORDS

- **hyperspherical color space**
- **image enhancement**
- **image fusion**
- **multimedia**
- **pansharpening**
- **satellite images**
- **wavelet resolution merge**

REFERENCES

1. Belward, A. S., & Jon, O. S., (2015). Who launched what, when, and why; trends in global land-cover observation capacity from civilian earth observation satellites. *ISPRS Journal of Photogrammetry and Remote Sensing*, *103*, 115–128.
2. Wilson, F., (2006). Recent advances in satellite propulsion and associated mission benefits. In: *24th AIAA International Communications Satellite Systems Conference* (p. 5306).
3. Wei-Cheng, L., Da-Yin, L., Chung-Yang, L., & Yong-Yao, L., (2005). Daily imaging scheduling of an earth observation satellite. *IEEE Transactions on Systems, Man, and Cybernetics-Part A: Systems and Humans, 35*(2), 213–223.
4. Toutin, T., (2011). Three-dimensional geometric correction of earth observation satellite data. *Advances in Environmental Remote Sensing* (pp. 173–217). CRC Press: Boca Raton, FL, USA.
5. Itten, K. I., & Peter, M., (1993). Geometric and radiometric correction of TM data of mountainous forested areas. *IEEE Transactions on Geoscience and Remote Sensing, 31*(4), 764–770.

6. Li, C., & Hao, X., (2017). A geometric and radiometric simultaneous correction model (GRSCM) framework for high-accuracy remotely sensed image preprocessing. *Photogrammetric Engineering & Remote Sensing, 83*(9), 621–632.
7. Dave, C. P., Rahul, J., & Srivastava, S. S., (2015). A survey on geometric correction of satellite imagery. *International Journal of Computer Applications, 116*(12).
8. Teillet, P. M., (1986). Image correction for radiometric effects in remote sensing. *International Journal of Remote Sensing, 7*(12), 1637–1651.
9. Wang, Z., Alan, C. B., Hamid, R. S., & Eero, P. S., (2004). Image quality assessment: From error visibility to structural similarity. *IEEE Transactions on Image Processing, 13*(4), 600–612.
10. Solanky, V., & Katiyar, S. K., (2016). Pixel-level image fusion techniques in remote sensing: A review. *Spatial Information Research, 24*(4), 475–483.
11. Intriligator, J., & Patrick, C., (2001). The spatial resolution of visual attention. *Cognitive Psychology, 43*(3), 171–216.
12. Vermeer, G. J. O., (1999). Factors affecting spatial resolution. *Geophysics, 64*(3), 942–953.
13. Welch, R., (1982). Spatial resolution requirements for urban studies. *International Journal of Remote Sensing, 3*(2), 139–146.
14. Rocchini, D., (2007). Effects of spatial and spectral resolution in estimating ecosystem α-diversity by satellite imagery. *Remote Sensing of Environment, 111*(4), 423–434.
15. Jin, X., & Curt, H. D., (2005). Automated building extraction from high-resolution satellite imagery in urban areas using structural, contextual, and spectral information. *EURASIP Journal on Advances in Signal Processing. 2005*(14), 1–11.
16. Canty, M. J., Allan, A. N., & Michael, S., (2004). Automatic radiometric normalization of multitemporal satellite imagery. *Remote Sensing of Environment, 91*(3, 4), 441–451.
17. Panteras, G., & Guido, C., (2018). Enhancing the temporal resolution of satellite-based flood extent generation using crowdsourced data for disaster monitoring. *International Journal of Remote Sensing 39*(5), 1459–1474.
18. Liu, H., & Qihao, W., (2012). Enhancing temporal resolution of satellite imagery for public health studies: A case study of West Nile Virus outbreak in Los Angeles in 2007. *Remote Sensing of Environment, 117*, 57–71.
19. Vermeer, G. J. O., (1999). Factors affecting spatial resolution. *Geophysics, 64*(3), 942–953.
20. Zhang, Y., (2004). Understanding image fusion. *Photogramm. Eng. Remote Sens., 70*(6), 657–661.
21. Connah, D., Mark, S. D., & Graham, D. F., (2014). Spectral edge image fusion: Theory and applications. In: *European Conference on Computer Vision* (pp. 65–80). Springer, Cham.
22. Pohl, C., & John, L. V. G., (1998). Review article multisensor image fusion in remote sensing: Concepts, methods, and applications. *International Journal of Remote Sensing, 19*(5), 823–854.
23. Schnebele, E., Tanyu, B. F., Cervone, G., & Waters, N., (2015). Review of remote sensing methodologies for pavement management and assessment. *European Transport Research Review, 7*(2), 1–19.
24. Shao, Z., Neema, S. S., Aleksei, P., Fanan, U., Walter, M., & Paulo, J. M., (2021). Urban sprawl and its impact on sustainable urban development: A combination of remote sensing and social media data. *Geo-spatial Information Science, 24*(2), 241–255.
25. Segarra, J., Maria, L. B., Jose, L. A., & Shawn, C. K., (2020). Remote sensing for precision agriculture: Sentinel-2 improved features and applications. *Agronomy, 10*(5), 641.

26. Govender, M., Kershani, C., & Hartley, B., (2007). A review of hyperspectral remote sensing and its application in vegetation and water resource studies. *Water Sa., 33*(2), 145–151.
27. Mhangara, P., Willard, M., & Naledzani, M., (2020). Comparison of image fusion techniques using satellite pour L' Observation de la Terre (SPOT) 6 satellite imagery. *Applied Sciences, 10*(5), 1881.
28. Wang, Z., Djemel, Z., Costas, A., Deren, L., & Qingquan, L., (2005). A comparative analysis of image fusion methods. *IEEE Transactions on Geoscience and Remote Sensing, 43*(6), 1391–1402.
29. Ruta, D., & Bogdan, G., (2000). An overview of classifier fusion methods. *Computing and Information Systems, 7*(1), 1–10.
30. Te-Ming, T., Shun-Chi, S., Hsuen-Chyun, S., & Ping, S. H., (2001). A new look at IHS-like image fusion methods. *Information Fusion, 2*(3), 177–186.
31. Gan, Q., & Chris, J. H., (2001). Comparison of two measurement fusion methods for Kalman-filter-based multisensor data fusion. *IEEE Transactions on Aerospace and Electronic Systems, 37*(1), 273–279.
32. Ghassemian, H., (2016). A review of remote sensing image fusion methods. *Information Fusion, 32*, 75–89.
33. Meher, B., Sanjay, A., Rutuparna, P., & Ajith, A., (2019). A survey on region-based image fusion methods. *Information Fusion, 48*, 119–132.
34. Ma, J., Yong, M., & Chang, L., (2019). Infrared and visible image fusion methods and applications: A survey. *Information Fusion, 45*, 153–178.
35. Zhang, J., (2010). Multi-source remote sensing data fusion: Status and trends. *International Journal of Image and Data Fusion, 1*(1), 5–24.
36. Belgiu, M., & Alfred, S., (2019). Spatiotemporal image fusion in remote sensing. *Remote Sensing, 11*(7), 818.
37. Crowson, M., Warren-Thomas, E., Jane, K. H., Bambang, H., Fahmuddin, A., Asmadi, S., Keith, C. H., et al., (2019). A comparison of satellite remote sensing data fusion methods to map peat swamp forest loss in Sumatra, Indonesia. *Remote Sensing in Ecology and Conservation, 5*(3), 247–258.
38. Zhu, X., Fangyi, C., Jiaqi, T., & Kay-Ann, W. T., (2018). Spatiotemporal fusion of multisource remote sensing data: Literature survey, taxonomy, principles, applications, and future directions. *Remote Sensing, 10*(4), 527.
39. Bretschneider, T., & Odej, K., (2000). Image fusion in remote sensing. In: *Proceedings of the 1st Online Symposium of Electronic Engineers* (pp. 1–8).
40. Achard, F., & Christine, E., (1995). Forest classification of Southeast Asia using NOAA AVHRR data. *Remote Sensing of Environment, 54*(3), 198–208.
41. Fernández, A., Pilar, I., & Jose, L. C., (1997). Automatic mapping of surfaces affected by forest fires in Spain using AVHRR NDVI composite image data. *Remote Sensing of Environment, 60*(2), 153–162.
42. Sidjak, R. W., (1999). Glacier mapping of the Illecillewaet icefield, British Columbia, Canada, using Landsat TM and digital elevation data. *International Journal of Remote Sensing, 20*(2), 273–284.
43. Hall, D. K., George, A. R., & Vincent, V. S., (1995). Development of methods for mapping global snow cover using moderate resolution imaging spectroradiometer data. *Remote Sensing of Environment, 54*(2), 127–140.
44. Dozier, J., (1989). Spectral signature of alpine snow cover from the Landsat thematic mapper. *Remote Sensing of Environment, 28*, 9–22.

45. Ben-Dor, E., Levin, N., & Saaroni, H., (2001). A spectral based recognition of the urban environment using the visible and near-infrared spectral region (0.4-1.1 μm). A case study over Tel-Aviv, Israel. *International Journal of Remote Sensing, 22*(11), 2193–2218.
46. McFeeters, S. K., (1996). The use of the normalized difference water index (NDWI) in the delineation of open water features. *International Journal of Remote Sensing, 17*(7), 1425–1432.
47. Bo-Cai, G., (1996). NDWI—A normalized difference water index for remote sensing of vegetation liquid water from space. *Remote Sensing of Environment, 58*(3), 257–266.
48. James, A. P., & Belur, V. D., (2014). Medical image fusion: A survey of the state of the art. *Information Fusion, 19*, 4–19.
49. Ahmad, H. A., Hui, J. Y., & Colin, G. M., (2014). Medical imaging modalities. In: *Medical Imaging in Clinical Trials* (pp. 3–26). Springer, London.
50. Marquering, H. A., Jouke, D., Quentin, J. A. B., Julien, P. M. D., Joanne, D. S., Jeroen, J. B., & Johan, H. C. R., (2008). Coronary CT angiography: IVUS image fusion for quantitative plaque and stenosis analyses. In: *Medical Imaging 2008: Visualization, Image-Guided Procedures, and Modeling* (Vol. 6918, p. 69181G). International Society for Optics and Photonics.
51. Schwein, A., Ponraj, C., Greg, B., Alan, B. L., Jean, B., & Carlos, F. B., (2018). Computed tomography angiography-fluoroscopy image fusion allows visceral vessel cannulation without angiography during fenestrated endovascular aneurysm repair. *Journal of Vascular Surgery, 68*(1), 2–11.
52. Hopp, T., Duric, N., & Nicole, V. R., (2015). Image fusion of ultrasound computer tomography volumes with X-ray mammograms using a biomechanical model based 2D/3D registration. *Computerized Medical Imaging and Graphics, 40*, 170–181.
53. Grier, S. J., Turner, A. S., & Alvis, M. R., (1996). The use of dual-energy x-ray absorptiometry in animals. *Investigative Radiology, 31*(1), 50–62.
54. Baco, E., Osamu, U., Erik, R., Ljiljana, V., Aud, S., Manju, A., Suzanne, P., et al., (2015). Magnetic resonance imaging–transectal ultrasound image-fusion biopsies accurately characterize the index tumor: Correlation with step-sectioned radical prostatectomy specimens in 135 patients. *European Urology, 67*(4), 787–794.
55. Kundur, D., & Dimitrios, H., (2004). Toward robust logo watermarking using multiresolution image fusion principles. *IEEE Transactions on Multimedia, 6*(1), 185–198.
56. Mehra, I., & Naveen, K. N., (2014). Image fusion using wavelet transform and its application to asymmetric cryptosystem and hiding. *Optics Express, 22*(5), 5474–5482.
57. Zhang, Y., & Rakesh, K. M., (2012). A review and comparison of commercially available pan-sharpening techniques for high resolution satellite image fusion. In: *2012 IEEE International Geoscience and Remote Sensing Symposium* (pp. 182–185). IEEE.
58. Palubinskas, G., (2013). Fast, simple, and good pan-sharpening method. *Journal of Applied Remote Sensing, 7*(1), 073526.
59. John, C. M., & Kavya, N., (2014). Integration of multispectral satellite and hyperspectral field data for aquatic macrophyte studies. *The International Archives of Photogrammetry, Remote Sensing and Spatial Information Sciences, 40*(8), 581.
60. Li, S., James, T. K., & Yaonan, W., (2002). Using the discrete wavelet frame transform to merge Landsat TM and SPOT panchromatic images. *Information Fusion, 3*(1), 17–23.
61. King, R. L., & Jianwen, W., (2001). A wavelet-based algorithm for pan sharpening Landsat 7 imagery. In: *IGARSS 2001; Scanning the Present and Resolving the Future:*

Proceedings IEEE 2001 International Geoscience and Remote Sensing Symposium (Cat. No. 01CH37217) (Vol. 2, pp. 849–851). IEEE.

62. Lemeshewsky, G. P., (2002). Multispectral image sharpening using a shift-invariant wavelet transform and adaptive processing of multiresolution edges. In: *Visual Information Processing XI* (Vol. 4736, pp. 189–200). International Society for Optics and Photonics.
63. Yocky, D. A., (1995). Image merging and data fusion by means of the discrete two-dimensional wavelet transform. *JOSA A*, *12*(9), 1834–1841.
64. Garguet-Duport, B., Jacky, G., Jean-Marc, C., & Guy, P., (1996). The use of multiresolution analysis and wavelets transform for merging SPOT panchromatic and multispectral image data. *Photogrammetric Engineering and Remote Sensing, 62*(9), 1057–1066.
65. Klonus, S., & Manfred, E., (2007). Image fusion using the Ehlers spectral characteristics preservation algorithm. *GI Science & Remote Sensing, 44*(2), 93–116.
66. Padwick, C., Michael, D., Fabio, P., & Scott, S., (2010). WorldView-2 pan-sharpening. In: *Proceedings of the ASPRS 2010 Annual Conference, San Diego, CA, USA* (Vol. 2630, pp. 1–14).
67. Lucien, W., (2002). *Data Fusion: Definitions and Architectures*. Les Presses de L'ecole des Mines.

CHAPTER 9

SPAM REMARK REVELATION UTILIZING SUPERVISED MACHINE LEARNING APPROACHES IN SECURITY ISSUES

SOHOM BHATTACHARYA,[1] PRIYANKA DHARA,[1] SHUBHAM BHATTACHARJEE,[1] and SUBIR GUPTA[2]

[1]*Department of Master of Computer Application, B. C. Roy Engineering College, Durgapur, West Bengal, India*

[2]*Department of Computer Science and Engineering (Artificial Intelligence and Machine Learning), Haldia Institute of Technology, West Bengal, India*

ABSTRACT

Since the beginning, the stunning turn of events and the conspicuousness of online gregarious affiliations restrict the primer of spamming, which generally prompts the uncontrolled dispersal of malware, unconventional advancements, phishing, and stunts. Online stages like YouTube, Facebook, Twitter, etc., are utilized in a sundry and set off to decrease individuals' optimistic visual view correspondingly as affiliations. Recorded as a printed duplicate, the supervised machine learning (ML) approach has been hugely epic to discern spam and spammers in online loosened up affiliations optically. Usually tardy, extraordinary supervised ML approaches propose for spammer openness tasks. This section expands on different well-kenned directed ML approaches in detaching the spam and ham comments.

9.1 INTRODUCTION

The effect of digital correspondence has turned it into the most fundamental part of the contemporary world. Moreover, it conjures some rudimentary issues

Multimedia Security: Tools, Techniques, and Applications. Bhaskar Mondal and Shyam Singh Rajput (Eds.)

within human society. Perhaps the most sizeable voluminous model is spam [1]. Spam remarks are generally spontaneous, unapproved, and customarily voluminous correspondence [2, 3]. Being a conveyor of malware, spam itself causes several processes. For example, the expansion of spontaneous advertisements, misrepresented plans, phished messages, unequivocal matter, advancements of cause, and so forth on a hierarchical front. Spam impacts include irritation to the singular client, less solid messages, loss of work usefulness, abuse of organization data transfer capacity, wastage of record worker extra room as well as computational puissance spread of infections in the form of worms and Trojan ponies, monetary swindling through phishing, Denial of Accommodation (DoS), catalog reaping assaults, etc. [4].

Resolving this issue has hugely become one of the paramount jobs for PC subject matter experts, for it is anything but a forthcoming examination region in the security space. Of numerous methodologies, machine learning (ML) is especially prominent because of its brood application [5]. ML is the examination of modernized computations that ameliorate mundanely through experience. And also by utilizing information utilizing cerebrating [6]. It might be used in a gregarious event of fields, including picture affirmation, material science, and exhibiting Data science and others. ML incorporates PCs discovering how they can perform endeavors without being explicitly adjusted to do overall [7–9]. A subset of ML is solidly cognate to computational experiences, which fixates making gauges utilizing likelihood thickness work. Program estimations prompt the machine to execute all methods expected to handle the recent concern, and the PC's part does not require learning. The goal of this book part is inclining to the spam remark issue utilizing ML.

The remaining chapter facilitates in the following way: In Section 9.2, we have demystified the fundamental concepts of ML. Section 9.3 has overviewed the computational part, like pseudo of the different estimations present in ML. Section 9.4 offers performance quantifications of ML calculations. Section 9.5 frameworks the logical examination of the regulated ML approaches to avoid spam remarks. Decisively, Section 9.6 culminates up about the entire vigilance.

9.2 CONCEPT OF MACHINE LEARNING (ML)

9.2.1 ORIGIN OF ML

The term 'machine learning' (ML) was first raised by an individual named [10, 11, 87]. It executes calculations and neural sodality models to oblige

PC structures in reliably altering their exhibits. ML assessments perfunctorily foster a numerical model utilizing test information in a certain way. It is ostensible as 'coordinating information to settle on culls without being autonomously adjusted to settle on those culls' [12]. An ML is a subfield of PC programming yet is frequently correspondingly implicatively suggested as feeling assessment or hunch exhibiting. Its objective and use are to construct beginning or conceivably sway remaining alive assessments to gain from information, amass expansive models that give exact sets, or discover plans solidly with nascent and covered equipollent news [13].

In 1997, Tom Mitchell gave a very much influenced numerical and convivial definition. The definition includes, "A computer program is verbalized to learn from experience E with veneration to some task T and some performance measure P, if its performance on T, as quantified by P, ameliorates with experience E" [14]. ML is the most recent word or expression coasting around. It has the right to, as it is possibly the fascinating subfields of Computer Science.

9.2.2 A PORTION OF THE REASONS WHY ML IS SO FUNDAMENTAL

1. **Increase in Data Generation:** Due to the unwise incite of information, it requires a technique that can habituate to structure, examine, and draw functional discernments from the report. It utilizes the knowledge to tackle scrapes and discover answers for the most involute undertakings visually concerned by sodalities.
2. **Amend Decision Making:** ML is used to acclimate to settle on better business culls by utilizing several calculations. For instance, ML is used to estimate deals, visualize managements in the securities exchange, and distinguish hazards, material science, information science, oddities, etc.
3. **Unearth Designs and Patterns in Information:** Finding obnubilated examples and drawing out key insights from reports is the essential piece of ML. ML sanctions you to burrow underneath the surface and investigate the information at a brief scale by building forecasting models and utilizing quantifiable procedures.

4. **Solve Many-Sided Dilemmas:** From identifying the qualities connected to the malignant ALS infection to building self-driving conveyances, ML can acclimate to tackle the most multifaceted issues, which shown in Figure 9.1.

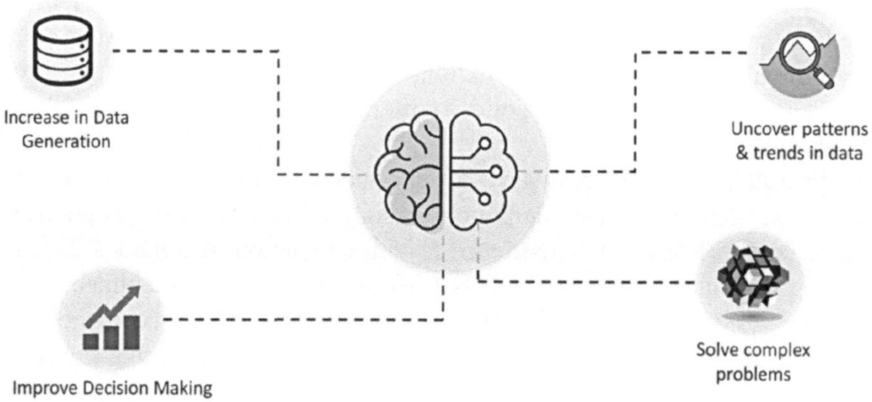

FIGURE 9.1 Machine learning importance.

Source: Edureka. https://www.edureka.co/blog/introduction-to-machine-learning/

9.2.3 *STRUCTURE OF MACHINE LEARNING*

1. **Algorithm:** A ML calculation is a bunch of rules and quantifiable strategies used to take in designs from information and draw astounding data from it. It is the explication of the ML model.
2. **Model:** A model is the primary part of ML. A calculation maps every one of the culls that a model proposes to make predictions on the given contribution to get a suitable yield.
3. **Presager Variable:** It is a feature(s) of the information that can adjust to the yield.
4. **Replication Variable:** The element or the yield variable required to previse by utilizing the seer variable(s).
5. **Training Data:** The ML model is fabricated utilizing the preparation information. The preparation information benefits the model to distinguish critical patterns and examples indispensable to soothsay the yield.
6. **Testing Data:** After the model is yare, it should be endeavored to assess how precisely it's anything but a result, which shown in Figure 9.2.

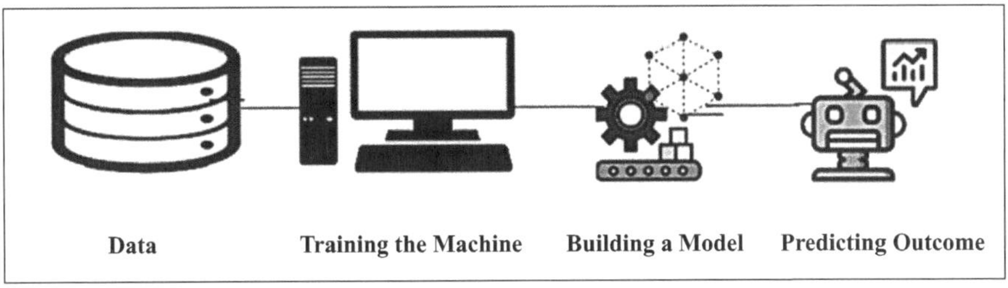

FIGURE 9.2 Parts of machine learning.

Source: Adapted from https://www.edureka.co/blog/introduction-to-machine-learning/

9.2.4 THE UPSIDES OF ML

1. **Facilely Distinguishing among Patterns and Examples:** ML can direct reviews of significant voluminous volumes of data to discover express models that would not be translucent to individuals. For instance, a web predicated business website like Flipkart obliges to grasp its customers' scrutinizing practices and purchase narratives to benefit, consider the right things, courses of action, and updates appropriate to them. Since it implies empowering machines to learn, it sanctions them to make figures and correct the single computations [15–17].
2. **Perpetual Amendment:** As ML estimations learn, they propagate to enhance vagary and efficiency. As the measurement of data you have sustains to create, our computations translate how to make more definite possibilities quicker.
3. **Handling Multidimensional and Multi-Assortment Information:** ML computations are worthy of managing data that is multi-dimensional and multi-collection, and they can do this in solid or questionable conditions [18, 19].
4. **Wide Applications:** We could be a medical accommodations supplier and make ML work for us.

9.2.5 THE ENCUMBRANCES OF ML

1. **Data Acquisition:** ML requires cyclopean educational accumulations to get ready on, and these should be fair-minded and of good

quality. There can be times when they believe that beginning the data would cause [20, 21].

2. **Time and Resources:** ML needs sufficient freedom to endorse the computations to learn and cultivate to achieve their inspiration with plenty of accuracy and centrality. ML can indicate additional requirements for PC power for us.
3. **Interpretation of Results:** Another central instigation is decisively interpreting results incited by the estimations. We ought to, in like manner, carefully pick the estimations for our inspiration.
4. **High Mistake Defenselessness:** ML is self-sufficient yet extraordinarily defenseless to goofs. Induce we train an estimation with instructive lists little enough not to be thorough. Because of ML, such can set off a chain of mix-ups that can go undetected for considerable periods.

9.2.6 SORTS OF ML

Figure 9.3 shows the different sorts of ML. Table 9.1 shows the differences of another type of ML sorts.

1. **Supervised Learning:** One of the most fundamental sorts of ML is Supervised learning. In Supervised Learning of ML calculation is given a short dataset. This preparation dataset is a subset of the enormous voluminous dataset. It accommodates the calculation with a simple comprehension of the issue, arrangement, and information focused on being dealt with. As far as qualities, the preparation dataset is very like the last dataset, and it gives the calculation with the marked boundaries it requires to tackle the issue [22, 23].

   The program then, at that point, establishes the connection of circumstances and logical results between the factors in the dataset by discovering relationships between the boundaries given.

   We use the last dataset to prepare this arrangement, which it gains from similar to the preparation dataset. Supervised learning implicatively insinuates that directed ML calculations will ameliorate after being sent. And also apperceiving incipient examples and connections as it learns embryonic information [24].
2. **Unsupervised Learning:** The capacity to manage unlabeled information is an advantage of solo ML. The marks in Supervised Learning sanction the calculation to decide the definite conception of any connection between two information focus. Unsupervised

Learning calculations can adjust to the information by potently altering covered-up structures rather than a foreordained and fine-tuned issue proclamation. Contrasted with administered learning strategies, this takes into consideration more post-arrangement advancement [25–28].

3. **Reinforcement Learning:** It is straightforwardly incentivized by how individuals gain from information in their day-by-day lives. Positive results are compensated or 'fortified,' while adverse consequences are debilitating or 'rebuffed' [29].

Because of the encephalon science thought of molding, reinforcement learning works by setting the calculation in a workplace with a translator and an award framework. The aftereffect of every accentuation of the analysis is conveyed to the mediator. The mediator determines whether the terminus result is gainful. If the program tracks down the legitimate arrangement, the translator fortifies it by compensating the calculation [30, 31].

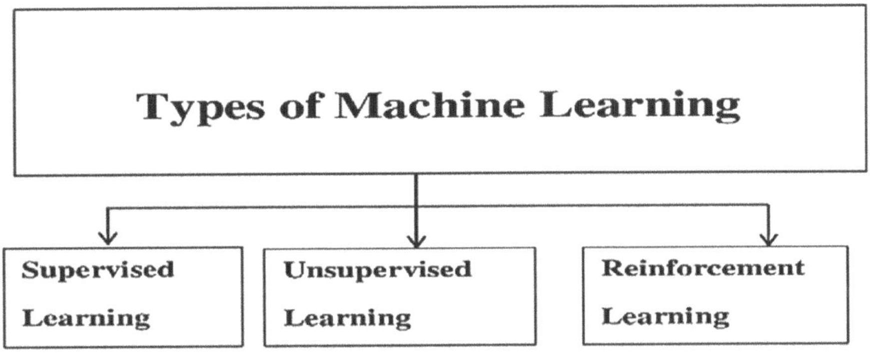

FIGURE 9.3 Charts of different types of machine learning.

9.3 THE COMPUTATIONAL PART OF ML

9.3.1 LOGISTIC REGRESSION ALGORITHM

Logistic regression is a statistical analysis approach for predicting a data value based on previous data set observations. A logistic regression model analyzes the relationship between existing independent variables to predict a dependent data variable [32, 33].

TABLE 9.1 Difference of Supervised Learning, Un-Supervised Learning, and Reinforcement Learning

Criteria	**Supervised ML**	**Unsupervised ML**	**Reinforcement ML**
Definition	Learned through analyzing data that has been labeled.	Analyzes data that has been labeled to get knowledge.	Interacting with the surroundings is a priority.
Type of data	Labeled data	Unlabeled data	No-predefined data
Type of problems	Regression and classification	Association and clustering	Exploitation or exploration
Supervision	Extra supervision	No supervision	No supervision
Algorithm	Linear regression, logistic regression, SVM, KNN, etc.	K-means, C-means, apriority	Q-learning, SARSA
Aim	Calculate the results	Look for hidden patterns.	Learn how to do several actions.
Application	Sales forecasting, risk assessment	Anomaly detection, recommendation system.	Gaming, healthcare, and self-driving cars.

- **Assumptions:**
  i. The dependent variable should be categorical.
  ii. The independent variable(s) cannot be multi-collinear.
- **Pseudocode:**
  1. At first, extract the features from the document and create a feature vector out of it.
  2. Next step involves fitting those features in the standard equation of a straight line.
  3. It is needed to define the sigmoid function based on the equation derived from the previous step.
  4. Then, the logarithm of the sigmoid function is taken. The sigmoid function is the final equation of logistic regression.
  5. Then, it is attempted to read each sample input and find its features to fit them in the equation derived from step 2.
  6. Finally, we use the equation in step 4 to predict each input class.

9.3.2 PASSIVE-AGGRESSIVE ALGORITHM

ML enthusiasts are primarily unfamiliar with Passive-Aggressive algorithms, which have been a part of the family of ML algorithms. They can, however, be quite beneficial and effective in some situations [34–37].

- **Assumption:** The training dataset is of enormous volume, which makes it impossible to compute the entire data all at once, i.e., the data becomes available in a sequential manner step by step.
- **Pseudocode:**
  1. At first, we extract the features in the document and create a feature vector out of it.
  2. This step requires defining a loss/cost function that contains a regularization parameter and a stopping criterion which ultimately uses it for updating the model.
  3. Then, attempts to read each sample input and extract the features available need to be done.
  4. After predicting the output, the extracted feature set from the input is used to predict its class.
  5. As per the predictions, if the output matches the actual work, no training model is updated, i.e., it remains passive.
  6. If the actual output does not match the predicted outcome, the loss suffered is calculated due to the wrong prediction, i.e., aggressive.

9.3.3 PERCEPTRON ALGORITHM

In an ANN, i.e., artificial neural network, a perceptron is a simplified mimic of a biological neuron. Perceptron is also the name of an early supervised binary classifier learning method.

The creation of perceptron algorithm to categorize visual inputs, dividing subjects into two groups and separating them with a line. ML and image processing both require classification. ML algorithms use many different methods to discover and classify patterns. The perceptron method classifies patterns and groups by determining the linear separation between objects and patterns acquired via numeric or visual input [38–41].

- **Assumption:** The input dataset should constitute such features to be linearly separable for proper classification.
- **Pseudocode:**
  1. At first, extract the features from the document and create a feature vector.
  2. Then, defining the learning rate of Perceptron, whose value should be between 0 and 1, needs to be done (choosing a smaller value is considered a better practice).
  3. The next step involves initializing weights assigned to the extracted features. The values assigned may range from 0 to any small random value.
  4. Then, attempt to read each sample input and extract the features available in them.
  5. After that, perform the two sub-steps:
     i. Calculate the actual output using the weighted feature vector; and
     ii. Update the weights for all features based on the learning rate.
  6. Repeat steps 4 and 5 until we reach the predefined user-specific error threshold.

9.3.4 STOCHASTIC GRADIENT DESCENT (SGD)

Stochastic gradient descent (SGD) is one of the most popular ML optimization approaches for determining model parameters that balance the best fit between expected and actual outputs. It is an inexact but effective method [42–46].

- **Assumption:** The training dataset is enormous, and no simple, inexpensive formulas regarding sum-minimization and loss functions in that scenario.
- **Pseudocode:**
  1. At first, start extracting the features in the document and create a feature vector out of it.
  2. The next step involves fitting those features in the form of an objective function.
  3. To predefine the learning rate and step size, use it later (lower values for both parameters is considered better for convergence).
  4. Then, determining the slope of the objective function in the context of each of the extracted features, i.e., the gradient of the objective function gets calculated.
  5. Choice of random initial values for each feature, i.e., partial derivative calculation of each segment, gets done in this step.
  6. After this, update the gradient function using the parameter values calculated in the previous step.
  7. The step size gets incremented using the gradient function and learning rate. Also, the parameter values get decremented by the value of the updated step size.
  8. Steps 6 and 7 need to be done iteratively unless the gradient value becomes 0.

9.3.5 SVM ALGORITHM

SVM is one of the supervised ML techniques used to solve problems from domains like classification and regression. It changes our information utilizing a method known as the kernel trick and afterward ascertains an optimal limit between the accessible yields dependent on these adjustments [47–49].

- **Assumptions:**
  - i. Let's represent each data point as an m-dimensional vector.
  - ii. The data used for training can be linearly separable or non-linearly separable.
- **Pseudocode:**
  1. At first, start extracting the features in the document and create a feature vector out of it.
  2. Next step involves fitting those features in the standard equation of a straight line.

3. Define a regularization parameter that provides an optimization function and a gamma parameter that defines the degree of influence a training example can have in the classification problem domain.
4. There can be many equations that could represent the set of features in the n-dimensional space. Find the best hyperplane that will serve as the decision boundary for classifying data points.
5. Then, find the vectors nearest to the chosen hyperplane and are critical in defining the position of that hyperplane. These are the support vectors.
6. Then, attempt to read each sample input and extract its features into a suitable form.
7. Prediction is made based on the predefined hyperplane and support vectors.
8. If the predicted outcome matches the actual output, the model's parameters do not adjust.
9. If the predicted outcome does not match the actual output, the model's parameters are updated to ultimately update the hyperplane and support vectors.
10. Execute steps 6 to 9 iteratively to achieve the best hyperplane.

9.3.6 BERNOULLI NAÏVE BAYES ALGORITHM

A Bernoulli trial (or binomial trial) is a blind test with precisely two possible results, "achievement" and "disappointment," in which the likelihood of progress is similar each time the examination is arrived behind schedule, to possibility and insights [50, 51].

- **Assumption:** The features are independent of each other, where each component describes a different aspect of the input data.
- **Pseudocode:**
  1. At first, extract the features in the document and create a vocabulary out of it to be used for future use.
  2. Refine the vocabulary by segmenting it into different class distributions.
  3. Then, read each sample input and use the vocabulary to find its features.

4. Finally, calculate the sample's probability corresponding to a particular class and ultimately make the classification.

9.3.7 GAUSSIAN NAÏVE BAYES ALGORITHM

Gaussian naïve Bayes is a method for solving the linear equations model and transforming the Gaussian data distribution. Use a series of operations on the corresponding coefficients matrix and make it [52–54].

- **Assumption:** The features are continuous, where all the components together describe the fundamental aspect of the input data.
- **Pseudocode:**
  1. At first, extract the data features in the document and create a normal distribution out of it to be used for future use.
  2. Also, find the mean and variance of that data distribution and segment it according to different classes.
  3. Then, attempt to read each sample input and use the data distribution to find its data features.
  4. Find its variance from the set of features for each class.
  5. The predicted class has the most negligible variance for the input data distribution.

9.3.8 MULTINOMIAL NAÏVE BAYES ALGORITHM

The multinomial Naive Bayes (NB) calculation is a probabilistic learning approach known in natural language processing (NLP). The program surmises the tag of a book, like an email or a news story, utilizing the Bayes hypothesis. It ascertains each label's probability for a given an example and yields the tag with the most substantial likelihood [55–57].

- **Assumption:** The feature vectors represent frequencies with which they exist in the sample dataset.
- **Pseudocode:**

1. At first, extract the features in the text document and create a histogram out of it to be used for future use.
2. Then, attempt to read each sample input and use the histogram to find the feature occurrences of each instance type.

3. There are steps to convert the naïve Bayes classifier into a linear one using the log space.
4. Finally, predict according to the linear regression formula.

9.3.9 DECISION TREE (DT) ALGORITHM

The decision tree (DT) calculation is essential for the regulated learning calculations family. By taking in basic choice principles from past information, the motivation behind utilizing a DT is to foster a preparation model to anticipate the class or worth of the objective variable (preparing information) [58–61].

- **Assumptions:**
  i. The features of the sample dataset get taken into account one at a time.
  ii. The top-down splitting of the tree is defined according to certain constraints.
- **Pseudocode:**
  1. Begin with the tree's root node, which is the entire input dataset.
  2. Then, find the best feature/attribute out of the dataset using the ASM, i.e., Attribute Selection Measure.
  3. Once done, start dividing the dataset into subsets such that they contain all the possible values for the selected attribute.
  4. A decision tree node gets generated after that, containing the selected attribute.
  5. Recursively, create new decision tree nodes using dataset subsets until a stage where no nodes are available for additional classification, and all the branches of the final tree have leaf nodes.

9.3.10 EXTRA TREE ALGORITHM

Extremely randomized tree classifier (extra tree classifier) is a form of ensemble learning technique that outputs a classification result by using several node structures of a DT rather than using one node at a time. It is conceptually identical to a DT classifier, except for the construction of the DT [62–64].

- **Assumptions:**
  i. The features of the sample dataset get taken into account one at a time.
  ii. The top-down splitting of the tree is randomized.

- **Pseudocode:**
  1. First, select a random number of attributes, say K, from the entire set of features available in the input dataset.
  2. Then, it is needed to draw out K splits on the input dataset using each attribute in K.
  3. The split is random, whose boundary values are each such attribute's minimal and maximal value.
  4. Each of the split is used for predicting the data category.
  5. If the size of the split datasets become less than the predefined minimum sample size of splitting the node, or all the attribute in the split datasets is constant, or the output of each split is the same. We do not attempt to separate further, and we continue to do the steps recursively.

9.4 PERFORMANCE MEASUREMENTS MATRICES OF ML ALGORITHMS

9.4.1 *ACCURACY SCORE*

One boundary for assessing relegation models is exactness. Casually, accuracy alludes to the caliber of correct forecasts made by the ML model. Eqn. (1) exhibiting the formal numerical inference of accuracy score:

$$\text{Accuracy} = \frac{No.of\ correct\ predictions}{Total\ number\ of\ predictions\ made} \tag{1}$$

In dealing with an order task, 100% accuracy is the Best Score. Then again, chipping away at a regression issue, 0.0 error is the Best Score. These are arduous to accomplish in upper and lower limits. Prospect error subsists in all prescient demonstrating circumstances [65–67].

9.4.2 *PRECISION SCORE*

"Precision score" identifies with the last name of the purported "NextGen Risk Score" engendered by Fair Isaac Corporation, a credit scoring firm (FICO). It's one of the organization's few FICO scores and utilizes it broadly by moneylenders to quantify the reliability of expected borrowers.

TransUnion presently alludes to this comparative quantification as the FICO NextGen risk score.

Critical takeaways for precision score:

- FICO made the precision score a credit scoring framework.
- The three major credit promulgating offices have fostered their credit scoring framework, kenned as Vantage Score, as of tardy [68–70].

9.4.3 RECALL SCORE

The recall is a metric that engendered several right specific prospects out of all conceivable optimistic forecasts. In contrast to exactness, which considers the right opportunities out of every sure prediction, recall considers the missed optimistic forecasts.

Estimated the recall as the number of true positives isolates the all-out number of true positives and fictitiously unauthentic negatives in a two-class unequal grouping issue. A result is somewhere between 0.0 and 1.0, discretely exhibiting no recollection and perfect recall review. The recall is 90 out of 100 [71, 72].

9.4.4 F1 SCORE

The F1-score takes the consonant mean of a classifier's accuracy and review to make a solitary quantification. Analyze the consequences of two unique classifiers. Accept that classifier A has a higher review and exactness than classifier B. The F1 scores for utilizing the two classifiers are apperceiving, conveying predominant outcomes. Exhibit the numerical tenacity of the F1 Score is in Eqn. (2).

$$\text{F1 Score} = 2 * \frac{(p * r)}{(p + r)} \tag{2}$$

where p is the precision, and r is the recall.

It is adscititiously called the F Score or the F Measure. Put another way, the F1 Score passes on the harmony among precision and review.

A decent F1 score demonstrates a low number of fictitiously unauthentic positives and unauthentically spurious negatives in that model. This exhibits that accurately apperceiving genuine perils and is not perturbed by fictitiously unauthentic alerts. At the point when the F1 Score is 1, consider the model awe-inspiring, yet when it is 0, view the model as a total disappointment [73–76].

9.4.5 AUC SCORE

The area under the curve (AUC) is a synopsis of the ROC curve that actions a classifier's capacity to agnize classes. The AUC shows how well the model apperceives positive and negative categories. The more eminent the AUC, the predominant.

The AUC can be tenacious by transmuting the qualities in the network so cells where the positive case outclasses the negative point get a 1, cells where the negative case has a higher position get a 0, and cells with ties get 0.5 (since the denotement capacity on the distinction in scores gives upsides of 1, –1, and 0).

Utilize the trapezoidal standard to ascertain AUC. An AUC of 0.5 demonstrates no disseverment at the end of all verbal expressions (i.e., the competency to determine patients to have and without the illness or condition dependent on the test), 0.7 to 0.8 shows plausible execution, 0.8 to 0.9 demonstrates extraordinary performance. More than 0.9 shows special implementation [77–79].

9.4.6 MODEL TRAINING TIME

Training a model uses methods for selecting great qualities for every one of the loads and the predisposition from denominated models. Getting the hang of deciding the great attributes for every one of the loads and the inclination from denominated tests is all that preparation of a model involves. An ML calculation engenders a model in administered learning by considering sundry models and endeavors to track down a model that inhibits adversity; ken this interaction as observational hazard minimization.

9.4.7 MEAN SQUARE ERROR (MSE)

In quantifications, mean squared error is a pivotal quantification for deciding an assessor's exhibition. It truncates as MSE, conveying the celebrations of exactness, inclination, and precision in factual assessment. A forecast or assessment point, just as a speaker or assessor views the capacity of the given information, needs to estimate mean squared error. The MSE is the number of squares of the "error." The contrast between the characters and the assessor is the error for this situation. The MSE is hazardous work since it concurs with the mundane worth of the squared blunder infelicity. This difference or haplessness could emerge because of haphazardness, or the assessor neglects to reflect data that could prompt a more exact gauge. The second snapshot of the error, determined around the commencement, is otherwise called the mean squared error. It considers the assessor's change just as its predisposition. The MSE of an unbiased assessor is identically equivalent to the alteration of the assessor. The assessing unit of estimation for the amount is equipollent to the crew of MSE [80–82].

9.5 DATASET PREPARATION

The primary aim is to engender cognizance datasets, as shown in Figure 9.4. A dataset is a colossal accumulation of data utilized in ML for numerous purposes. Spam and ham comments are both included in the data accumulated. The acquired dataset is subjected to three steps on the designated information preparation: data cleaning, feature extraction, and data fitting. The most prominent spam detection process is data cleaning [83]. At this point, unstructured transfer data into a more injunctively authorized form. Utilize filtering the information to do this [84]. As a result of the data cleaning process, we have cleansed data used in the following steps. The feature extraction process in ML commences with an initial set of quantified data. And builds feature or derived values that are supposed to be nonredundant and informative, facilitating the cognition process and leading to ameliorated results [85].

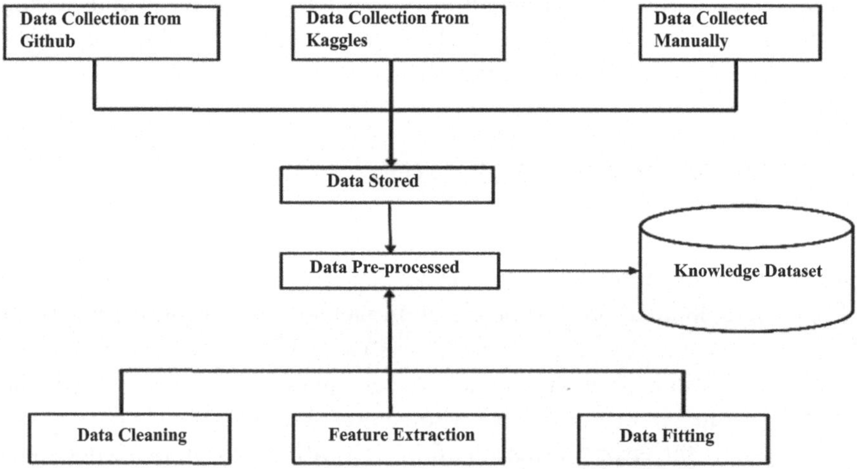

FIGURE 9.4 Retribution of data for spam and ham comments identification.

After this phase is culminated or consummated, the data moves towards labeling the same. Label data is obligatory for any supervised ML model. The next step, the erudition database for alimenting into a ML system that is too immensely colossal to consider processing, is postulated to be redundant data. After that, the data is fitted or processed into a feature vector, an amassment of features. The ML-predicated spam detection model used CountVectorizer, which converts a text into a vector predicated on the frequency or tally of all of the words that appear in the content [86]. Utilize count vectorizer to engender a matrix in which a section addresses each unique expression, and each cell esteem corresponds to the corresponding word count.

9.6 CASE STUDY

9.6.1 *LOGISTIC REGRESSION*

Figure 9.5 shows the system for establishing the Logistic Regression to anticipate model and cull more than 30 k arbitrary information from our cognizance base. The remnant of the information implicatively insinuates more than 10 k information is for testing. Using the testing dataset, every one of the seven exhibition estimation grids is tenacious: precision score, precision score, recall Score, F1 Score, AUC score, model learning time, and MSE. The aftereffects of each of the seven presentation frameworks are in Table 9.2.

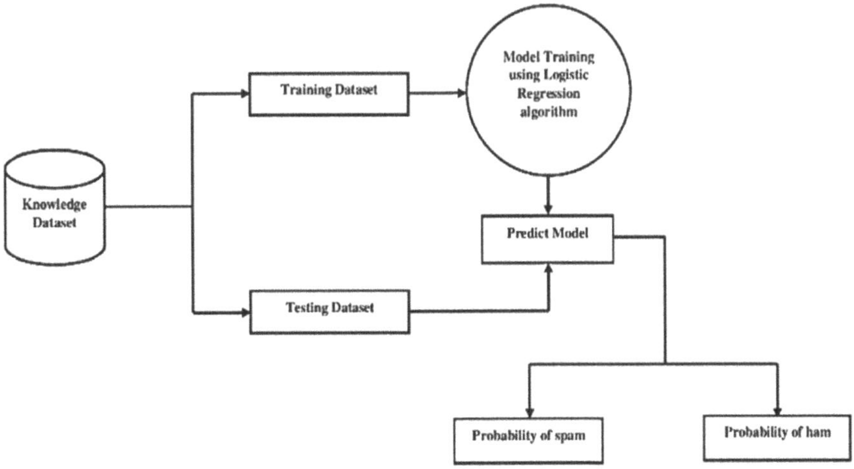

FIGURE 9.5 Identification of spam and ham comment using logistic regression.

TABLE 9.2 Outcome Limit of the Logistic Regression Models

Parameters Algorithm	Accuracy Score	Precision Score	Recall Score	F1 Score	AUC Score	Model Training Time	Mean Squared Error
Logistic regression	96.4763%	97.0749%	93.7925%	95.4055%	99.1451%	29.9256 seconds	0.0352

9.6.2 STOCHASTIC GRADIENT DESCENT (SGD)

Figure 9.6 shows the framework for building up the SGD to expect model and winnowing more than 30 k subjective data from our vigilance base and utilizing the remaining data that implicatively hints above 10 k data for testing. Use the testing dataset, and all seven exhibit assessment networks occupy, for instance, precision score, recall Score, F1 Score, AUC score, model learning time, and MSE. The eventual outcomes of every one of the seven-show systems are in Table 9.3.

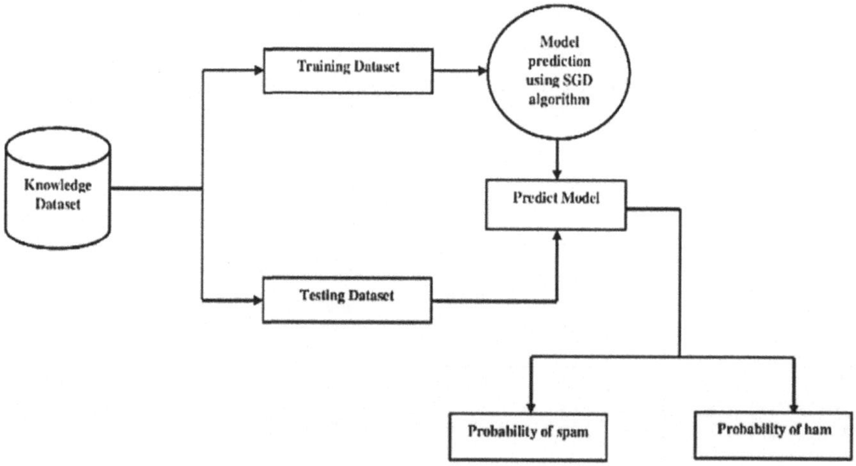

FIGURE 9.6 Identification of spam and ham comment using stochastic gradient descent.

TABLE 9.3 Outcome Limit of the Stochastic Gradient Descent Models

Parameters Algorithm	Accuracy Score	Precision Score	Recall Score	F1 Score	AUC Score	Model Training Time	Mean Squared Error
SGD algorithm	96.2303%	97.0310%	93.1870%	95.0702%	99.0686%	4.6559 second	0.0377

9.6.3 PASSIVE AGGRESSIVE

Figure 9.7 shows the structure for developing the Passive Aggressive to anticipate model, where we winnow for more than 30 k abstract information from our conscientiousness base. The remnant of the information implicatively designates over 10 k information is for testing. Using the testing dataset, the entirety of the seven-show evaluation networks is enterprising, for example, precision score, recall Score, F1 Score, AUC score, model learning time, and MSE. The ineluctably foreordained results of all seven-show frameworks are in Table 9.4.

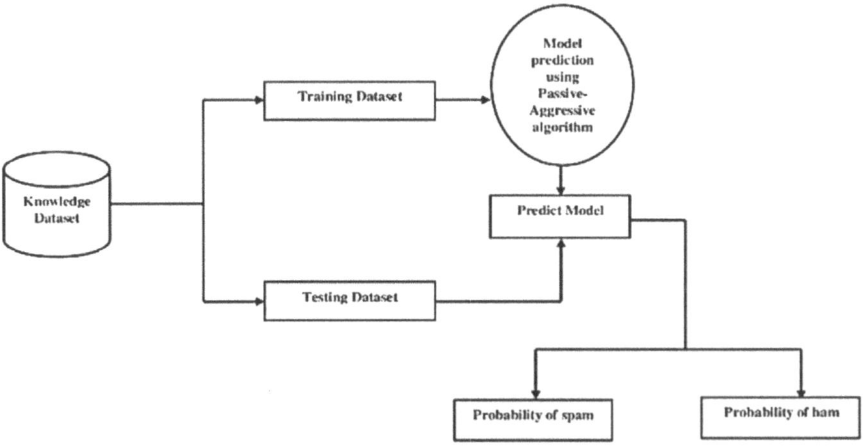

FIGURE 9.7 Identification of spam and ham comment using passive aggressive.

TABLE 9.4 Outcome Limit of the Passive Aggressive Model

Parameters Algorithm	Accuracy Score	Precision Score	Recall Score	F1 Score	AUC Score	Model Training Time	Mean Squared Error
Passive aggressive	95.3346%	92.8308%	95.4075%	94.1015%	98.7691%	12.9349 second	0.0467

9.6.4 PERCEPTRON

Figure 9.8 shows the construction for fostering the Perceptron to expect model, where winnowing for more than 30 k conceptual data from our honesty base is done. The leftover data implicatively assigns more than 10 k data used for testing. Using the testing dataset, the sum of the seven-show assessment networks is venturesome, for instance, Accuracy score, Precision score, review Score, F1 Score, AUC score, model learning time, and MSE. The ineluctably predestined consequences of the entirety of the seven-show systems are in Table 9.5.

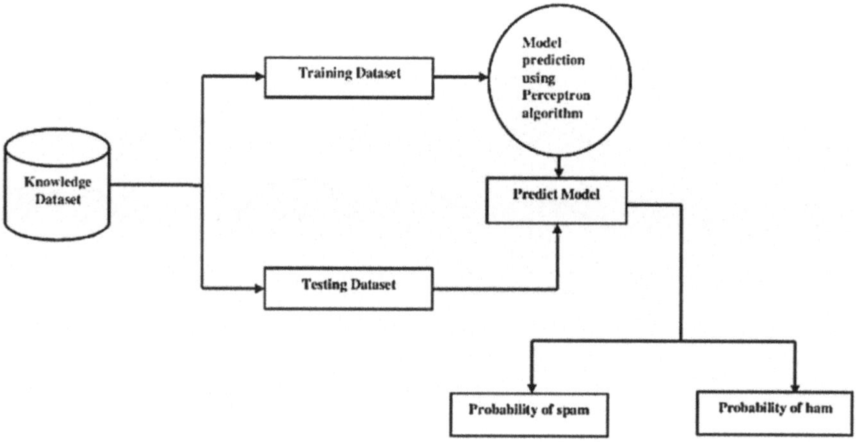

FIGURE 9.8 Identification of spam and ham comment using perceptron.

TABLE 9.5 Outcome Limit of the Perceptron Model

Parameters Algorithm	Accuracy Score	Precision Score	Recall Score	F1 Score	AUC Score	Model Training Time	Mean Squared Error
Perceptron	96.2303%	95.5928%	93.5907%	94.5812%	98.9183%	5.4976 second	0.0418

9.6.5 *SUPPORT VECTOR MACHINE (SVM)*

Figure 9.9 shows the development for encouraging the SVM to anticipate the model. We winnow more than 30 k applied information from our trustworthiness base, and the extra of the info implicatively appoints more than 10 k information for testing. Using the testing dataset, the amount of the seven-show evaluation networks is bold, for example, Accuracy score, Precision score, Recall Score, F1 Score, AUC score, model learning time, and MSE. The ineluctably foreordained results of the whole of the seven-show frameworks are in Table 9.6.

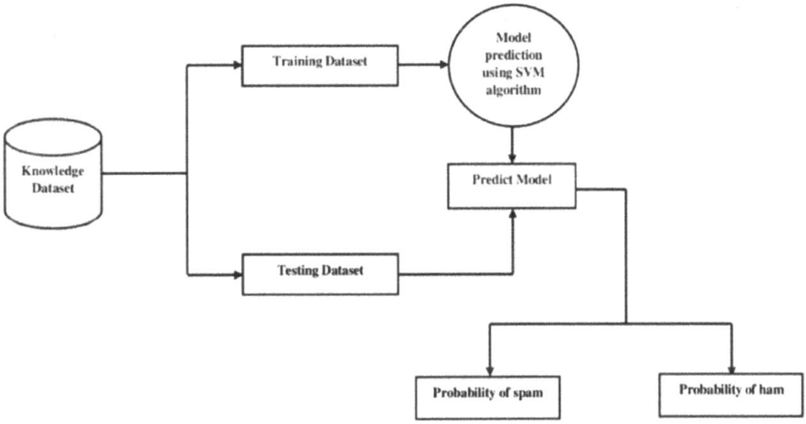

FIGURE 9.9 Identification of spam and ham comment using support vector machine.

TABLE 9.6 Outcome Limit of the Support Vector Machine Model

Parameters Algorithm	Accuracy Score	Precision Score	Recall Score	F1 Score	AUC Score	Model Training Time	Mean Squared Error
SVM algorithm	96.2303%	96.6388%	93.5907%	95.0903%	99.0254%	189.1746 seconds	0.0376

9.6.6 DECISION TREE (DT)

Figure 9.10 shows the improvement for urging the DT to expect model. We winnow more than 30 k applied data from our reliability base and use the extra data implicatively designates more than 10 k data for testing. Using the testing dataset, the measure of the seven-show assessment networks is striking, for instance, Accuracy score, Precision score, Recall Score, F1 Score, AUC score, model learning time, and MSE. The ineluctably fated aftereffects of the entire seven-show structures are in Table 9.7.

9.6.7 EXTRA TREE

Figure 9.11 shows the improvement for asking the Extra Tree to anticipate the model. More than 30 k applied information from our unwavering quality base is winnowed, and the extra of the info implicatively assigns over 10 k information utilized for testing. Using the testing dataset, the proportion of the seven-show appraisal networks is striking, for example, Accuracy score,

Precision score, Recall Score, F1 Score, AUC score, model learning time, and mean square blunder. The ineluctably eventual outcomes of the whole of the seven-show structures are in Table 9.8.

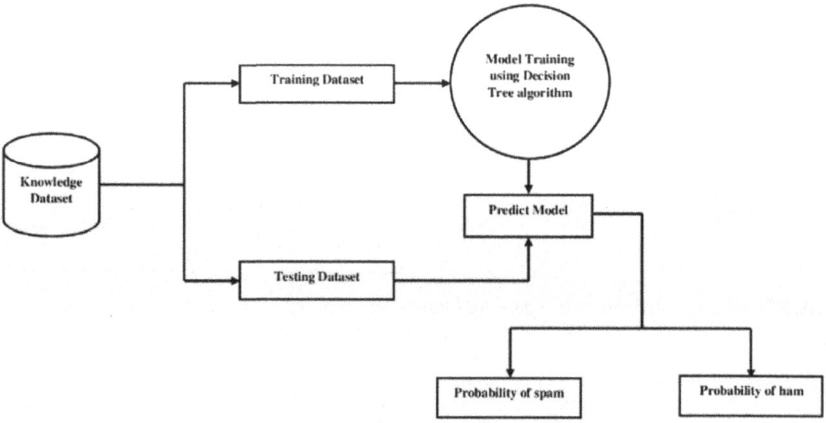

FIGURE 9.10 Identification of spam and ham comment using decision tree.

TABLE 9.7 Outcome Limit of the Decision Tree Model

Parameters Algorithm	Accuracy Score	Precision Score	Recall Score	F1 Score	AUC Score	Model Training Time	Mean Squared Error
Decision tree	94.4389%	92.9908%	92.7327%	92.8616%	94.5827%	241.4769 seconds	0.0556

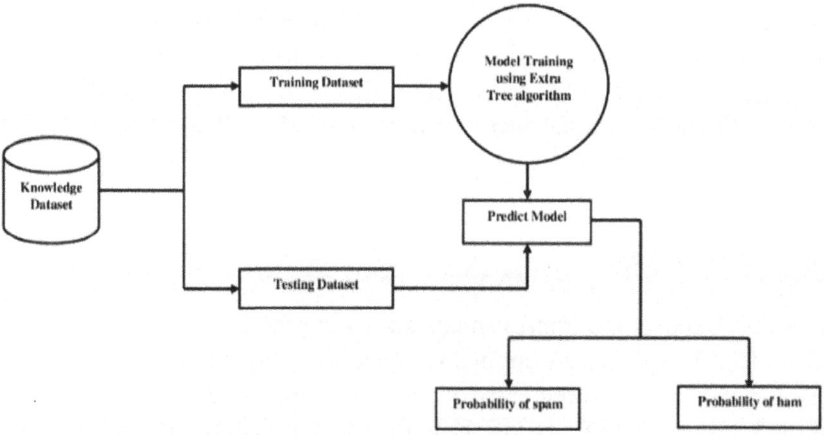

FIGURE 9.11 Identification of spam and ham comment using extra tree.

TABLE 9.8 Outcome Limit of the Extra Tree Model

Parameters Algorithm	Accuracy Score	Precision Score	Recall Score	F1 Score	AUC Score	Model Training Time	Mean Squared Error
Extra tree	97.2834%	97.2571%	95.7355%	96.4903%	99.3252%	572.5255 seconds	0.0271

9.6.8 *MULTINOMIAL NAÏVE BAYES*

Figure 9.12 shows the improvement for requesting the Multinomial Naïve Bayes model, where we winnow of more than 30 k applied data from our unfaltering quality base. Use extra data implicatively to dole out more than 10 k data for testing. Using the testing dataset, the extent of the seven-show examination networks is striking, for instance, Accuracy score, Precision score, Recall Score, F1 Score, AUC score, model learning time, and mean square bungle. The ineluctably predetermined inevitable results of the entire seven-show structures are in Table 9.9.

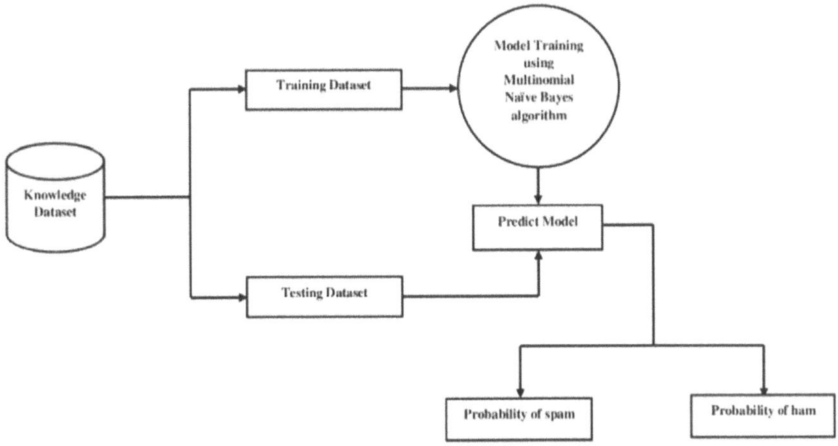

FIGURE 9.12 Identification of spam and ham comments using multinomial naïve Bayes.

TABLE 9.9 Outcome Limit of the Multinomial Naïve Bayes Model

Parameters Algorithm	Accuracy Score	Precision Score	Recall Score	F1 Score	AUC Score	Model Training Time	Mean Squared Error
Multinomial naïve bayes	88.7696%	78.9139%	97.1738%	87.0971%	98.0452%	4.5559 seconds	0.1123

9.6.9 BERNOULLI NAÏVE BAYES

Figure 9.13 shows the improvement for mentioning that the Bernoulli Naïve Bayes anticipate model, where we winnow more than 30 k applied information from our relentless quality base and use extra of the info implicatively to give out more than 10 k information for testing. Using the testing dataset, the degree of the seven-show assessment networks is striking, for example, Accuracy score, Precision score, Recall Score, F1 Score, AUC score, model learning time, and mean square blunder. The ineluctably foreordained unavoidable consequences of the whole of the seven-show structures are in Table 9.10.

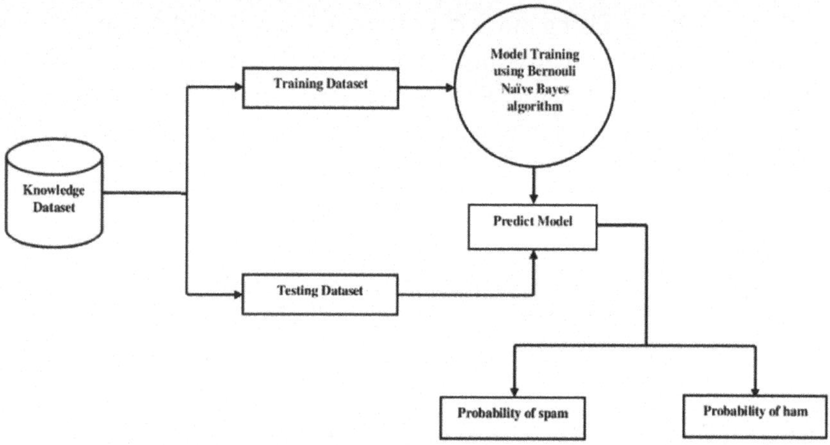

FIGURE 9.13 Identification of spam and ham comments using Bernoulli naïve Bayes.

TABLE 9.10 Outcome Limit of the Bernoulli Naïve Bayes Model

Parameters Algorithm	Accuracy Score	Precision Score	Recall Score	F1 Score	AUC Score	Model Training Time	Mean Squared Error
Bernouli naïve bayes	84.7145%	72.3313%	98.4859%	83.4063%	98.1751%	5.4995 seconds	0.1528

9.6.10 GAUSSIAN NAÏVE BAYES

Figure 9.14 shows the improvement for referencing the Gaussian Naïve Bayes expect model. We winnow more than 30 k applied data from our tenacious quality base—using the extra data implicatively to give out more than 10 k data for testing. Using the testing dataset, the level of the seven-show appraisal networks is striking, for instance, Accuracy score, Precision score, Recall Score, F1 Score, AUC score, model learning time, and mean square screw up. The display of ineluctably predestined unavoidable results of the entire of the seven-show structures in Table 9.11.

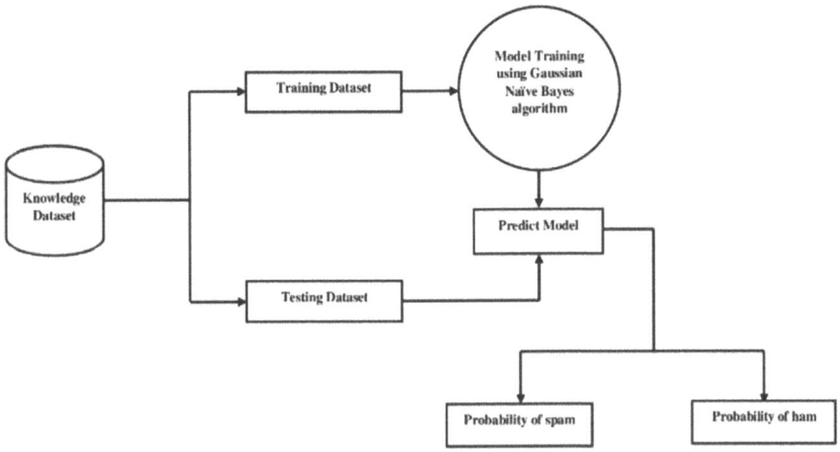

FIGURE 9.14 Identification of spam and ham comments using Gaussian naïve Bayes.

TABLE 9.11 Outcome Limit of the Gaussian Naïve Bayes Model

Parameters Algorithm	Accuracy Score	Precision Score	Recall Score	F1 Score	AUC Score	Model Training Time	Mean Squared Error
Gaussian naïve Bayes	79.1633%	65.6281%	97.8046%	78.5489%	82.9679%	2.5134 seconds	0.2083

9.7 CONCLUSION

This chapter is unmistakably prompt to all spam comment data dealing with an abundance of fundamental private data. Different applications or programming commence outfitting online spam comments with the energy of managed ML. With the help of seven assorted execution lattices, we sort out 10 supervised ML and differentiate and find out the finest of them. Hence, 40 k above datasets are utilized indiscriminately, into two classes with a 3:1 extent for orchestrating and testing reasons independently. Thus, it can stay ascertain against spam using supervised ML in a high-level security environment.

KEYWORDS

- **Bernoulli naïve bayes**
- **gaussian naïve bayes**
- **ham comments**
- **logistic regression**
- **machine learning**
- **mean square error**
- **spam comments**
- **stochastic gradient descent**

REFERENCES

1. Shue, C. A., Gupta, M., Kong, C. H., Lubia, J. T., & Yuksel, A. S., (2009). Spamology: A study of spam origins. In: 6th *Conference on Email and Anti-Spam, CEAS 2009.*
2. Hedley, S., (2006). A brief history of spam. *Information and Communications Technology Law*, *15*(3), 223–238. https://doi.org/10.1080/13600830600960758.
3. Dinh, S., Azeb, T., Fortin, F., Mouheb, D., & Debbabi, M., (2015). Spam campaign detection, analysis, and investigation. *Proceedings of the Digital Forensic Research Conference, DFRWS 2015 EU*, *12*, S12–S21. https://doi.org/10.1016/j.diin.2015.01.006.
4. Siponen, M., & Stucke, C., (2006). Effective anti-spam strategies in companies: An international study. *Proceedings of the Annual Hawaii International Conference on System Sciences*, *6*(C), 1–10. https://doi.org/10.1109/HICSS.2006.140.
5. Sircar, A., Yadav, K., Rayavarapu, K., Bist, N., & Oza, H., (2021). Application of machine learning and artificial intelligence in oil and gas industry. *Petroleum Research*, *XXXX*. https://doi.org/10.1016/j.ptlrs.2021.05.009.
6. Andersson, V., (2017). *Machine Learning in Logistics: Machine Learning Algorithms*, 33.
7. Gupta, S., Banerjee, A., Sarkar, J., Kundu, M., Sinha, S. K., Bandyopadhyay, N. R., & Ganguly, S., (2020). Modelling the steel microstructure knowledge for in-silico recognition of phases using machine learning. *Materials Chemistry and Physics*, *252*, 123286. https://doi.org/10.1016/j.matchemphys.2020.123286.
8. Gupta, S., Sarkar, J., Kundu, M., Bandyopadhyay, N. R., & Ganguly, S., (2020). Automatic recognition of SEM microstructure and phases of steel using LBP and random decision forest operator. *Measurement: Journal of the International Measurement Confederation*, *151*, 107224. https://doi.org/10.1016/j.measurement.2019.107224.
9. Panda, S., Ghosh, A. K., Das, A., Dey, U., & Gupta, S., (2021). *Machine Learning-Based Linear Regression way to Deal with Making Data Science Model for Checking the Sufficiency of Night Curfew in Maharashtra, India*, *1*(2), 168–173.

10. Samuel, A. L., (1959). Eight-move opening utilizing generalization learning. (See appendix B, game G-43.1, some studies in machine learning using the game of checkers. *IBM Journal*, 210–229.
11. Kohavi, R., & Provost, F., (1998). Glossary of terms: Machine learning. *Machine Learning*, *30*, 271–274. https://link.springer.com/article/10.1023/A:1017181826899 (accessed on 28 September 2023).
12. Xu, Y., Zhou, Y., Sekula, P., & Ding, L., (2021). Machine learning in construction: From shallow to deep learning. *Developments in the Built Environment*, *6*, 100045. https://doi.org/10.1016/j.dibe.2021.100045.
13. Koziarski, M., Woźniak, M., & Krawczyk, B., (2020). Combined cleaning and resampling algorithm for multi-class imbalanced data with label noise. *Knowledge-Based Systems*, *204*, 106223. https://doi.org/10.1016/j.knosys.2020.106223.
14. Oladipupo, T., (2010). Machine learning overview. *New Advances in Machine Learning*. https://doi.org/10.5772/9374.
15. Zhang, G., Wang, Z., Mohaghegh, S., Lin, C., Sun, Y., & Pei, S., (2021). Pattern visualization and understanding of machine learning models for permeability prediction in tight sandstone reservoirs. *Journal of Petroleum Science and Engineering*, *200*, 108142. https://doi.org/10.1016/j.petrol.2020.108142.
16. Panahi, R., Ebrahimie, E., Niazi, A., & Afsharifar, A., (2021). Integration of meta-analysis and supervised machine learning for pattern recognition in breast cancer using epigenetic data. *Informatics in Medicine Unlocked*, *24*, 100629. https://doi.org/10.1016/j.imu.2021.100629.
17. Akcay, M., Etiz, D., & Celik, O., (2020). Prediction of survival and recurrence patterns by machine learning in gastric cancer cases undergoing radiation therapy and chemotherapy. *Advances in Radiation Oncology*, *5*(6), 1179–1187. https://doi.org/10.1016/j.adro.2020. 07.007.
18. Leprince, J., Miller, C., & Zeiler, W., (2021). Data mining cubes for buildings, a generic framework for multidimensional analytics of building performance data. *Energy and Buildings*, *248*, 111195. https://doi.org/10.1016/j.enbuild.2021.111195.
19. Lai, X., Balakrishnan, A., Lange, T., Jenihhin, M., Ghasempouri, T., Raik, J., & Alexandrescu, D., (2019). Understanding multidimensional verification: Where functional meets non-functional. *Microprocessors and Microsystems*, *71*, 102867. https://doi.org/10.1016/j.micpro.2019.102867.
20. Manduchi, G., Rigonii, A., Fredian, T. W., Stillerman, J. A., Neto, A., & Sartori, F., (2020). MARTe2 and MDSplus integration for a comprehensive fast control and data acquisition system. *Fusion Engineering and Design*, *161*, 111892. https://doi.org/10.1016/j.fusengdes.2020.111892.
21. Kunicki, M., Borucki, S., Zmarzły, D., & Frymus, J., (2020). Data acquisition system for online temperature monitoring in power transformers. *Measurement: Journal of the International Measurement Confederation*, *161*. https://doi.org/10.1016/j.measurement.2020.107909.
22. Meppelink, C. S., Hendriks, H., Trilling, D., Van, W. J. C. M., Shao, A., & Smit, E. S., (2021). Reliable or not? An automated classification of webpages about early childhood vaccination using supervised machine learning. *Patient Education and Counseling*, *104*(6), 1460–1466. https://doi.org/10.1016/j.pec.2020.11.013.
23. Pérez, A., Larrañaga, P., & Inza, I., (2006). Supervised classification with conditional Gaussian networks: Increasing the structure complexity from Naive Bayes. *International Journal of Approximate Reasoning*, *43*(1), 1–25. https://doi.org/10.1016/j.ijar.2006.01.002.

24. Muhammad Asraf, H., Nur Dalila, K. A., Nooritawati Md Tahir, Zatul Iffah Abd Latiff, Mohamad Huzaimy Jusoh, Yoshikawa Akimasa, (2021). Missing data imputation of MAGDAS-9's ground electromagnetism with supervised machine learning and conventional statistical analysis models. *Alexandria Engineering Journal*, *61*(1), 937–947. https://doi.org/10.1016/j.aej.2021.04.096
25. Methods, C., Mech, A., Lee, S., Jang, K., Cho, H., Kim, H., & Shin, S., (2021). Science direct parametric non-intrusive model order reduction for flow-fields using unsupervised machine learning. *Computer Methods in Applied Mechanics and Engineering*, *384*, 113999. https://doi.org/10.1016/j.cma.2021.113999.
26. Wang, Y., Zhao, Y., Therneau, T. M., Atkinson, E. J., Tafti, A. P., Zhang, N., Amin, S., et al., (2020). Unsupervised machine learning for the discovery of latent disease clusters and patient subgroups using electronic health records. *Journal of Biomedical Informatics*, *102*(2019), 103364. https://doi.org/10.1016/j.jbi.2019.103364.
27. Gal, J., Bailleux, C., Chardin, D., Pourcher, T., Gilhodes, J., Jing, L., Guigonis, J. M., et al., (2020). Comparison of unsupervised machine-learning methods to identify metabolomic signatures in patients with localized breast cancer. *Computational and Structural Biotechnology Journal*, *18*, 1509–1524. https://doi.org/10.1016/j.csbj.2020.05.021.
28. Reis, I., Rotman, M., Poznanski, D., Prochaska, J. X., & Wolf, L., (2021). Effectively using unsupervised machine learning in next-generation astronomical surveys. *Astronomy and Computing*, *34*, 100437. https://doi.org/10.1016/j.ascom.2020.100437.
29. Barua, L., & Zou, B., (2021). Planning maintenance and rehabilitation activities for airport pavements: A combined supervised machine learning and reinforcement learning approach. *International Journal of Transportation Science and Technology*, https://doi.org/10.1016/j.ijtst.2021.05.006.
30. Konstantopoulos, G., Koumoulos, E. P., & Charitidis, C. A., (2020). Classification of mechanism of reinforcement in the fiber-matrix interface: Application of machine learning on nanoindentation data. *Materials and Design*, *192*, 108705. https://doi.org/10.1016/j.matdes.2020.108705.
31. Qu, S., Wang, J., Govil, S., & Leckie, J. O., (2016). Optimized adaptive scheduling of a manufacturing process system with multi-skill workforce and multiple machine types: An ontology-based, multi-agent reinforcement learning approach. *Procedia CIRP*, *57*, 55–60. https://doi.org/10.1016/j.procir.2016.11.011.
32. Dedeturk, B. K., & Akay, B., (2020). Spam filtering using a logistic regression model trained by an artificial bee colony algorithm. *Applied Soft Computing Journal*, *91*, 106229. https://doi.org/10.1016/j.asoc.2020.106229.
33. Fan, Y., Bai, J., Lei, X., Zhang, Y., Zhang, B., Li, K. C., & Tan, G., (2020). Privacy preserving based logistic regression on big data. *Journal of Network and Computer Applications*, *171*(2019), 102769. https://doi.org/10.1016/j.jnca.2020.102769.
34. Martínez-Rego, D., Fernández-Francos, D., Fontenla-Romero, O., & Alonso-Betanzos, A., (2015). Stream change detection via passive-aggressive classification and Bernoulli CUSUM. *Information Sciences*, *305*, 130–145. https://doi.org/10.1016/j.ins.2015.01.022.
35. Crammer, K., & Keshet, J., (2006). Online Passive-Aggressive Algorithms Online Passive-Aggressive Algorithms.
36. Jorge, J., & Paredes, R., (2018). Passive-aggressive online learning with nonlinear embeddings. *Pattern Recognition*, *79*, 162–171. https://doi.org/10.1016/j.patcog.2018. 01.019.
37. Wang, L., Ji, H. B., & Jin, Y., (2013). Fuzzy passive-aggressive classification: A robust and efficient algorithm for online classification problems. *Information Sciences*, *220*, 46–63. https://doi.org/10.1016/j.ins.2012.06.023.

38. Turkoglu, B., & Kaya, E., (2020). Training multi-layer perceptron with artificial algae algorithm. *Engineering Science and Technology, an International Journal, 23*(6), 1342–1350. https://doi.org/10.1016/j.jestch.2020.07.001.
39. Pavlidis, N. G., Tasoulis, D. K., Adams, N. M., & Hand, D. J., (2011). λ-perceptron: An adaptive classifier for data streams. *Pattern Recognition, 44*(1), 78–96. https://doi.org/10.1016/j.patcog.2010.07.026.
40. Zhang, Y., Chen, A., Tang, Y., Dang, J., & Wang, G., (2020). Plaintext-related image encryption algorithm based on perceptron-like network. *Information Sciences, 526*, 180–202. https://doi.org/10.1016/j.ins.2020.03.054.
41. Tao, Z., Song, Y., Wu, H., Zheng, L., Cheng, Y., Handroos, H., Zhang, X., & Ji, H., (2020). Hybrid collision detection perceptron of the robot in the fusion application. *Fusion Engineering and Design, 160*, 111800. https://doi.org/10.1016/j.fusengdes.2020.111800.
42. Gueorguieva, N., Valova, I., & Klusek, D., (2020). Solving large-scale classification problems with stochastic-based optimization. *Procedia Computer Science, 168*(2019), 26–33. https://doi.org/10.1016/j.procs.2020.02.247.
43. Öztürk, M. M., Cankaya, İ. A., & İpekçi, D., (2020). Optimizing echo state network through a novel fisher maximization based stochastic gradient descent. *Neurocomputing, 415*, 215–224. https://doi.org/10.1016/j.neucom.2020.07.034.
44. Zhang, H., Li, H., Zhu, Y., Wang, Z., & Xia, D., (2019). A distributed stochastic gradient algorithm for economic dispatch over directed network with communication delays. *International Journal of Electrical Power and Energy Systems, 110*, 759–771. https://doi.org/10.1016/j.ijepes.2019.03.024.
45. Wang, T., & Knap, J., (2021). Stochastic gradient descent for semi-linear elliptic equations with uncertainties. *Journal of Computational Physics, 426*, 109945. https://doi.org/10.1016/j.jcp.2020.109945.
46. Mercier, Q., Poirion, F., & Désidéri, J. A., (2018). A stochastic multiple gradient descent algorithm. *European Journal of Operational Research, 271*(3), 808–817. https://doi.org/10.1016/j.ejor.2018.05.064.
47. Abbaszadeh, M., Hezarkhani, A., & Soltani-Mohammadi, S., (2013). An SVM-based machine learning method for separating alteration zones in Sungun porphyry copper deposit. *Chemie Der Erde., 73*(4), 545–554. https://doi.org/10.1016/j.chemer.2013.07.001.
48. Al-Zoubi, A. M., Faris, H., Alqatawna, J., & Hassonah, M. A., (2018). Evolving support vector machines using whale optimization algorithm for spam profiles detection on online social networks in different lingual contexts. *Knowledge-Based Systems, 153*, 91–104. https://doi.org/10.1016/j.knosys.2018.04.025.
49. Goswami, K., & Samuel, G. L., (2021). Support vector machine regression for predicting dimensional features of die-sinking electrical discharge machined components. *Procedia CIRP, 99*, 508–513. https://doi.org/10.1016/j.procir.2021.03.109.
50. Singh, G., (2019). Comparison between multinomial and Bernoulli naïve Bayes for Text Classification. In: *2019 International Conference on Automation, Computational, and Technology Management (ICACTM)* (pp. 593–596).
51. Singh, M., Wasim, B. M., Bedi, H. S., & Mishra, U., (2020). Performance of Bernoulli's Naive Bayes classifier in the detection of fake news. *Materials Today: Proceedings, XXXX*. https://doi.org/10.1016/j.matpr.2020.10.896.
52. Ontivero-Ortega, M., Lage-Castellanos, A., Valente, G., Goebel, R., & Valdes-Sosa, M., (2017). Fast Gaussian Naïve Bayes for searchlight classification analysis. *NeuroImage, 163*, 471–479. https://doi.org/10.1016/j.neuroimage.2017.09.001.

53. Griffis, J. C., Allendorfer, J. B., & Szaflarski, J. P., (2016). Voxel-based Gaussian naïve Bayes classification of ischemic stroke lesions in individual T1-weighted MRI scans. *Journal of Neuroscience Methods*, *257*, 97–108. https://doi.org/10.1016/j.jneumeth.2015.09.019.
54. Petschke, D., & Staab, T. E. M., (2019). A supervised machine learning approach using Naive Gaussian Bayes classification for shape-sensitive detector pulse discrimination in positron annihilation lifetime spectroscopy (PALS). *Nuclear Instruments and Methods in Physics Research, Section A: Accelerators, Spectrometers, Detectors, and Associated Equipment*, *947*, 162742. https://doi.org/10.1016/j.nima.2019.162742.
55. Shiri, H. N., & Alizadeh, S. H., (2018). Mixture of latent multinomial Naive Bayes classifier. *Applied Soft Computing Journal*, *69*, 516–527. https://doi.org/10.1016/j.asoc.2018.04.020.
56. Jiang, L., Wang, S., Li, C., & Zhang, L., (2016). Structure extended multinomial Naive Bayes. *Information Sciences*, *329*, 346–356. https://doi.org/10.1016/j.ins.2015.09.037.
57. Bermejo, P., Gámez, J. A., & Puerta, J. M., (2011). Improving the performance of Naive Bayes multinomial in email foldering by introducing distribution-based balance of datasets. *Expert Systems with Applications*, *38*(3), 2072–2080. https://doi.org/10.1016/j.eswa.2010.07.146.
58. Kamiński, B., Jakubczyk, M., & Szufel, P., (2018). A framework for sensitivity analysis of decision trees. *Central European Journal of Operations Research*, *26*(1), 135–159. https://doi.org/10.1007/s10100-017-0479-6.
59. Quinlan, J. R., (1999). Simplifying decision trees. *International Journal of Human Computer Studies*, *51*(2), 497–510. https://doi.org/10.1006/ijhc.1987.0321.
60. Moshkov, M., (2021). On the depth of decision trees over infinite 1-homogeneous binary information systems. *Array*, *10*(2020), 100060. https://doi.org/10.1016/j.array.2021.100060.
61. Panhalkar, A. R., & Doye, D. D., (2021). Optimization of decision trees using modified African buffalo algorithm. *Journal of King Saud University – Computer and Information Sciences*, *XXXX*. https://doi.org/10.1016/j.jksuci.2021.01.011.
62. Naik, B., (2020). Extra-tree learning based socio-economic factor analysis and multi-class adaptive boosting meta-estimator for prediction of agricultural productivity. *Indian Journal of Science and Technology*, *13*(29), 2081–2101. https://doi.org/10.17485/ijst/v13i29.839.
63. Okoro, E. E., Obomanu, T., Sanni, S. E., Olatunji, D. I., & Igbinedion, P., (2021). Application of artificial intelligence in predicting the dynamics of bottom hole pressure for under-balanced drilling: Extra tree compared with feed forward neural network model. *Petroleum*, *XXXX*. https://doi.org/10.1016/j.petlm.2021.03.001.
64. Maier, O., Wilms, M., Von, D. G. J., Krämer, U. M., Münte, T. F., & Handels, H., (2015). Extra tree forests for sub-acute ischemic stroke lesion segmentation in MR sequences. *Journal of Neuroscience Methods*, *240*, 89–100. https://doi.org/10.1016/j.jneumeth.2014.11.011.
65. Valverde-Albacete, F. J., & Peláez-Moreno, C., (2014). 100% classification accuracy considered harmful: The normalized information transfer factor explains the accuracy paradox. *PLoS One*, *9*(1). https://doi.org/10.1371/journal.pone.0084217.
66. Hassan, Z., Burhamah, W., Alabdulmuhsen, S., Al Saffar, A., Oroszlányová, M., & Aziz, H., (2021). The analysis and accuracy of mortality prediction scores in burn patients admitted to the intensive care burn unit (ICBU). *Annals of Medicine and Surgery*, *65*, 1–4. https://doi.org/10.1016/j.amsu.2021.102249.

67. Imam, Y. Z., Ahmedullah, H., Chandra, P., Almaslamani, M., Alkhal, A., & Deleu, D., (2020). Accuracy of clinical scoring systems for the diagnosis of tuberculosis meningitis in a case mix of meningitides a retrospective cohort study. *Journal of the Neurological Sciences*, *416*, 116979. https://doi.org/10.1016/j.jns.2020.116979.
68. Ozenne, B., Subtil, F., & Maucort-Boulch, D., (2015). The precision-recall curve overcame the optimism of the receiver operating characteristic curve in rare diseases. *Journal of Clinical Epidemiology*, *68*(8), 855–859. https://doi.org/10.1016/j.jclinepi. 2015.02.010.
69. Zhang, X., Feng, X., Xiao, P., He, G., & Zhu, L., (2015). Segmentation quality evaluation using region-based precision and recall measures for remote sensing images. *ISPRS Journal of Photogrammetry and Remote Sensing*, *102*, 73–84. https://doi.org/10.1016/j. isprsjprs.2015.01.009.
70. Berger, A., & Guda, S., (2020). Threshold optimization for F measure of macro-averaged precision and recall. *Pattern Recognition*, *102*. https://doi.org/10.1016/j. patcog.2020.107250.
71. Easton, R. M., Bendinelli, C., Sisak, K., Enninghorst, N., Regan, D., Evans, J., & Balogh, Z. J., (2012). Recalled pain scores are not reliable after acute trauma. *Injury*, *43*(7), 1029–1032. https://doi.org/10.1016/j.injury.2011.12.020.
72. Lowe, J. T., Li, X., Fasulo, S. M., Testa, E. J., & Jawa, A., (2017). Patients recall worse preoperative pain after shoulder arthroplasty than originally reported: A study of recall accuracy using the American shoulder and elbow surgeons score. *Journal of Shoulder and Elbow Surgery*, *26*(3), 506–511. https://doi.org/10.1016/j.jse.2016.09.004.
73. Gimeno, R., Lobán, L., & Vicente, L., (2020). A neural approach to the value investing tool F-Score. *Finance Research Letters*, *37*(2019), 101367. https://doi.org/10.1016/j.frl. 2019.101367.
74. Liang, P., Yang, W., Chen, X., Long, C., Zheng, L., Li, H., & Zuo, Y., (2020). machine learning of single-cell transcriptome highly identifies mRNA signature by comparing F-score selection with DGE analysis. *Molecular Therapy–Nucleic Acids*, *20*, 155–163. https://doi.org/10.1016/j.omtn.2020.02.004.
75. Guns, R., Lioma, C., & Larsen, B., (2012). The tipping point: F-score as a function of the number of retrieved items. *Information Processing and Management*, *48*(6), 1171–1180. https://doi.org/10.1016/j.ipm.2012.02.009.
76. Sasaki, Y., (2007). *The Truth of the F-Measure* (pp. 1–5). Teach Tutor Mater. http://www.cs.odu.edu/~mukka/cs795sum09dm/Lecturenotes/Day3/F-measure-YS-26Oct07.pdf (accessed on 28 September 2023).
77. Zhang, X., Li, X., Feng, Y., & Liu, Z., (2015). The use of ROC and AUC in the validation of objective image fusion evaluation metrics. *Signal Processing*, *115*, 38–48. https://doi.org/10.1016/j.sigpro.2015.03.007.
78. Mas-Fuster, M. I., Ramon-Lopez, A., Lacueva, J., Más-Serrano, P., & Nalda-Molina, R., (2018). Suitability of the AUC ratio as an indicator of the pharmacokinetic advantage in HIPEC. *Journal of Pharmaceutical Sciences*, *107*(2), 764–769. https://doi.org/10.1016/j. xphs.2017.10.008.
79. Van, G. N., He, Y., Zwinderman, A. H., & Putter, H., (2021). Estimation of incident dynamic AUC in practice. *Computational Statistics and Data Analysis*, *154*. https://doi.org/10.1016/j.csda.2020.107095.
80. Guo, M., & Ghosh, M., (2012). Mean squared error of James-Stein estimators for measurement error models. *Statistics and Probability Letters*, *82*(11), 2033–2043. https://doi.org/10.1016/j.spl.2012.06.019.

81. Li, G., Zhao, S., Wu, J., Li, C., & Liu, Y., (2019). DV-hop localization algorithm based on minimum mean square error in internet of things. *Procedia Computer Science, 147*, 458–462. https://doi.org/10.1016/j.procs.2019.01.272.
82. Nicolson, A., & Paliwal, K. K., (2019). Deep learning for minimum mean-square error approaches to speech enhancement. *Speech Communication, 111*(2018), 44–55. https://doi.org/10.1016/j.specom.2019.06.002.
83. Lattar, H., Salem, B. A., & Ghezala, H. H. B., (2020). Does data cleaning improve heart disease prediction? *Procedia Computer Science, 176*, 1131–1140. https://doi.org/10.1016/j.procs.2020.09.109.
84. Huang, Y., Milani, M., & Chiang, F., (2020). Privacy-aware data cleaning-as-a-service. *Information Systems, 94*, 101608. https://doi.org/10.1016/j.is.2020.101608.
85. Yeomans, C. M., Shail, R. K., Grebby, S., Nykänen, V., Middleton, M., & Lusty, P. A. J., (2020). A machine learning approach to tungsten prospectivity modelling using knowledge-driven feature extraction and model confidence. *Geoscience Frontiers, 11*(6), 2067–2081. https://doi.org/10.1016/j.gsf.2020.05.016.
86. Wang, J., & Wang, M., (2021). Review of the emotional feature extraction and classification using EEG signals. *Cognitive Robotics, 1*(2020), 29–40. https://doi.org/ 10.1016/j.cogr.2021.04.001.
87. Samuel, A., (1959). "Some Studies in Machine Learning Using the Game of Checkers." *IBM Journal of Research and Development. 3*(3), 210–229.

CHAPTER 10

SECURE CHAOS-BASED IMAGE ENCRYPTION TECHNIQUES BASED ON FEEDBACK CONVOLUTION METHOD

R. VIDHYA[1] and M. BRINDHA[2]

[1]*School of Computer Science and Engineering (SCOPE), Department of IoT, Vellore Institute of Technology, Katpadi, Vellore, Tamil Nadu, India*

[2]*Department of Computer Science and Engineering, National Institute of Technology, Tiruchirappalli, Tamil Nadu, India*

ABSTRACT

For the protection of images, plenty of image encryption algorithms have been developed in the present years. Cryptanalysis of the existing papers solely depends on the keys used in the encryption process, whereas differential attacks also have equal importance. Two measures commonly used for analyzing differential attacks in a system are NPCR and UACI. From the study of existing papers, even though high NPCR and UACI are obtained, few of the values lie in the critical values of α-level significance, which in turn increases the possibility of differential attacks. To overcome this, an efficient encryption scheme is proposed based on the convolution operation and applied on the spatial information of the plain image for generating the keys of the encryption process. The one-time key is also utilized for achieving Shannon's perfect secrecy. The proposed chaotic feedback convolution is proved against various test cases using simulations. Both numerical and theoretical analyses are performed to prove that the proposed encryption method is resistant to differential attacks.

Multimedia Security: Tools, Techniques, and Applications. Bhaskar Mondal and Shyam Singh Rajput (Eds.)

10.1 INTRODUCTION

Internet protocol security (IPsec) is the most challenging protocol in the network layer. To strengthen the security level of digital transmission system, it consists of three phases: key management, reliability, and confidentiality. Among these, key management is connected with random number generation (RNG) and secure key distribution to send the generated keys between the sender and receiver before transferring the information (plain texts). Reliability of information system is related to two features: (i) integrity – a verification procedure to determine no information about data/message has been altered or modified by unauthorized persons; and (ii) authentication – a verification about the originality of the plain images. Confidentiality is ensured by the encryption–decryption method in both transmitter and receiver side. In general, for any information processing system, confidentiality, integrity, and non-repudiation features are achieved by cryptographic mechanizm.

As a result of the rapid advancement of multimedia and internet technologies, multimedia information such as images, videos, and other types of media are shared online. Therefore, information confidentiality, i.e., security of information, has become more important. In this chapter, a brief introduction to cryptographic techniques, image encryption and challenges in image encryption are described.

10.1.1 CRYPTOGRAPHIC TECHNIQUES

Cryptography is a science and technique to safeguard information/messages. It is the study and procedure of concealing messages. It seeks to make sure that the message is accessible only to the authorized person. The motivation is that if an unauthorized person acquires the information without the secret key, he won't be able to make sense of it. In cryptography, the knowledge of the secret key is agreed upon and shared between the sender and receiver side, and this classification comes under the symmetric-key mechanizm, whereas the complement is asymmetric key or public key cryptography. Moreover, cryptography is one type of security techniques to retain messages from unintended eyes. In order to achieve information security, different techniques are used. Security can be accomplished using standard and internationally approved symmetric encryption techniques such as AES, DES [1] and asymmetric algorithms like RSA, Blow fish, IDEA, etc. Essentially, cryptography has unique constraints: confusion, diffusion, and demands on keys. Existing encryption methods such as DES and AES are time-consuming and also, other public key methods such as RSA (Rivest Shamir Adleman) are time consuming which leads to high computational complexity. This appears to be a significant impediment to secure image transmission in the real world, where time is of the essence, therefore these techniques are solely recommended for text [2].

The inability of traditional encryption approaches to adapt to images is due to factors such as large data sizes, high neighboring pixel correlation, and high inter-pixel redundancy, among others, resulting in security leakage. In order to overcome these problems, chaos is an attractive theory which lies under non-linear systems in the domain of mathematics. This theory is utilized in various laboratories, universal systems and rigorous research areas such as fractals, weather, organizational behavior [3]. As a result, chaos theory has a large body of literature, and one possible application of chaos is security. Additionally, chaos theory is considered for the generation of pseudo-random numbers with impressive properties such as highly sensitive initial conditions, topological mixing, ambiguity, etc. Because of these unique features, many researchers are attracted towards chaos for the generation of keys to encryption mechanizm. In general, chaotic maps [2] are extremely sensitive to initial conditions, which means that their performance over time diverges greatly from the initial conditions. Furthermore, chaotic maps provide control parameters and initial conditions, which are required for iterating the maps and which promise a large key space too. It also assures that the generated key sequences have a high degree of randomness.

10.1.2 IMAGE ENCRYPTION

An encryption or decryption system is named as cryptosystem, or a cipher. Consequently, the encryption mechanizm is called encipher, and the decryption mechanizm is called decipher. The data used for encryption is described as the plaintext, whereas the enciphered data is named as cipher text. The plain text and cipher text are considered as P and C respectively. The procedure for encryption is $C = E_{k_e}(P)$ where $E(.)$ is the encryption function and k_e is the key for encryption. Similarly, the decryption procedure is given as $P = D_{k_d}(C)$, where $D(.)$ is the decryption function and k_d is the key for decryption. Meanwhile, if $k_e = k_d$, then the cipher is named as symmetric or private key cipher, here the keys are transferred to the receiver side from sender through a secure channel. If $k_e \neq k_d$, then this cipher is represented as asymmetric or public key cipher, here the key used for encryption k_e is distributed but the decryption key k_d is private so that there is no need for additional secure channel for transferring the keys. The overview of image encryption and decryption scenario is shown in Figure 10.1.

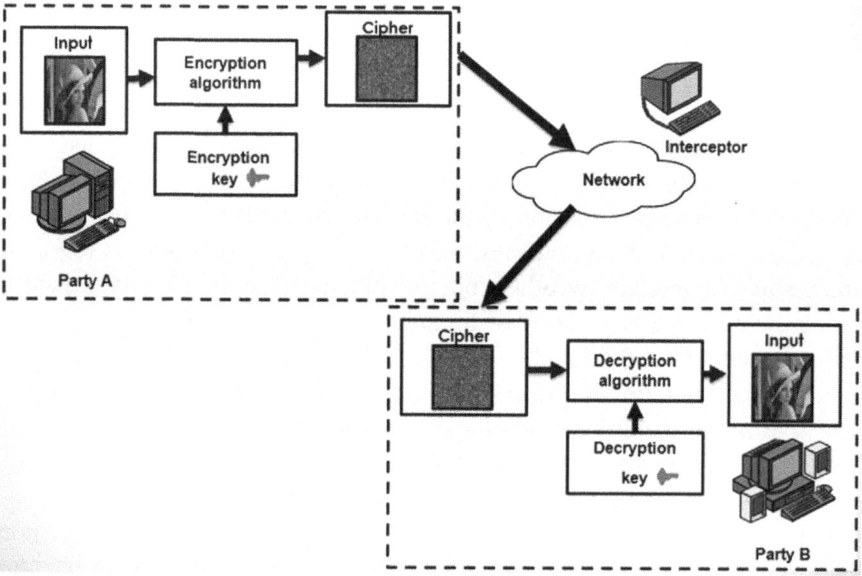

FIGURE 10.1 Image encryption/decryption scenario.

It has been utilized in various applications such as military, personal image, telemedicine, biometric system, video conference, biometric systems, digital rights management (DRM), multimedia storage, signal processing in encrypted domain, real-time interaction, Secure surveillance framework for IoT, wireless/Mobile communication, etc., Every application listed above requires an image encryption mechanizm.

10.1.3 TYPICAL ARCHITECTURE FOR CHAOS-BASED IMAGE CRYPTOSYSTEM (CIC)

Any encryption technique is based on two phases in general. The first is permutation, while the second is diffusion. In permutation, the pixels' locations are modified, whereas in diffusion, the gray value is changed pixel-by-pixel or block-by-block. Mathews [4] created the first chaotic function and demonstrated that it is suitable for cryptography. Later, in order to meet

Shannon's requirements [5], the first common architecture for chaos-based image encryption is presented, as shown in Figure 10.2, with two steps – (i) permutation; and (ii) substitution by Fridrich [6]. Permutation is the process of moving pixels around in an image, whereas substitution is the process of changing pixel values.

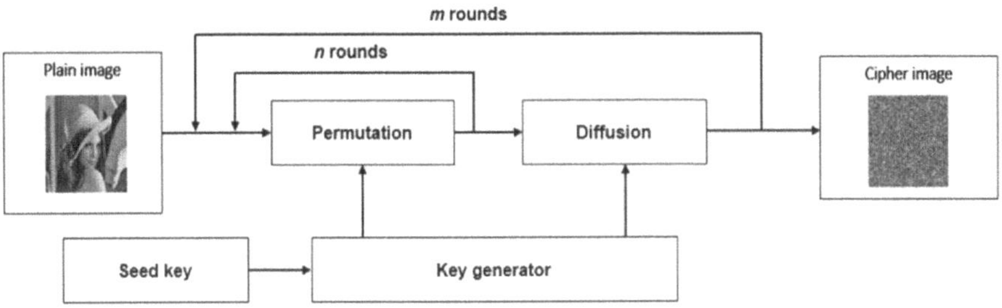

FIGURE 10.2 Typical architecture of chaos-based image cryptosystem.

Moreover, the principle of the diffusion step is not only to modify the values of pixel but also to change many pixels in an image for a small change in one pixel. To disrupt the association among adjacent pixels, the confusion stage is performed n times, and diffusion is replicated for m times. Both m and n are replicated more than one time, to get adequate level of security.

10.2 CHAOS-BASED IMAGE ENCRYPTION AGAINST DIFFERENTIAL ATTACKS

In this section, the detailed study of the differential attacks, the problem of chaos-based image encryption techniques, and the challenges are discussed.

10.2.1 DIFFERENTIAL ATTACKS

The differential assault is assessed using two metrics: NPCR (Number of Pixels Change Rate) and UACI (Unified Average Changing Intensity). Yue et al. [7] developed a mathematical design for NPCR and UACI and used these values to control NPCR and UACI's statistical belief. Few values are selected as critical values based on this statistical hypothesis, with three α-level significance levels of 0.01, 0.05, and 0.001. All existing encryption methods, on the other hand, ignore differential attack observations. The researchers are concentrating on encryption methods with lower computational complexity and shorter encryption durations, but they are not looking to differential attack prevention. Differential attacks are the most vulnerable among the many possible attacks in image encryption. Furthermore, because permutation-only processes are easy to cryptanalyze, the security of any encryption method is strongly dependent on the diffusion process. Enayatifar et al. [8] have created image encryption using DNA sequences, and the NPCR and UACI values of various sample images are indeed theoretically critical in the null hypothesis test at the 0.05 level proposed in Ref. [7]. Teng & Wang [9], used self-adaptive chaotic image encryption for bit-level encryption and decryption, with an NPCR of 93.6768% and a UACI of 33.3364%. The cryptosystem is subject to differential attack based on these parameters. Using an extended Arnold cat and a Bernoulli shift map, Ye & Huang [10] proposed an effective permutation-substitution for image encryption approach. Even if the keys are correlated with plain image, the resultant NPCR and UACI values have theoretical critical values, indicating that differential assaults are ineffective. Similarly, Liu, Tong, & Ma [11] suggested a novel hyper chaos system that relies on linear feedback shift register (LFSR)-based initial key generation to defeat selected plain text attacks. However, the NPCR and UACI values for the conventional lena image are 99.27% and 33.25%, respectively, while the NPCR and UACI values for other evaluated images are similarly less and potentially critical, resulting in differential attacks.

Wang et al. [12] presented a hybrid image compression and encryption scheme that employs arithmetic coding to improve compression ratios and chaotic sequences for encryption. Even if the keys are related to plain images, the collected NPCR and UACI values for a few sample images have theoretically critical values, increasing the likelihood of differential attacks. According to the Ref. [13] technique, block level image encryption is achieved using the Baker map and several one-dimensional chaotic maps. However, despite the high NPCR values, the UACI values in this scheme are only 27.88%. Synchronizing process is used in Ref. [14] scheme to improve the security of chaotic systems, but the UACI value is as low as 26.27%, which is lower than the optimal value. Color image encryption with one-dimensional chaotic maps using complete plain image features is used in Ref. [15] method but the UACI value is 33.36% and the key space is only around 10^{38}. The method [16] which is based on an upgraded cat map and a dynamic random growth technique, produces an optimal NPCR value, however, the UACI value is 26.5869% for some images, i.e., the UACI values are theoretically critical, and differential attacks may result.

The technique [17] proposes a new chaotic image encryption based on a novel hyperchaotic system, but the NPCR and UACI values are 99.24% and 33.13%, respectively. The [18] encryption process is only practical after many encryption rounds, and the value of UACI grows with the number of rounds, reaches a peak, and then slowly decreases. The correlation coefficients for the final cipher image reach as high as 0.0541.

The UACI score for few images using Ref. [19], which uses the parametric switching chaotic system (PSCS) for image encryption is only 33.14%, despite having strong NPCR values. The key space is likewise just 10^{68} characters long. Chen et al. [20] suggested an effective image encryption based on gray-code shuffling, and pixel-related substitution is performed to defeat chosen plain text attacks. However, after only a few encryption cycles, the values of NPCR and UACI reach optimal levels, and the key space is as small as 10^{47}. For gray images with a 128-bit size, Pareek, Patidar, & Sud [21] suggest a novel key-dependent diffusion-substitution. In this approach, the key space is 10^{38}, the entropy is 7.9952, and the UACI values (on average) are 32.11.

10.2.2 RESEARCH ISSUES

Even after several years of research in this area, chaotic image encryption remains an open research problem. Today, developing a strong, unique, and high-performance strategy remains a major difficulty. The level of difficulty of this problem is mostly determined by the image features. In an image encryption, the main problems are: (i) security level; (ii) computational time; (iii) huge data size; (iv) higher redundancy; and (v) high correlation among adjacent pixels. Moreover, the most important challenge is to reduce encryption time without compromising the level of security in real time scenarios.

The main challenges in image encryption are:

1. **Security Level:** This level of most information security applications is determined by computational uncertainty, for which the system provides strong randomness. In addition, security is assessed at the practical level of cryptography by examining its resistance to various types of assaults such as differential attacks, brute force attacks, pattern attacks, and so on.
2. **Huge Data Size:** Multimedia files are large and consume lots of computations in image encryption.

3. **Higher Redundancy:** In a plain image, the pixels are highly redundant to other pixels. It is difficult to perform encryption without redundancy. Handling such redundant pixels in input image is a challenging task.
4. **Computational Time:** If the computations involved in the encryption mechanizm are complex, then computation time is also a challenging task.
5. **High Correlation Among Adjacent Pixels:** In essence, image data has a built-in feature: adjacent pixels are highly connected. The chance of an adversary breaking the cryptosystem is great unless the relationship between adjacent pixels is changed. To account for this, a larger effort is required to separate the entire relationship between each and every adjacent pixel.

10.2.3 PROBLEM STATEMENT

Surveillance video processing, military, social media, medical diagnostics, robotics, industries, space, and other applications that employ images extensively use chaos-based image encryption techniques. Image encryption techniques frequently face a range of obstacles in the real world due to environmental complexity. Section 10.2.2 goes over it in detail. The cryptosystem that requires less time without compromising security are preferable for real-time image encryption. Even if it had stronger security features, an encryption technique that ran slowly would be of no use for real-time computing. Because traditional techniques demand more computational power and time, they are not suited for digital image encryption.

Chaos-based algorithms, on the other hand, provide a fair balance of speed, confusion, and powerful diffusion. Some chaos-based image encryption techniques, on the other hand, are broken due to their limited key space and encryption scheme vulnerability. As a result, the focus of this research is on the development of chaos-based cryptosystems for digital image encryption. The security properties of the chaos-based cipher are further investigated in relation to various sorts of assaults. When cryptanalyzing a cryptosystem, a general assumption is that the cryptanalyst can obtain information about the design and operation of the investigated cryptosystem, except for the secret keys for encryption and decryption, according to Kerchoff's principle [22]. As a result, four typical attacks [23] are based on the aforesaid idea. There are several types of attacks that can be used, including cipher text alone attacks, known-plain text attacks, chosen plain text attacks, and chosen-cipher text attacks, with difficulty levels ranging from the most difficult to the easiest.

All sorts of cryptanalytic assaults should be resistant to image encryption. The attacker only knows the cipher text in a cipher text attack. Because it is the most difficult attack, any cryptosystem should be able to withstand it. In a known-plain text attack, a cryptanalyst has access to a plaintext and the associated encrypted data and attempts to derive a correlation between the two (plain and cipher text) in order to gain the key, then uses the same key to decrypt the cipher text. A cryptanalyst with access to the encryption algorithm obtains the cipher text for the chosen plaintext in a chosen-plaintext attack. To decrypt encrypted message, the association between these two is examined. In a chosen-cipher text attack, a cryptanalyst looks for a plain text that matches the cipher text chosen at random. As a result, a good cryptanalysis research [24–26] is required. To demonstrate the cipher's resistance to known assaults, it must be included.

In addition, Jakimoski & Kocarev [27] introduced a novel block cipher based on a one-dimensional chaotic map that is tested against two attacks – (i) differential analysis (chosen/known-plaintext attacks); and (ii) linear cryptanalysis (use of linear expressions used in the cryptosystem to reveal the weakness in the cipher image). The differential attack is the most vulnerable of these conceivable attacks, because attackers focus on plain image-related attacks, and any encryption scheme should be able to withstand this type of attack. Furthermore, a heuristic analysis of chaos-based video and image encryption algorithms is performed in Ref. [28], and the following difficulties are discovered as a result of this research: (i) computational burden; and (ii) security gains are doubtful in a few known schemes. Among these, the second is the most vulnerable of these concerns. Less key space, less plain image sensitivity, statistical assaults, differential attacks (chosen/known plain text), and non-randomness of system are all security challenges with chaos-based encryption. Few security concerns are only quantitatively analyzed in some existing methodologies and declared secure in these. Furthermore, the majority of current methods are quite sensitive to differential attacks. The main reason for this is that encryption methods do not achieve strong diffusion properties, and plain images are not dissipated using the keys used in the encryption process.

Furthermore, present algorithms in the field of chaos-based image encryption have one or more drawbacks:

- Researchers who concentrate on encryption methods with low computational complexity and short encryption times ignore the need for differential attack resistance. Differential attacks are the most vulnerable of the many possible attacks in image encryption.

- Furthermore, the security of any encryption method is strongly dependent on the diffusion process, as permutation alone is trivial to cryptanalyze.
- Fixed keys are used, and key does not correlate with plain and cipher image.
- Proper information about the precision of the keys is not specified.
- Direct visualization of the encrypted image itself tells the information about the plain image (statistical attack).
- System fully relies on one-time key (OTK) or shuffling.

10.2.4 CURRENT RESEARCH DEVELOPMENTS IN CHAOTIC ENCRYPTION

The goal of this chapter is to address the concerns mentioned above, which are heavily influenced by existing chaotic image encryption techniques. Few solutions overcome the aforementioned concerns, but they do not attain all of the qualities of a good encryption scheme, such as strong replacement, plain image related (PIR), and an extra level of protection with a OTK. As seen in a few methods, 2D and 3D chaotic maps can provide a respectable amount of security while reducing computing load. As a result, selecting the optimal chaotic map is one aspect for raising the level of security, but it does not play a significant influence completely. In addition, choosing the first seeds of any chaotic map without knowing the plain image can lead to attacks.

In the field of chaos-based image encryption, various techniques have been developed. However, no traditional method can solve all of the issues outlined in Section 10.2.2 at the same time. As a result, research is being conducted to better understand and address these concerns.

10.2.5 OBJECTIVES OF THIS CHAPTER

This study presents the development of image ciphers based on chaos with efficient permutation and diffusion properties. For such ciphers, the observations and investigations are concentrated particularly on the ability to withstand different kinds of attacks. Various chaos-based ciphers are proposed to achieve each of these goals, and a cryptanalysis study of these ciphers is provided. Moreover, considering the limitations listed in Section 10.1.5, the goals of this research are:

- Developing chaos-based image cryptosystems with efficient permutation and diffusion characteristics.
- Developing a chaos-based cipher that has good encryption speed without compromising security.
- Investigating the security characteristics of the developed chaos-based ciphers with respect to various kinds of attacks.

10.3 CHALLENGES IN CHAOTIC IMAGE ENCRYPTION

Detailed survey of current literature in this domain is needed to develop good encryption schemes for handling images. This chapter reviews the traditional and existing chaos-based image encryption approaches. Chaotic encryption algorithms are suffered from different problems like limited security, high computational time, a large amount of data, considerable redundancy, and a high correlation between adjacent pixels. This survey is performed in four essential dimensions for achieving the objectives of this chapter. The first category addresses basic chaotic ciphers, the second one deals with pixel and bit-wise permutation, i.e., permutation only architectures along with encryption metrics, the third one covers the PIR and non-plain image related (n-PIR) concepts, and the fourth one addresses OTK related concepts.

10.3.1 CHAOTIC CIPHERS

Various types of image encryption algorithms are presented with chaos as it is one of the most appealing approaches for encrypting images due to its complicated behavior. The reason for using chaos in image encryption is that it has two basic properties: ergodicity and high sensitivity to initial vectors, which can be generalized to the 'confusion' and 'diffusion' features of cryptography to develop encryption schemes [25] as suggested by Shannon. Arnold & Avez [29] introduced image encryption using chaotic maps in 1967, and the subject of chaos-related image encryption developed into a robust research topic. Following this research, Mathews [4] developed a chaotic function for image encryption to solve the drawbacks of the one-time pad technique. Due to their ease of implementation, a variety of chaotic maps such as the Henon map, Logistic map, Chebyshev's map, Arnold's map, Tent map, Skew map, and others are commonly employed at first. Images have a two-dimensional space, and one-dimensional chaotic maps may result in weak security and limited key space for which two-dimensional chaotic maps, three-dimensional chaotic maps, and hyper chaos are used in image encryption, despite the one-dimensional chaotic maps' simplicity. Following that, Fridrich [6] proposed the first chaotic encryption architecture, consisting of two stages – (i) confusion and (ii) diffusion and has been generalized and adopted by the majority of researchers. Confusion shuffles the values of an image, whereas diffusion uses operations to change the intensity values of each pixel. Most of the cryptosystems proposed after Lian, Sun, & Wang [31]; Ye [32] have adopted the aforementioned architecture.

10.3.2 PERMUTATION ONLY CIPHERS

In the encryption of images, permutation techniques are crucial. Permutation is always the first stage of a cryptosystem, and its goal is to modify the positions of all pixels. Pixel level permutation and bit level permutation are two types of permutation based on the movement of pixels or bits within a pixel. In general, there are two types of pixel level permutations: (i) look up table-based (S-box); and (ii) sorting-based. The position of pixels is shuffled using a permutation mapping matrix in existing general shuffling algorithms (look up table-based shuffle or sorting).

From the described architecture, the confusion stage is performed by different ways – (i) pixel values are shuffled in the image [6, 31]; (ii) position of the pixel values are shuffled [32]; and (iii) bit by bit level permutation is done at some algorithms [9].

Further, few image encryptions totally depend on the confusion stage itself [18, 33]. Moreover, a novel permutation only hierarchical chaotic image encryption is proposed comprising of intra and inter block level permutation, which is designed by VLSI architecture, and the key components are realized in FPGA to validate the features of the chaotic method [35]. Subsequently, in Ref. [36], it is found that O (log r ($M.N/K$)) chosen or known plain images are enough to break this approach, i.e., chosen plain text attack is possible. As a result, it can be deduced that the scheme is entirely based on permutation and that cipher text and known/chosen plain text assaults may easily break it.

10.3.3 PLAIN IMAGE RELATED (PIR) AND NON-PLAIN IMAGE RELATED (N-PIR)

Any image encryption algorithm's security is entirely dependent on the diffusion process. Keys play an important function in modifying pixel values during the substitution stage. The fundamental reason for this is that encryption methods do not achieve strong diffusion properties, and the plain picture is not dissipated with the encryption keys used. Some encryption keys are generated independently from the plain image [31], while others are generated precisely by correlating with the plain image [38, 39]. If the substitution is based on a correlation with a plain image, it produces strong security and can also withstand chosen/known plain text attack.

10.3.3.1 NON-PLAIN IMAGE RELATED (N-PIR)

In Ref. [40], a piece-wise linear map is used to construct two levels of permutations and three levels of diffusion in the aspects of row, column, and block. The keys are not connected with the input (plain) image, which could lead to differential (chosen/known plain text) attacks. While the system [41] uses simultaneous symmetric encryption to reduce computing time, it is vulnerable to differential attacks because it does not relate to plain images.

A novel chaotic system [42] based on the cosine transform is presented to produce high complex behavior in generated key streams in addition to these strategies. Effective permutation and random based diffusion are proposed in this encryption technique to raise the security level and to obtain high NPCR and good UACI without knowing the plain image. In Refs. [43, 44], a novel cyclic group-based sequence is created for image encryption and there is no substantial spread in these systems since the keys are not associated with the plain image. As a result, plain image-related attacks are possible in both methods. In Ref. [45], row and column wise shift operations are used for pixel permutation, and simple XOR operation based diffusion is used for image encryption. But this scheme is cracked by Ref. [46], and this scheme has the following security concerns: (i) weak keys; (ii) fixed keys; and (iii) encryption provides information about the plain image.

10.3.3.2 PLAIN IMAGE RELATED (PIR)

This category includes the following schemes: [39] – hash functions such as SHA-512, SHA-256, and MD5 are used (ii) [46] – some mathematical operations, such as sum of plain image for initial seed creation, and (ii) other ways, such as Ref. [47], in which first seeds of chaotic map used in encryption procedure are formed using plain image. Zhu, Zhao, & Zhang [48] suggested a lossless image compression and encryption based cryptosystem with a 4:1

compression ratio, but [49] have broken this cryptosystem by employing moduli as the sub-key in the confusion stage, which results in low sensitivity to changes in plain text. Brindha & Gounden [38] presented a hash table-based cryptosystem with compression based on the Chinese remainder theorem. A 5:1 compression ratio is achieved. The same authors have also presented a novel image encryption based on three key matrices and cryptanalyzed their work to show that it is not subject to chosen/known plain text attacks [39].

10.3.4 ONE-TIME KEY (OTK) RELATED CONCEPTS

Shannon's theory [5] states that good mixing of transformations is required for perfect secrecy of the cryptosystem, and that the one-time pad technique is very secure because the key is changed for each encryption. OTKs are created and used in encryption scenarios based on this [37]. Liu et al. [30] has presented a coupled chaotic network model for image encryption that does not use crucial values and instead uses a one-time pad to ensure security. Dong [34] proposes an image encryption strategy based on a coupled chaotic system with OTKs, average NPCR and UACI values of 99.61233% and 33.45033%, respectively, demonstrating that all NPCR and UACI are non-critical due to OTKs.

Referring to Ref. [37], this scheme uses many impulse noise signals from environmental noise, i.e., OTK is utilized for initial vectors of chaotic map to generate the random sequences applied in image encryption.

The above-mentioned techniques rely exclusively on OTK to provide high plain image sensitivity, but one drawback with OTKs is that if the OTK is not totally random for each encryption, differential attacks are possible.

10.4 NOVEL APPROACH FOR ENCRYPTION OF IMAGES AGAINST DIFFERENTIAL ATTACKS

To overcome the differential attacks, a novel image encryption, which is based on convolution function and OTK, is utilized in the proposed model.

10.4.1 TRADITIONAL CONVOLUTION APPROACH

In image processing, each pixel value is changed by a linear combination of its neighbors using kernel/mask or filter. Linear filter is mainly applied for smoothening, edge detection and the kernel is always fixed. Figure 10.3 clearly shows the convolution applied for a sample plain image.

The following are the steps involved in general convolution:

- To execute convolution, the kernel matrix is shifted horizontally and vertically in the plain image of each pixel A(i,j).
- The adjacent pixels values of A(i,j) are multiplied with the kernel matrix, and the multiplied values are then added to get A'(i,j), which is then stored in the same place.

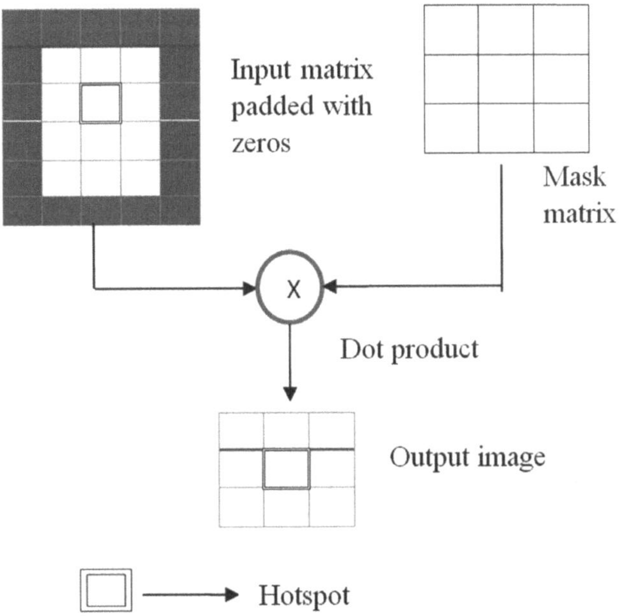

FIGURE 10.3 General framework of convolution.

10.4.2 ONE TIME ENCRYPTION

A OTK is used for encryption to avoid differential attacks or known plaintext attacks. The initial seed in the chaotic map is generated as follows:

- A random number R is chosen such that $1 \leq R \leq m \times n$, where m and n are the height and width of the image.
- The random number is transformed into Binary digits.
- The binary numbers are first transformed to binary fractions, and then back to decimal fractions.

- One such value is employed as the logistic map's seed.
- Consider an example for generating the initial seed for logistic map:
- Take a random number R = 156.
- The equivalent binary number is B = 0111100010.
- The fraction part of Binary number is B = 0.0111100010 and the corresponding decimal fraction is F = 0.472656250000000.
- This F value is used as the initial seed for the logistic map $x_0 = 0.472656250000000$.

10.4.3 PROPOSED FRAMEWORK

In this section, the proposed methodology and the detail description about the framework which is introduced in chaos-based image encryption are explained.

10.4.3.1 CHAOTIC-BASED FEEDBACK CONVOLUTION

For the proposed work, the kernel is varied for each plain pixel and related to the previously computed value for achieving the concept of feedback mechanizm. The convolution theorem is modified and applied to image encryption in the key generation.

Let $A(i, j)$ denote the plain pixel values, $1 \leq i \leq m \text{ and } 1 \leq j \leq n$, where m and n are the no of rows and columns in the image.

Let $H(x, y)$ denotes the kernel or mask matrix values applied in the plain image pixels which is described as $-\infty \leq x \leq \infty \text{ and } -\infty \leq y \leq \infty$.

The generic convolution formula, written as following, can be used in any discrete two-dimensional space.

$$A'(i,j) = \sum_{x=-\infty}^{\infty} \sum_{y=-\infty}^{\infty} A(i-x, j-y) . H(x,y) \tag{1}$$

From that, the proposed convolution operation for each pixel is calculated as:

$$A'(i,j) = \sum_{x=-\infty}^{\infty} \sum_{y=-\infty}^{\infty} A(i-x, j-y) . H_k(x,y); 1 \leq k \leq m \times n \tag{2}$$

For the intermediate output, initially, all values are zeros and this matrix is described as:

$$O_1 = \begin{pmatrix} o_{11} & \cdots & o_{1m} \\ \vdots & \ddots & \vdots \\ o_{n1} & \cdots & o_{mn} \end{pmatrix} \tag{3}$$

the first kernel is represented as:

$$H_1(x,y)=\begin{pmatrix} H_1(-1,1) & H_1(0,1) & H_1(1,1) \\ H_1(-1,0) & H_1(0,0) & H_1(1,0) \\ H_1(-1,-1) & H_1(0,-1) & H_1(1,-1) \end{pmatrix} \tag{4}$$

After that, apply the value of $H_1(x, y)$ to $A(1,1)$ then compute the value $O_1(1,1)$ and store into matrix O_1.

From the second pixel onwards, the kernel matrix is represented as:

$$H_{2-mn}(x,y)=\begin{pmatrix} H_2(-1,1) & H_2(0,1) & H_2(1,1) \\ H_2(-1,0) & O_1(i-1,j-1) & H_2(1,0) \\ H_2(-1,-1) & H_2(0,-1) & H_2(1,-1) \end{pmatrix} \tag{5}$$

After computing intermediate values in the forward direction from the first to the last pixel using plain image, convolution is applied in the reverse direction from the last to the first pixel for these values. This can be written as:

$$O'(i,j)=\sum_{x=-\infty}^{\infty}\sum_{y=-\infty}^{\infty} O(i-x,j-y).\,Q_k(x,y);\,1\le k\le m\times n \tag{6}$$

The output matrix is represented as:

$$O_2=\begin{pmatrix} o_{11} & \cdots & o_{1m} \\ \vdots & \ddots & \vdots \\ o_{n1} & \cdots & o_{mn} \end{pmatrix}$$

Now the mask is applied for the last pixel:

$$Q_1(x,y)=\begin{pmatrix} Q_1(-1,1) & Q_1(0,1) & Q_1(1,1) \\ Q_1(-1,0) & Q_1(0,0) & Q_1(1,0) \\ Q_1(-1,-1) & Q_1(0,-1) & Q_1(1,-1) \end{pmatrix} \tag{7}$$

The feed given for the mask middle is illustrated in Eqn. (8), and then the mask is applied in the last pixel of intermediate output matrix O_1.

$$Q_1(x,y)=\begin{pmatrix} Q_1(-1,1) & Q_1(0,1) & Q_1(1,1) \\ Q_1(-1,0) & O_1(1,1) & Q_1(1,0) \\ Q_1(-1,-1) & Q_1(0,-1) & Q_1(1,-1) \end{pmatrix} \tag{8}$$

Now the computed value of last pixel is stored in output matrix O_2. The kernel value is applied for the pixel $O_1(m-1, n-1)$ to $O_1(1,1)$ and is represented as follows:

$$Q_{(m-1,n-1)-(1,1)}(x,y)=\begin{pmatrix} Q_1(-1,1) & Q_1(0,1) & Q_1(1,1) \\ Q_1(-1,0) & O_2(i+1,j+1) & Q_1(1,0) \\ Q_1(-1,-1) & Q_1(0,-1) & Q_1(1,-1) \end{pmatrix} \quad (9)$$

The computed output matrix O_2 is used as Key matrix for the image encryption. This key matrix is applied to the encryption of images for achieving high level of security because keys are thoroughly correlated with the plain image and one time pad technique is included in the proposed encryption. This is clearly illustrated in Figure 10.4.

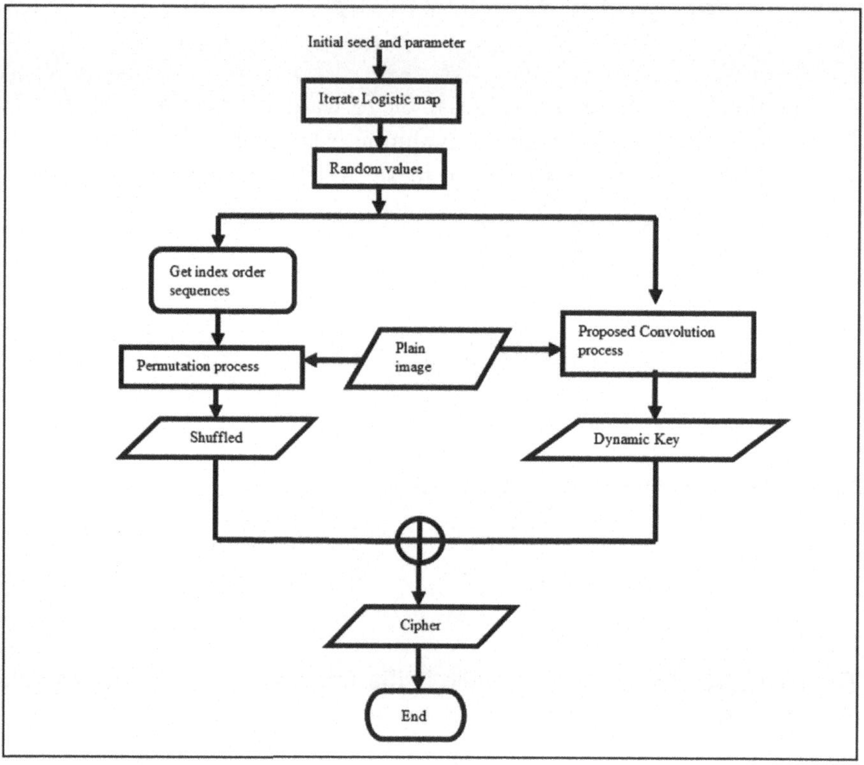

FIGURE 10.4 Flow chart for encryption scenario.

10.4.3.2 ENCRYPTION SCHEME

The encryption of images is divided into two stages – (i) permutation and (ii) diffusion. It is explained here.

10.4.3.2.1 Permutation Phase

- **Step 1:** A one-dimensional array $P = \{p_1, p_2, \ldots, p_{m \times n}\}$, is created by taking a plain image of size $I = m \times n$ and is converted into a one-dimensional array $P = \{p_1, p_2, \ldots, p_{m \times n}\}$, where m and n are the height and width of the plain image, respectively.
- **Step 2:** For logistic map, put the initial value x_0, iterate $m \times n$ times, and the produced sequence is $X = \{x_1, x_2, \ldots, x_{m \times n}\}$. The index order sequence $F_X = \{F_{x_1}, F_{x_2}, \ldots, F_{x_{m \times n}}\}$ is created by sorting the sequence X in ascending order.
- **Step 3:** From the index order sequence, the plain image is shuffled, the shuffled values are stored into an array $S = \{s_1, s_2, \ldots, s_{m \times n}\}$ and then the array is reshaped into a matrix $C = \begin{pmatrix} c_{11} & \cdots & c_{1m} \\ \vdots & \ddots & \vdots \\ c_{n1} & \cdots & c_{mn} \end{pmatrix}$ to form a shuffle image.

10.4.3.2.2 Key Generation

- **Step 1:** Set and iterate the logistic map for $2 \times ((m \times n) \times 9) = q$ times and the generated random values are stored in an array $A = \{a_1, a_2, \ldots, a_q\}$.
- **Step 2:** This array is splitted into two arrays based on the middle value and the arrays are represented as $A_1 = \{a_1, a_2, \ldots, a_{mid}\}$ and $A_2 = \{a_{mid+1}, a_{mid+2}, \ldots, a_q\}$.
- **Step 3:** The arrays A_1 and A_2 are used for the key generation and these two arrays are reshaped into matrices M_1 and M_2 of size $1 \le M_1 \le m \times n$ Rows and $1 \le M_1 \le 9$ columns, same as for matrix M_2.
- **Step 4:** In the matrix M_1, each row is extracted and reshaped into a 3×3 matrix to change the intensity of the plain image in the forward direction, which means from the first pixel to the last pixel as described in Refs. [2–5].
- **Step 5:** Now the computed intermediate matrix is named as O_1 and this matrix is input to the generation of key matrix. Now using matrix O_1 and M_2, the convolution is applied in the reverse manner which means from last pixel to first pixel as illustrated in (7–9).
- **Step 6:** Now the computed output matrix O_2 is used as key matrix $K = \begin{pmatrix} k_{11} & \cdots & k_{1m} \\ \vdots & \ddots & \vdots \\ k_{n1} & \cdots & k_{mn} \end{pmatrix}$ and it will change for every encryption.

- **Step 7:** Ex-OR the key matrix with permuted matrix to finally get the encrypted image. Figure 10.5 shows the plain image and encrypted image after encryption process.

$$\begin{pmatrix} E_{11} & \cdots & E_{1m} \\ \vdots & \ddots & \vdots \\ E_{n1} & \cdots & E_{mn} \end{pmatrix} = \begin{pmatrix} k_{11} & \cdots & k_{1m} \\ \vdots & \ddots & \vdots \\ k_{n1} & \cdots & k_{mn} \end{pmatrix} \oplus \begin{pmatrix} c_{11} & \cdots & c_{1m} \\ \vdots & \ddots & \vdots \\ c_{n1} & \cdots & c_{mn} \end{pmatrix} \quad (10)$$

(a) (b)

FIGURE 10.5 Plain image and encrypted image.

Source: Adapted from Ref. [50]. https://creativecommons.org/licenses/by/4.0/

10.4.3.2.3 Decryption Procedure

The main feature of the proposed encryption mechanizm is dynamic keys that are used for every encryption to withstand differential attacks. Dynamic keys means that one time pad technique is applied here but the limitations behind that is the need for sending the key together with the cipher text. Due to this, protection of the keys from the attacker becomes uncertain and transmission load gets increased. For this purpose, keys are obtained from the following steps:

- Set and iterate the logistic map into $m \times n$ times, and these values are stored in a one dimensional array named W_{pq}. $1 \leq W \leq p \times q$.
- The key matrix is generated from the plain image K_{pq} $1 \leq K \leq p \times q$ and this is reshaped into an array Z_{pq}.
- These two arrays are EX-ORed in Eqn. (11) to generate another matrix named as D_{pq}, and the initial condition for a logistic map with

matrix D is sent along with the cipher image. Using this, key matrix is generated in the decryption side.

$$W_{pq'} \oplus Z_{pq'} = D_{pq} \tag{11}$$

From Eqn. (11), using matrix D_{pq}, the logistic map is iterated to get the sequence for retrieving the key matrix. From the key matrix, the shuffled image is retrieved using Eqn. (12):

$$\begin{pmatrix} c_{11} & \cdots & c_{1m} \\ \vdots & \ddots & \vdots \\ c_{n1} & \cdots & c_{mn} \end{pmatrix} = \begin{pmatrix} k_{11} & \cdots & k_{1m} \\ \vdots & \ddots & \vdots \\ k_{n1} & \cdots & k_{mn} \end{pmatrix} \blacklozenge \begin{pmatrix} E_{11} & \cdots & E_{1m} \\ \vdots & \ddots & \vdots \\ E_{n1} & \cdots & E_{mn} \end{pmatrix} \tag{12}$$

Once the shuffled image is obtained, reverse permutation is performed using generated random values to get the plain image.

10.5 ANALYSIS OF PROPOSED FEEDBACK MODEL FOR KEY GENERATION

If the encryption is designed to give greater security, a single pixel change in the plain image should result in completely different cipher images. Various test situations are investigated for the key generation of the encryption scheme because the proposed system is designed to attain excellent security. Following the analysis of the test instances, it is found that feedback applied to the center of the mask to change the intensity of the plain image in forward and reverse directions results in improved security. The test instances are analyzed using two measures – (i) average changed intensity values; and (ii) number of pixel change rates (NPCRs).

10.5.1 CASE 1: WITHOUT FEEDBACK

i. **Mask Applied to Individual Plain Pixel Without One Time Key:** For the input matrix, the convolution operation is applied to each plain pixel with individual mask but without one time key and feedback. The cipher matrix C_1 is compared with the cipher image named C_2, which is obtained from the same plain image with one pixel change. After analyzing these two-cipher matrices, it is found that one pixel change affects only the neighbor pixels for the changed pixel value. Also, the NPCR and UACI value is very low.

From Figures 10.7(a) and 10.8(a), it is shown clearly that individual mask applied to individual pixel for changing the intensity of the pixel doesn't give much security.

ii. **Mask Applied to Individual Pixel with One Time Key:** For the same input matrix, mask is applied to individual pixel but in this case, one time key is used for every encryption. Because of the onetime key, the NPCR and UACI values are nearer to >99% and > 33%. In Ref. [17], theoretically critical values of NPCR and UACI are given; from this, few of the obtained values are found to be critical values, so it insists the differential attacks. Theoretically, critical values are highlighted in Figures 10.7(b) and 10.8(b). One pixel change in plain image is done at random positions, and for these changes, NPCR and UACI are found, which are given in Tables 10.1 and 10.2, respectively.

10.5.2 CASE 2: FEEDBACK

Without feedback convolution, individual pixel doesn't affect other pixels in case 1. If one pixel change in case 1 affects the neighbor pixels, then convolution is actually dependent on the neighbor pixels; otherwise, the same pixel only will get affected after one pixel change. But, from assumption 1, to overcome the differential attack, the previously computed value is applied to the next pixel. Since spatial information of each pixel is used for computation, so there is a need to analyze whether the given feed is applied in the middle of the mask or the neighborhood pixels of each pixel.

i. **Feedback Given in Middle:** If the feedback is given in the middle of the mask, it will increase the intensity value and gives much efficiency in the edge and corner pixels. For the convolution operation of the filtering method, the impulse function is high in the middle compared to the neighboring pixels. In the proposed method inte,nsity change plain pixels are sliding at the middle of the mask; hence, to perform convolution, the middle position is good compared to eight neighbors. Figures 10.7(c) and 10.8(c) show that the NPCR and UACI values are high due to middle replacement done at the mask and the values are given in Tables 10.1 and 10.2.

ii. **Feedback Applied in the Eight Neighbors:** Feedback of kernel is done at N_8 pixels based on modular operation with plain image but it doesn't give much effect in the edge and corner pixels. Some information is missed. The reason for that is the feedback is multiplied with padded zeros which impacts on the NPCR and the average changed intensity.

The dashed line shows the replacement done at corner and edge of mask. For example, if the feedback, i.e., previously computed pixel value is given to the edge position, the effect is negligible because of the multiplication with padded zeros. Then, the mask matrix is applied to the second plain pixel value. As the effect of feedback is negligible in the corner and edge pixel values, middle replacement is good and effective. This is clearly illustrated in Figure 10.6.

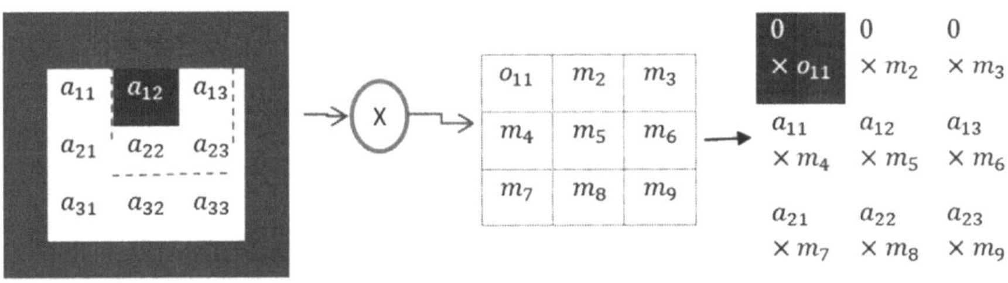

FIGURE 10.6 Replacement at corner pixel.

Figures 10.7(d)–(e) and 10.8(d)–(e), show the replacement done at corner pixel that is in the first position or N_8 position of the kernel pixel, respectively. From this, it is inferred that few values are theoretically critical values given in Ref. [17]. The NPCR and UACI values are given in Tables 10.1 and 10.2, in this the critical values are present in the few test cases which clearly shows that it leads to differential attack.

10.5.3 CASE 3: FEEDBACK APPLIED IN FORWARD DIRECTION

i. **Mask Applied to Individual Pixel Without One Time Key:** The same input matrix is taken for analysing the feedback without onetime key. Feedback is given as independent pixel operation is also a factor for the attacks in the encrypted image. The dependency of the previous computed value reduces the attacks at the same time it will increase the NPCR. If one pixel changes, the analysis begins with the neighboring pixels and progresses to the remaining pixels. However, the primary flaw in this strategy may be shown in the following cases. Example (i) If the pixel at position (1, 1) changes, all pixels in the encrypted images C_2 are changed because a change begins with the first pixel. However, sometimes less intensity change leads to a lower pixel change rate. The pixel value remains in the same position due to less intensity change, and hence the remaining pixels are not changed. Example (ii) if the pixel at position (8, 8) changes then in the cipher image C_2 only 10 pixels are changed so the NPCR rate will decrease. Also, in this case, feedback is given in forward direction only. If the last pixel changes then the pixel change rate and change of intensity is very less and is clearly illustrated in Figures 10.7(f) and 10.8(f) and the values are given in Tables 10.1 and 10.2.

ii. **Mask Applied to Individual Pixel with One Time Key:** In the previous case 6(i), the NPCR and UACI values are very less because of the same key, so one time key is used for generation of mask value applied in the convolution. Because of the onetime key, the two cipher images C_1 and C_2 have totally different pixels. So the NPCR and UACI values are nearer to 100 and >33% respectively. But in this case few, values are theoretically critical values, and it leads to differential attacks. So the convolution is applied in the computed value of the plain image in the reverse direction from last pixel to first pixel. In Figures 10.7(e) and 10.8(e), the critical values are clearly noticed and the values are given in Tables 10.1 and 10.2.

10.5.4 CASE 4: FEEDBACK APPLIED IN FORWARD AND REVERSE DIRECTION

i. **Mask Applied to Individual Pixel Without One Time Key:** In the above two cases, onetime key is the factor for achieving high NPCR and UACI values at the same time some values are proved to be theoretically critical values from Ref. [17]. So, convolution is applied in both directions, and the NPCR and UACI values are analyzed. From the analyzes, it is noticed that without the onetime key itself, it will give high NPCR and UACI values. At the same time, the limitation is the NPCR and UACI values are not ideal. Changed position and its corresponding intensity change are the important factors for not having ideal values. The NPCR and UACI values are high for the changed position (1, 1) and (256, 256) and are illustrated in Figures 10.7(h) and 10.8(h), respectively.

ii. **Mask Applied to Individual Pixel with One Time Key:** Above Case 3 (i) achieves good NPCR and UACI values only in changed positions (1, 1) and (256, 256) but the remaining positions depends on the change of intensity in the pixels. If they are very low, then they will induce less pixel change rate. So the onetime key plays an important role for achieving the good NPCR and UACI values without any theoretically critical values for the resistance of differential attack. Figures 10.7(f) and 10.8(f) clearly show that the good values are achieved if the mask is applied to individual pixel with one time key. In that, first, the convolution is performed in the plain image with feedback from the first pixel to the last pixel, and the convolution is applied in the computed value of the plain image from the last pixel to the first pixel, and it is given in Refs. [10–13]. So, this is good for key generation and it is utilized in the proposed encryption method. The NPCR and UACI values of the proposed method are given in Tables 10.1 and 10.2.

10.6 NUMERIC RESULTS

As the earlier section, the various cases are analyzed, and for these, simulations are performed using the input matrix of size 8 × 8 for all test cases. The output matrix C_1 is analyzed with one pixel change in the same input matrix and the resultant matrix C_2.

The analyzes clearly explains the change of pixel rate and changed intensity value. The simulation for all test cases using the sample matrix of size 8 × 8 and its variant (i.e., one-bit change in the location of (1,4)) is performed and the illustrations of each case is given in Figure 10.9.

10.6.1 INTENSITY CHANGE AVERAGE

The difference of the two cipher images is $\Delta = C_1 - C_2$ and the average of the difference matrix is used for analyzing the proposed work. Because of the use of the difference matrix, the attackers easily reveal the key matrix and the information about the intermediate parameters used in the encryption algorithm. So, the difference should be high between two cipher images to resist the differential attacks. From these test cases, the best one should be found, and it should be integrated into the key generation of the proposed encryption.

TABLE 10.1 NPCR Values for Various Test Cases

SL. No.	Location Changed	NPCR Values							
		Case 1 (i)	Case 1 (ii)	Case 2 (i)	Case 2 (ii)	Case 3 (i)	Case 3 (ii)	Case 4 (i)	Case 4 (ii)
1.	1, 1	0.097656	99.5346	99.62616	99.55444	1.0376	99.57581	99.56207	99.62616
2.	88, 150	0.183105	99.5743	99.58496	99.5697	1.037598	99.58496	4.554749	99.58496
3.	193, 66	0.190735	99.5773	99.60175	99.60175	1.025391	99.57581	2.661133	99.60175
4.	179, 229	0.097656	99.5453	99.6048	99.56055	0.567627	99.56207	71.53778	99.6048
5.	141, 36	0.09613	99.5804	99.61548	99.5636	0.662231	99.5575	1.046753	99.61548
6.	66, 216	0.190735	99.5361	99.56207	99.58191	0.662231	99.53168	27.95715	99.56207
7.	209, 63	0.190735	99.5193	99.5636	99.53613	1.405334	99.58344	2.758789	99.5636
8.	158, 122	0.091553	99.6002	99.56055	99.5697	0.762939	99.5438	2.206421	99.56055
9.	213, 150	0.093079	99.5331	99.58191	99.50256	0.570679	99.56207	2.304077	99.58191
10.	235, 74	0.093079	99.559	99.63529	99.58038	0.570679	99.58191	1.435852	99.63529
11.	193, 98	0.18158	99.5392	99.59717	99.58496	0.944519	99.56818	3.215027	99.59717
12.	246, 168	0.382996	99.54071	99.57886	99.59869	1.045227	99.57123	2.952576	99.57886
13.	10, 218	0.094604	99.5697	99.62616	99.57428	0.668335	99.58954	6.176758	99.62616
14.	128, 128	0.187683	99.60785	99.61853	99.54987	1.127625	99.61548	3.985596	99.61853
15.	256, 256	0.011986	99.61243	99.6109	99.54987	0.65918	99.52553	99.57886	99.6109

TABLE 10.2 UACI Values for Various Test Cases

SL. No.	Location Changed	UACI Values							
		Case 1 (i)	Case 1 (ii)	Case 2 (i)	Case 2 (ii)	Case 3 (i)	Case 3 (ii)	Case 4 (i)	Case 4 (ii)
1.	1, 1	0.005379	33.46317	33.64274	33.67729	0.0734	33.52945	33.52383	33.64274
2.	88, 150	0.012644	33.7252	33.51695	33.77807	0.093569	33.70976	0.612183	33.51695
3.	193, 66	0.012404	33.6729	33.6424	33.75439	0.09488	33.59035	0.357528	33.6424
4.	179, 229	0.009544	33.6869	33.55024	33.52577	0.077084	33.6482	23.74878	33.55024
5.	141, 36	0.009155	33.6054	33.63213	33.93215	0.08141	33.65812	0.219056	33.63213
6.	66, 216	0.010879	33.3368	33.51941	33.55119	0.091678	33.71139	8.845257	33.51941
7.	209, 63	0.01171	33.4241	33.63187	33.57729	0.125056	33.48833	0.374313	33.63187
8.	158, 122	0.00964	33.6029	33.53319	33.5253	0.078753	33.46313	0.458589	33.53319
9.	213, 150	0.00809	33.6532	33.63902	33.64414	0.074918	33.69428	0.520319	33.63902
10.	235, 74	0.00605	33.4969	33.63113	33.66101	0.075181	33.58312	0.306528	33.63113
11.	193, 98	0.008647	33.5422	33.63537	33.56995	0.084677	33.58389	0.426971	33.63537
12.	246, 168	0.019441	33.76274	33.46369	33.64131	0.125128	33.27718	0.651251	33.46369
13.	10, 218	0.00593	33.6601	33.51016	33.5623	0.084719	33.68605	1.637543	33.51016
14.	128, 128	0.011986	33.42669	33.61079	33.74874	0.094814	33.66702	0.547043	33.61079
15.	256, 256	0.007617	33.6594	33.63104	33.67178	0.044801	33.49678	33.36045	33.63104

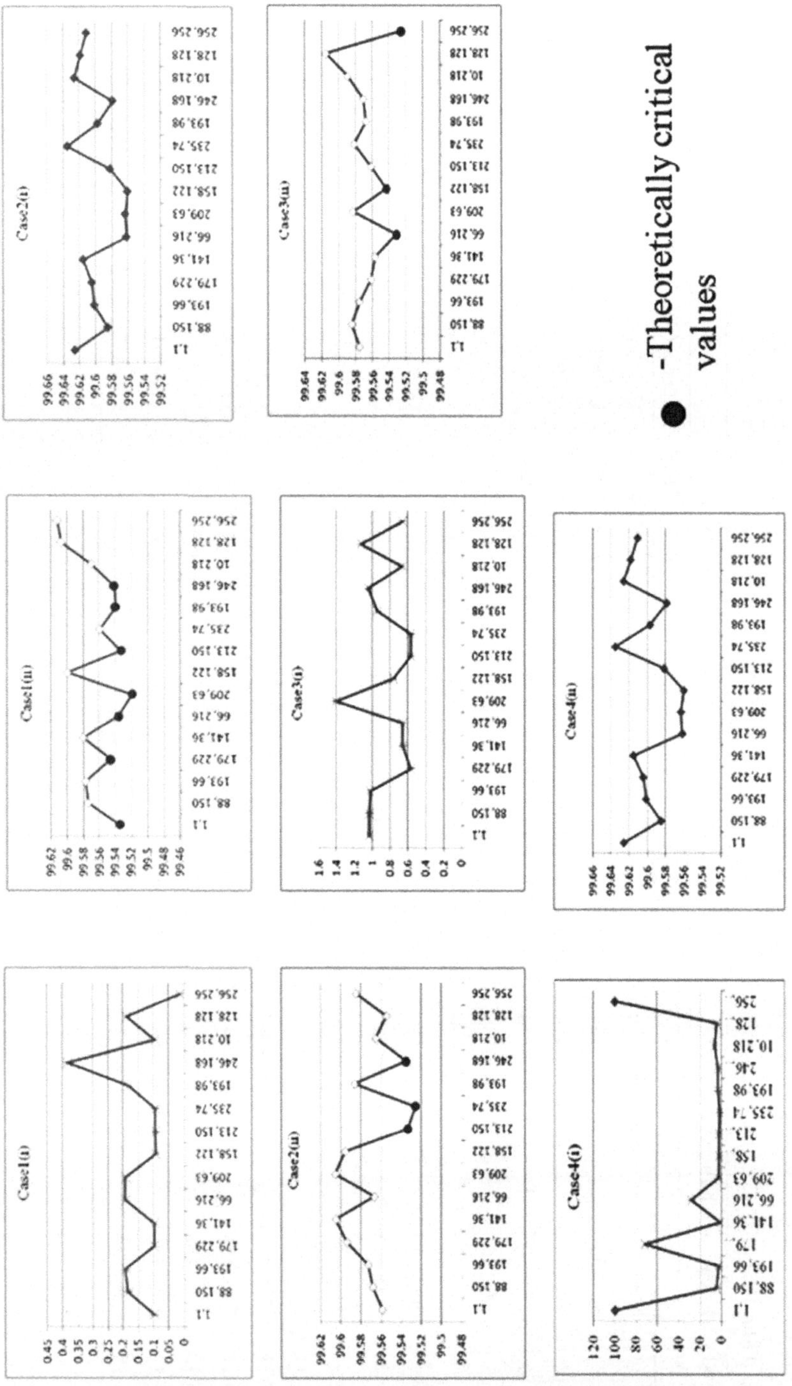

FIGURE 10.7 NPCR analysis of different test cases.

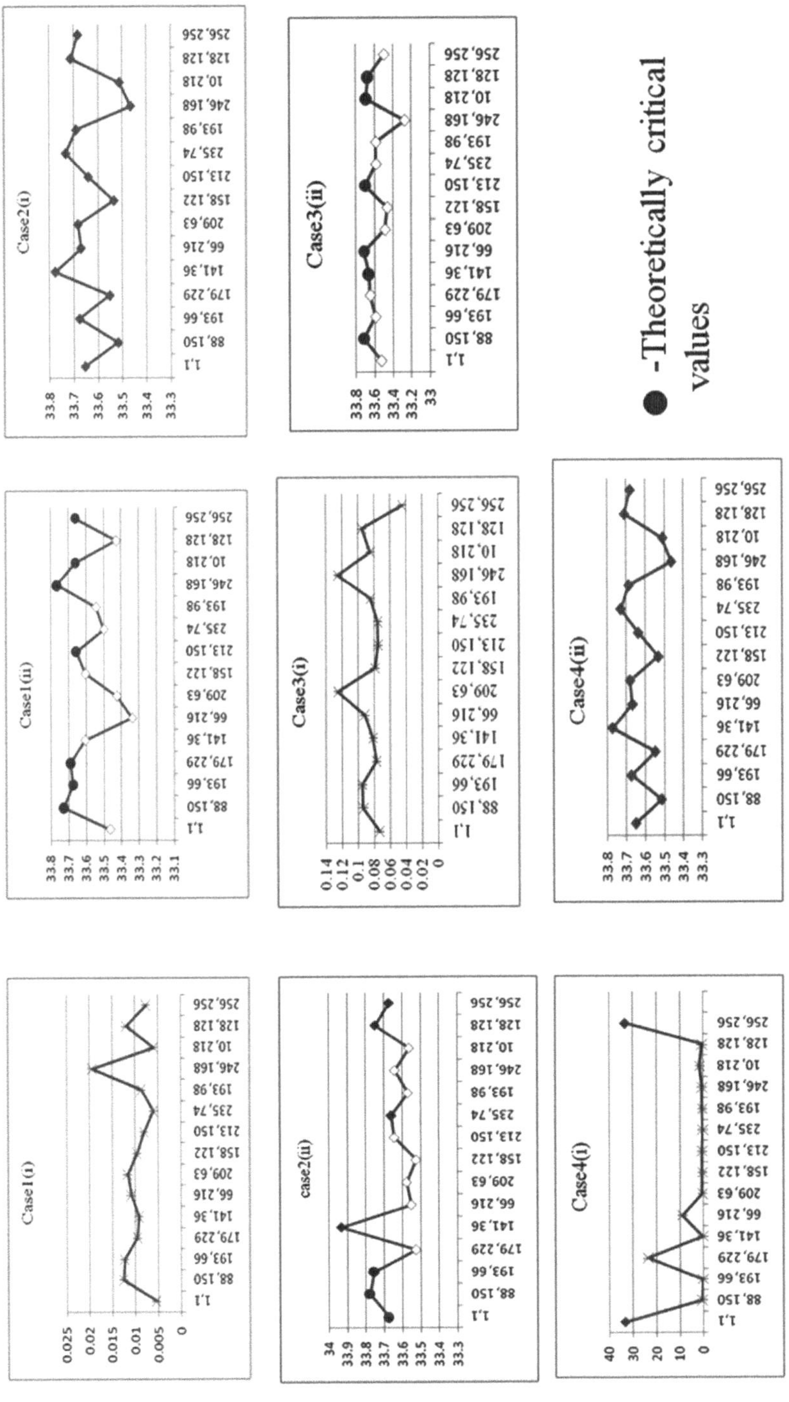

FIGURE 10.8 UACI analysis of different test cases.

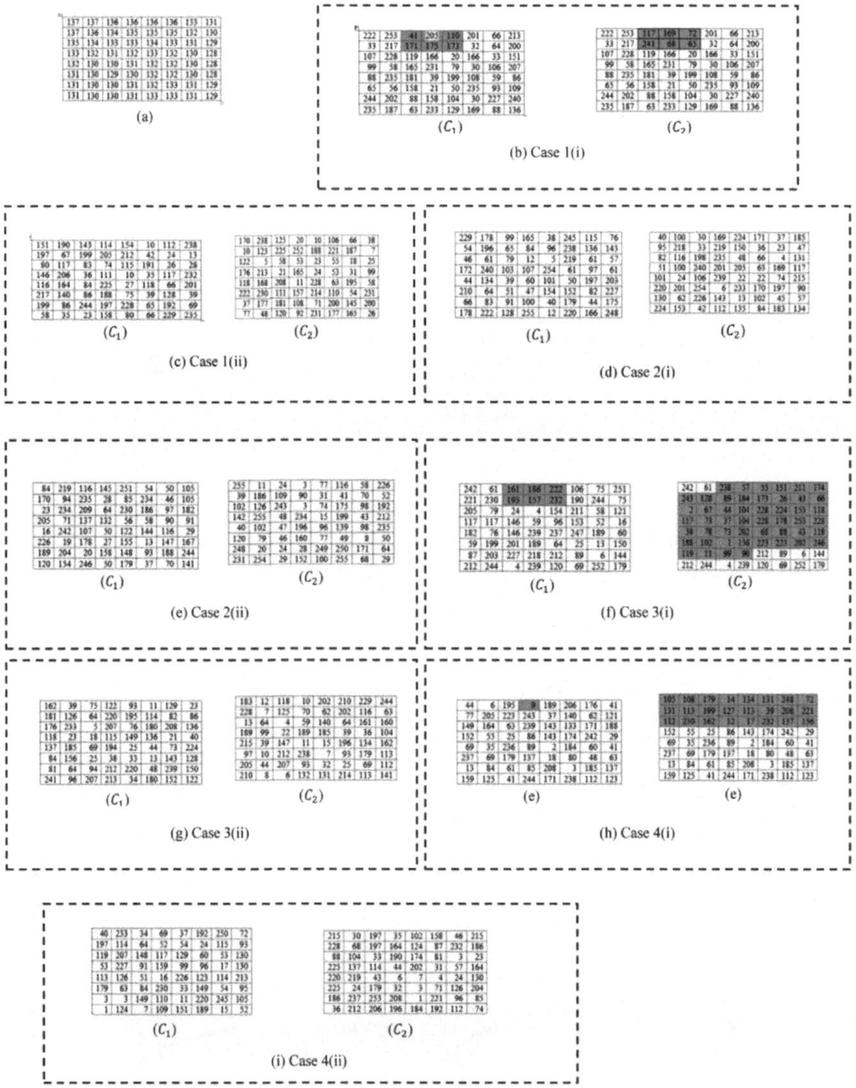

FIGURE 10.9 Numerical results for all test cases.

For the above all cases, the average of the difference matrix are represented in Figure 10.10(a) and (b). It shows clearly that case 4 (ii) has high average compared to the other cases which means that the intensity change rate is good compared to other cases.

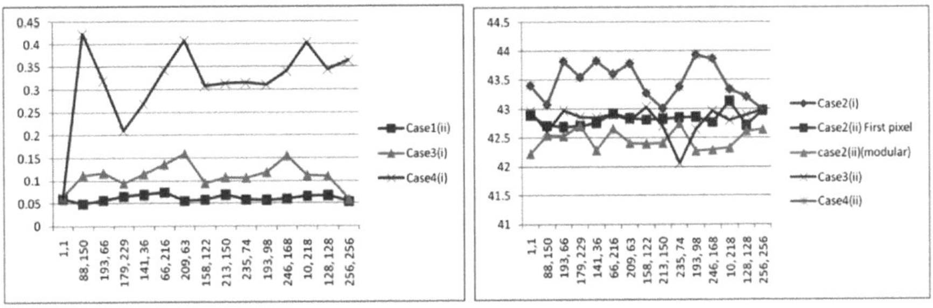

(a) (b)

FIGURE 10.10 Average for difference of two cipher matrix.

10.7 RESULTS AND DISCUSSIONS

The security of the proposed algorithm needs to be analyzed by the following parameters–(i) key space analysis, (ii) key sensitivity analysis, (iii) correlation among adjacent pixels, and (iv) distribution of gray values in the cipher image.

10.7.1 *KEY SPACE ANALYSIS*

The total number of keys used in the encryption algorithm is essential to overcome the Brute force attack. If the number of keys is high, it takes a huge amount of time to break the cryptosystem.

In the proposed work, the secret key and parameter of the logistic map is set to the IEEE standard precision of 10^{15} and the overall key space is 10^{30} which is equivalent to 2^{98}. From the analysis, the key space is high for resistance of Brute force attacks.

10.7.2 *KEY SENSITIVITY ANALYSIS*

The effect of a minor adjustment in the encryption and decryption keys is examined in this section. The following three examples are examined in this regard. (i) The encryption/decryption key is either changed from x_0 into $x_0 + 1$ or $x_0 - 1$. (ii) The key's magnitude is modified from 10^{15} to 10^7. (iii) The value of the parameter μ is altered from $\mu + 1$ or $\mu - 1$. The simulation is done and displayed in Figure 10.11 for these three situations. The proposed image encryption is sensitive to the secret keys, and it will withstand the differential attack, according to this discussion.

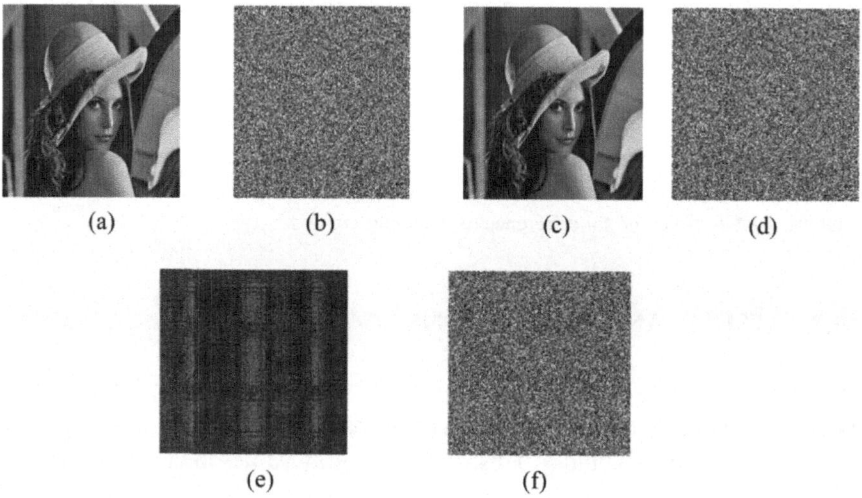

FIGURE 10.11 Key sensitivity analysis.

10.7.3 CORRELATION AMONG ADJACENT PIXELS

The statistical assault will be induced in the plain image because the adjacent pixels correlation is quite high. So it's necessary to disguise the image's statistical behavior, which means that if the correlation of neighboring pixels is close to zero, the intruder won't be able to learn anything about the image. For each image, a simulation is run to examine the correlation of neighboring vertical and horizontal pixels. For each image, w = 6000 pairs of random samples are collected from the horizontal, vertical, and diagonal positions, and the correlation between adjacent pixels is computed using the formulas below:

$$R_{ij} = \frac{|cov(i,j)|}{\sqrt{D(i)} \times \sqrt{D(j)}} \tag{13}$$

in which:

$$cov(i,j) = \frac{1}{w} \sum_{t=1}^{w} (i_t - E(i))(j_t - E(j)) \tag{14}$$

$$E(i)=\frac{1}{w}\sum_{t=1}^{w} i_t, D(i)=\frac{1}{w}\sum_{t=1}^{w}\left(i_t-E(i)\right)^2 \quad (15)$$

where; i_t and j_t are the chosen t^{th} pair of adjacent pixel gray values, respectively. $E(i)$ and $D(i)$ are the expectation and variance of the pixels. The correlation between the adjacent pixels of the plain and encrypted image is given in Table 10.3. Figure 10.12 clearly shows the uniform distribution among the pixels in vertical and horizontal directions.

TABLE 10.3 Correlation Co-Efficient among Adjacent Pixels for the Cipher Image

SL. No.	Sample Images	Correlation Co-Efficient		
		Horizontal	Vertical	Diagonal
1.	Lena	–0.02062	0.003685	0.024912
2.	Baboon	–0.00589	–0.01884	–0.00967
3.	Cameraman	–0.02874	–0.0445	0.025938
4.	Boat	–0.02261	–0.01625	0.02075
5.	peppers	–0.00434	–0.00446	0.019522

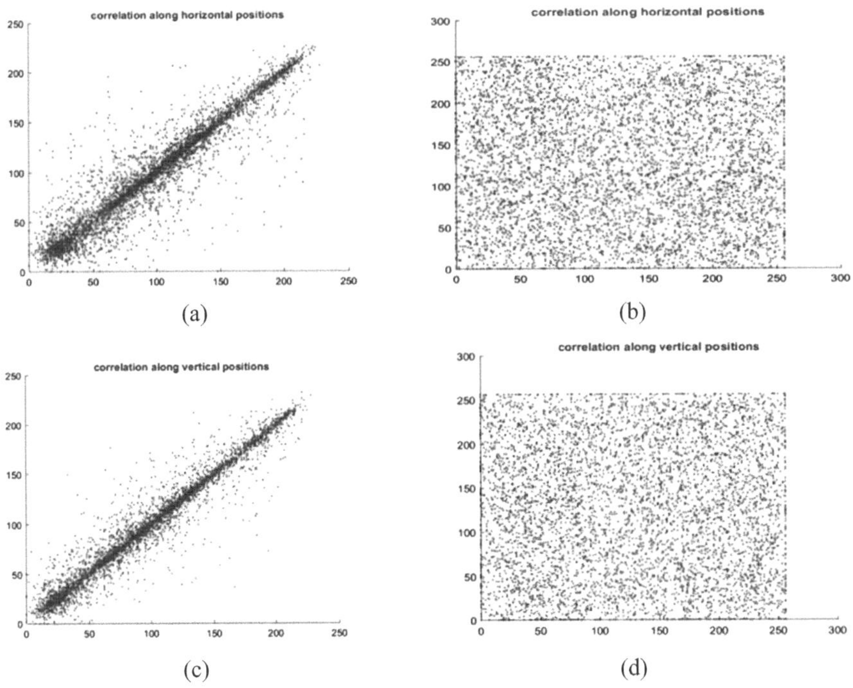

FIGURE 10.12 Correlation between adjacent pixels in horizontal and vertical directions.

10.7.4 HISTOGRAM ANALYSIS

The histograms indicate the statistical properties of the image. If the gray values are not evenly distributed, then it makes easier for attackers to extract information from the histogram analysis. The gray values in the image are evenly distributed if the encryption algorithm is good, which indicates that all pixel values have the same probability. Figure 10.13 shows the histograms of simple and enciphered images, which reveals that obtaining an image via histogram analysis is challenging.

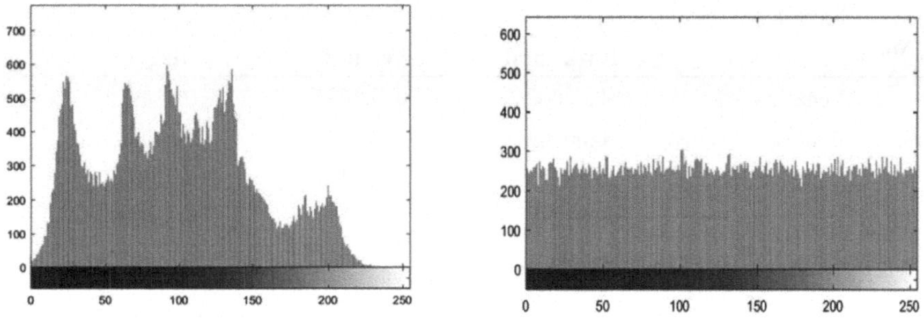

(a) (b)

FIGURE 10.13 Histogram of plain image and cipher image.

10.7.5 ANALYSIS OF KEY GENERATION USING NUMERICAL RESULTS WITH DIFFERENT TEST CASES

In the proposed key generation, the key matrix generated is analyzed in the following test cases. Consider for Case 1: The first 100 pixel levels are taken. Case 2: Last 100 pixel levels are taken. Case 3: Middle 100 pixel levels are taken Case 4: Randomly chosen 100 pixel levels are taken. For all these cases, find the probabilities of all pixel levels and then analyze all pixel levels for uniform distribution. The pixel value ranges from 0 to 255.

➢ **Case 1:** Pixel levels are considered as $p_0, p_1, p_2, \ldots, p_{99}$,
The sum of the 100 pixel levels is represented as:

$$sum_{P_{1-99}} = \sum_{0}^{99} p_0, p_1, p_2, \ldots, p_{99}$$

$= 0.0021+0.0041+0.0040+\ldots\ldots +0.00412 + 0.003311 + 0.004028$
$= 0.3909$

- **Case 2:** Pixel levels are considered as $p_{256}, p_{255}, p_{254}, \ldots, p_{156}$
  The sum of the 100 pixel levels is represented as:

$$sum_{p_{256-156}} = \sum_{256}^{156} p_{256}, p_{255}, p_{254}, \ldots, p_{156}$$

  = 0.003479+0.003845 + 0.004089 + …… + 0.004013 + 0.003387 + 0.003616 = 0.390052
- **Case 3:** Pixel levels are considered as $p_{78}, p_{79}, p_{80}, \ldots, p_{178}$
  The sum of the 100 pixel levels is represented as:

$$sum_{p_{78-178}} = \sum_{78}^{178} p_{78}, p_{79}, p_{80}, \ldots, p_{178}$$

  = 0.004028 + 0.004135 + 0.00415 + ……. + 0.004044 + 0.003937 + 0.003708 = 0.390106
- **Case 4:** Pixel levels are considered as random.
  The sum of the 100 randomly chosen pixel levels is represented as:

$$sum_{R_{1-100}} = \sum_{1}^{100} r_1, r_2, r_3, \ldots, r_{100}$$

  = 0.003784 + 0.003799 + 0.003693 + ……. + 0.003754 + 0.003983 + 0.003769 = 0.390305

Figure 10.14 clearly shows that the proposed work in the generation of key has uniform distribution and it indicates that the cryptosystem achieves a perfect secrecy.

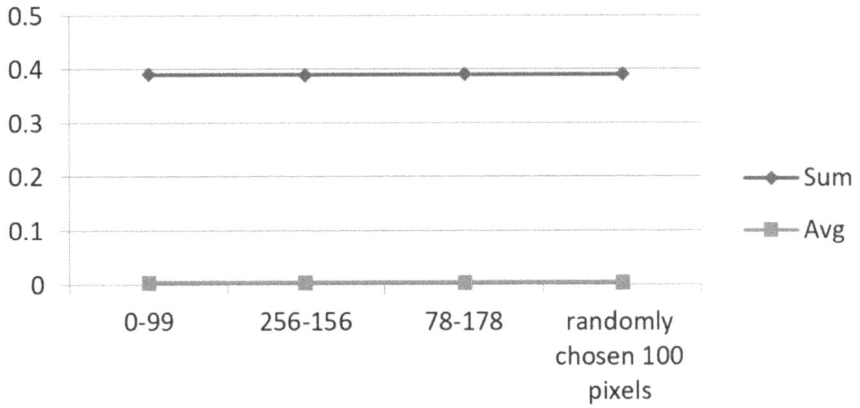

FIGURE 10.14 Probability of key matrix distribution.

10.8 CONCLUSION

An efficient dynamic key is generated from the proposed chaotic feedback convolution method using one dimensional chaotic map. OTKs are proposed to overcome the differential attacks with given α level significance compared to the exiting papers discussed. The proposed chaotic feedback convolution is proved with different cases using simulations. A good NPCR and UACI are obtained from the proposed key generation with low correlation, uniform distributions in the generated keys and cipher for achieving better randomness and high sensitivity keys. The simulation results are investigated numerically, and it is shown that the proposed approach is effective at creating dynamic keys. The proposed encryption system's absolute secrecy is theoretically proven.

KEYWORDS

- **chaos**
- **confusion**
- **differential attacks**
- **diffusion**
- **image encryption**
- **key generation**
- **key sensitivity analysis**
- **one time encryption**
- **permutation phase**

REFERENCES

1. Stallings, W., (2017). *Cryptography and Network Security: Principles and Practice.* Pearson Upper Saddle River.
2. Kocarev, L., (2001). Chaos-based cryptography: A brief overview. *IEEE Circuits and Systems Magazine, 1*(3), 6–21.
3. Aihara, K., (2012). Chaos and its applications. *Procedia IUTAM, 5*, 199–203. ISSN 2210-9838. doi: https://doi.org/10.1016/j.piutam.2012.06.027. IUTAM Symposium on 50 Years of Chaos: Applied and Theoretical.

4. Mathews, R., (1989). On the derivation of a "chaotic" encryption algorithm. *Cryptologia, 13*(1), 29–42.
5. Shannon, C. E., (1949). Communication theory of secrecy systems. *Bell System Technical Journal, 28*(4), 656–715.
6. Fridrich, J., (1998). Symmetric ciphers based on two-dimensional chaotic maps. *International Journal of Bifurcation and Chaos, 8*(06), 1259–1284.
7. Wu, Y., Noonan, J. P., & Agaian, S., (2011). NPCR and UACI randomness tests for image encryption. Cyber journals: Multidisciplinary journals in science and technology. *Journal of Selected Areas in Telecommunications (JSAT), 1*(2), 31–38.
8. Enayatifar, R., Abdullah, A. H., Isnin, I. F., Altameem, A., & Lee, M., (2017). Image encryption using a synchronous permutation-diffusion technique. *Optics and Lasers in Engineering, 90*, 146–154.
9. Teng, L., & Wang, X., (2012). A bit-level image encryption algorithm based on spatiotemporal chaotic system and self-adaptive. *Optics Communications, 285*(20), 4048–4054.
10. Ye, G., & Huang, X., (2016). An image encryption algorithm based on autoblocking and electrocardiography. *IEEE Multimedia, 23*(2), 64–71.
11. Liu, Y., Tong, X., & Ma, J., (2016b). Image encryption algorithm based on hyper-chaotic system and dynamic S-box. *Multimedia Tools and Applications, 75*(13), 7739–7759.
12. Wang, B., Zheng, X., Zhou, S., Zhou, C., Wei, X., Zhang, Q., & Che, C., (2014). Encrypting the compressed image by chaotic map and arithmetic coding. *Optik-International Journal for Light and Electron Optics, 125*(20), 6117–6122.
13. Wang, X. Y., Chen, F., & Wang, T., (2010). A new compound mode of confusion and diffusion for block encryption of image based on chaos. *Communications in Nonlinear Science and Numerical Simulation, 15*(9), 2479–2485.
14. Xiao, D., & Shih, F. Y., (2010). Using the self-synchronizing method to improve security of the multi chaotic systems-based image encryption. *Optics Communications, 283*(15), 3030–3036.
15. Murillo-Escobar, M., Cruz-Hernández, C., Abundiz-Pérez, F., López-Gutiérrez, R. M., & Del, C. O. A., (2015). A RGB image encryption algorithm based on total plain image characteristics and chaos. *Signal Processing, 109*, 119–131.
16. Wang, X., Liu, L., & Zhang, Y., (2015). A novel chaotic block image encryption algorithm based on dynamic random growth technique. *Optics and Lasers in Engineering, 66*, 10–18.
17. Boriga, R., Dăscălescu, A. C., & Priescu, I., (2014). A new hyperchaotic map and its application in an image encryption scheme. *Signal Processing: Image Communication, 29*(8), 887–901.
18. Zhang, G., & Liu, Q., (2011). A novel image encryption method based on total shuffling scheme. *Optics Communications, 284*(12), 2775–2780. doi: 10.1016/j.optcom.2011.02.039.
19. Zhou, Y., Bao, L., & Chen, C. P., (2013). Image encryption using a new parametric switching chaotic system. *Signal Processing, 93*(11), 3039–3052.
20. Chen, J. X., Zhu, Z. L., Fu, C., Yu, H., & Zhang, L. B., (2015a). An efficient image encryption scheme using gray code-based permutation approach. *Optics and Lasers in Engineering, 67*, 191–204.
21. Pareek, N. K., Patidar, V., & Sud, K. K., (2013). Diffusion–substitution based gray image encryption scheme. *Digital Signal Processing, 23*(3), 894–901.

22. Menezes, A. J., Katz, J., Van, O. P. C., & Vanstone, S. A., (1996). Handbook of applied cryptography. CRC press.
23. Stinson, D. R., (1995). Classical cryptography. In: Kenneth, H., (ed.), *Cryptography, Theory, and Practice* (2nd edn., pp. 1–20). Chapman & Hall/CRC.
24. Alvarez, G., J. M. Amigó, Arroyo, D., & Li, S., (2011). Lessons learnt from the cryptanalysis of chaos-based ciphers. In: *Chaos-Based Cryptography* (pp. 257–295). Springer.
25. Alvarez, G., & Li, S., (2006). Some basic cryptographic requirements for chaos-based cryptosystems. *International Journal of Bifurcation and Chaos, 16*(08), 2129–2151.
26. Solak, E., Rhouma, R., & Belghith, S., (2010). Cryptanalysis of a multi-chaotic systems-based image cryptosystem. *Optics Communications, 283*(2), 232–236.
27. Jakimoski, G., & Kocarev, L., (2001). Chaos and cryptography: Block encryption ciphers based on chaotic maps. *IEEE Transactions on Circuits and Systems I: Fundamental Theory and Applications, 48*(2), 163–169.
28. Preishuber, M., Hutter, T., Katzenbeisser, S., & Uhl, A., (2018). Depreciating motivation and empirical security analysis of chaos-based image and video encryption. *IEEE Transactions on Information Forensics and Security, 13*(9), 2137–2150.
29. Arnol'd, V. I., & Avez, A., (1967). *Problèmes Ergodiques de la mÉcanique Classique, 159*(3821), 1344.
30. Liu, H., Wan, H., Chi, K. T., & Lu, J., (2016a). An encryption scheme based on synchronization of two-layered complex dynamical networks. *IEEE Transactions on Circuits and Systems I: Regular Papers, 63*(11), 2010–2021.
31. Lian, S., Sun, J., & Wang, Z., (2005a). A block cipher based on a suitable use of the chaotic standard map. *Chaos, Solitons & Fractals, 26*(1), 117–129.
32. Ye, R., (2011). A novel chaos-based image encryption scheme with an efficient permutation-diffusion mechanism. *Optics Communications, 284*(22), 5290–5298.
33. Belkhouche, F., & Qidwai, U., (2003). Binary image encoding using 1D chaotic maps. In: *IEEE Region 5, 2003 Annual Technical Conference* (pp. 39–43). IEEE. doi: 10.1109/REG5.2003.1199708.
34. Dong, C., (2014). Color image encryption using one-time keys and coupled chaotic systems. *Signal Processing: Image Communication, 29*(5), 628–640.
35. Yen, J. C., & Guo, J. I., (2000). Efficient hierarchical chaotic image encryption algorithm and its VLSI realization. *IEE Proceedings-Vision, Image, and Signal Processing, 147*(2), 167–175.
36. Li, C., (2016). Cracking a hierarchical chaotic image encryption algorithm based on permutation. *Signal Processing, 118*, 203–210.
37. Kadir, A., Aili, M., & Sattar, M., (2017). Color image encryption scheme using coupled hyper chaotic system with multiple impulse injections. *Optik, 129*, 231–238.
38. Brindha, M., & Gounden, N. A., (2016). A chaos-based image encryption and lossless compression algorithm using hash table and Chinese remainder theorem. *Applied Soft Computing, 40*, 379–390. doi: 10.1016/j.asoc.2015.09.055.
39. Murugan, B., & Gounder, A. G. N., (2016). Image encryption scheme based on block-based confusion and multiple levels of diffusion. *IET Computer Vision, 10*(6), 593–602. doi: 10.1049/iet-cvi.2015.0344.
40. Patro, K. A. K., & Acharya, B., (2019). An efficient color image encryption scheme based on 1-D chaotic maps. *Journal of Information Security and Applications, 46*, 23–41.
41. Pak, C., & Huang, L., (2017). A new color image encryption using combination of the 1D chaotic map. *Signal Processing, 138*, 129–137.

42. Hua, Z., Zhou, Y., & Huang, H., (2019). Cosine-transform-based chaotic system for image encryption. *Information Sciences, 480*, 403–419.
43. Kandar, S., Chaudhuri, D., Bhattacharjee, A., & Dhara, B. C., (2019). Image encryption using sequence generated by cyclic group. *Journal of Information Security and Applications, 44*, 117–129.
44. Luo, Y., Lin, J., Liu, J., Wei, D., Cao, L., Zhou, R., Cao, Y., & Ding, X., (2019a). A robust image encryption algorithm based on Chua's circuit and compressive sensing. *Signal Processing, 161*, 227–247.
45. Parvin, Z., Seyedarabi, H., & Shamsi, M., (2016). A new secure and sensitive image encryption scheme based on new substitution with chaotic function. *Multimedia Tools and Applications, 75*(17), 10631–10648.
46. Norouzi, B., & Mirzakuchaki, S., (2016). Breaking an image encryption algorithm based on the new substitution stage with chaotic functions. *Optik, 127*(14), 5695–5701.
47. Ping, P., Xu, F., Mao, Y., & Wang, Z., (2018). Designing permutation–substitution image encryption networks with Henon map. *Neurocomputing, 283*, 53–63.
48. Zhu, H., Zhao, C., & Zhang, X., (2013). A novel image encryption–compression scheme using hyperchaos and Chinese remainder theorem. *Signal Processing: Image Communication, 28*(6), 670–680. doi: 10.1016/j.image.2013.02.004.
49. Li, C., Liu, Y., Zhang, L. Y., & Wong, K. W., (2014). Cryptanalyzing a class of image encryption schemes based on Chinese remainder theorem. *Signal Processing: Image Communication, 29*(8), 914–920. 173.
50. Li, X. Y., Zhou, X. B., Zhou, Q. L., Han, S. J., & Liu, Z. (2020). High-Capacity Reversible Data Hiding in Encrypted Images by Information Preprocessing. *Complexity,* 2020(1), 6989452.

INDEX

A

Accuracy, 5–8, 93–96, 100–103, 108–110, 124, 143, 193, 234, 240, 241, 248, 249, 260, 269, 274–281
  color space, 7
  model, 8
Acquired templates, 119
Actuators, 179, 183, 200
AdaBoost algorithm, 135
Adaptable, 35, 143, 145, 179
  houses, 183
Adjacent
  neighbor pixel, 30
  pixels, 223, 293, 295, 296, 299, 303, 319–321
Administered learning strategies, 261
Advanced encryption standard (AES), 17, 20–23, 159–161, 173, 203, 290, 291
Adversary, 115, 118, 119, 146, 296
Advertisements, 88, 89, 256
Alignment-free technique, 135, 141
Alteration, 5, 8, 10, 12, 14, 15, 23, 25, 26, 28, 32, 36, 40, 43–45, 49, 121, 257, 261, 272
  audio, 43
  image, 4, 6, 41, 42
  media, 4–6, 11, 36, 41
Ambiguity, 291
  attack, 38
Amend decision making, 257
Amplitude, 34, 43, 66
  modulation, 34
Analytical hierarchy process (AHP), 95
Android, 47, 93–95, 100–103, 106, 188, 189, 199, 200
  application, 94, 101, 102
    source code, 100
  malware, 94, 100, 102, 103
  virtualized environments, 106
Angiography, 234
Angle correlation attack, 135
Anticipate
  model, 273, 275, 280
  threats, 188
Anti-tampering features, 47
Anti-virus scanning, 92
Appraisal networks, 277, 281
Approximation, 32, 161
Arbitrary information, 273
Arduino, 188, 189, 199
  microcontroller, 189
Area under curve (AUC), 271, 273–281
Arithmetic coding, 32
Arnold
  cat map, 70, 207
  scrambling, 42, 155, 159, 160, 162–165, 167, 168, 173, 174
  transforms, 42
Arrays, 307, 308
Artificial
  characteristics, 234
  intelligence (AI), 87, 255
  minutiae, 135
  neural network (ANN), 92, 94, 95, 135, 264
    architecture, 92
Assaults, 186, 294–297, 300
Assembly code, 92, 93, 106
Assessment
  networks, 274, 276, 277, 280
  records, 194
Assessor, 271, 272
Associated key, 27
Assumption, 263–267
Astounding data, 258
Asymmetric, 195
  algorithm, 22, 290
  cryptography, 24
  cryptosystems, 234, 235
  key, 15, 16, 22, 290
    cryptography, 15, 16, 22–24
  matrix, 43

Attribution system, 94
Audio
scrambling, 43
signal, 28, 33, 34, 157, 211
stenography, 27, 28, 33
Authentication, 2, 14, 16, 22, 44, 45, 47, 50, 117, 133, 134, 136–140, 144–147, 156, 159, 160, 162, 193, 205, 206, 211–213, 290
Automated, 179, 183, 188, 189, 200
detection, 91, 108, 109, 111
component, 109
fingerprint identification systems (AFISs), 143
system, 188
Auto-sklearn library, 100
Auxiliary value, 22
Availability, 206

B

Bandwidth, 195, 197, 231
Barcodes, 45
Bashlite, 93
Bayes hypothesis, 267
Behavioral
biometric traits, 116, 117
characteristics, 117, 146
semantics, 100
Benign, 93, 94, 96, 100, 102, 109, 110
program, 93, 100
samples, 94, 96
Berne convention, 48
Bernoulli
Naive Bayes, 266, 280, 282
algorithm, 266
trial, 266
Big
commercial companies, 4
firms, 35, 47
Binary, 27, 28, 38, 87, 91–94, 99, 106, 107, 126, 129–132, 135, 136, 140, 141, 143, 144, 160, 162, 165, 210, 219, 264, 303, 304
biometric data, 132
code, 99, 140, 143
digits, 303
equivalent, 130
sequence, 28, 160, 165, 210, 219
string, 129–131, 141, 143, 144
template, 131, 136
vector, 129–132, 143, 219
Binomial trial, 266
BioCodes, 140
Bio-cryptosystem, 131, 135
BioHash, 140
methodology, 131
Biological neuron, 264
Biometric, 115–133, 135–142, 144–147, 187, 292
authentication, 117, 128, 147
systems, 115, 117
cryptosystem, 115, 119, 120, 125–128, 132, 133, 136, 144, 145, 147
data, 115–119, 121, 123–127, 129, 130, 132, 133, 135–138, 145, 146
encryptionTM, 126–128
recognition systems, 117
salting, 140
system, 115, 118, 119, 121, 123, 124, 129, 136, 137, 139, 140, 145, 146, 292
template, 132, 147
protection (BTP), 115–117, 119, 120, 122, 123, 125, 146
security, 115–118, 146, 147
technique, 115–117, 119, 120, 122, 123, 125, 146
traits, 117, 129, 136, 146
BioPhasor, 140
Bioscrypt, 127
Birthday attack, 124
Bit, 11, 12, 16–18, 20, 21, 25, 27, 29, 30, 33, 34, 40, 58, 129–132, 160, 161, 300
error rate (BER), 39, 40
information, 41
key, 18, 19, 21
level permutation, 300
plane, 30
Bitcoin, 24, 178
Blind detection, 37
Block, 17, 22, 24, 73, 165, 166, 217
chain (BC), 155, 178–182, 187, 188, 191, 192, 194, 195, 197, 200, 201
technology, 178, 181, 191, 194, 200
cipher, 16, 17, 213, 297
data, 182
diagram, 72, 126, 128, 129, 133, 134, 137, 139, 165, 166
original plain text bits, 17

Blowfish, 17, 20
algorithm, 20
Bluetooth, 199
Bogdanov map, 71
Bot, 89
malware, 108
Botnet, 89
Brand reputation damage, 88
Breached, 2, 185
Brightness, 7
Broadcast monitoring, 34
Brovey transform, 240, 243–245
Brute force attack, 17, 73, 83, 123, 146, 295
Bunch, 258
Burdens, 185
Burglar attempts, 188
Burrow, 257
Business intelligence, 88
Bypass, 89
Byte, 21, 93

C

Cable television signal transmission, 41
Call sequences, 100, 101
Cancelable biometrics, 115, 119, 120, 138, 139, 145, 147
non-invertible transformation techniques, 141
salting techniques, 139
Cartosat-1 satellite, 234
Cat map, 207, 220–222, 294
Catalog reaping assaults, 256
Cerber, 95
Chaff points, 133
Channel models, 9
Chaotic
cryptography, 206
encryption
algorithms, 299
architecture, 300
image encryption
challenges, 299
chaotic ciphers, 299
non-plain image related (N-PIR), 300, 301
one-time key (OTK), 302
permutation only ciphers, 300
plain image related (PIR), 301
maps, 57, 69, 72, 73, 204, 207, 222, 291, 294, 298, 299
sequences, 219, 294
system, 72, 83, 132, 206, 218, 219, 294, 301, 302
Chatroom, 108, 109
Cheques, 45
Chinese remainder theorem, 213, 302
Chunk loop, 24
Cipher, 15, 17–20, 59, 61–68, 74–79, 81–83, 222, 224, 291, 295–300, 308, 309, 312, 313, 319, 321, 322, 324
images, 68, 75–77, 83, 222, 224, 309, 312, 313
text, 14–19, 21, 23, 291, 296, 297, 300, 308
attack, 296, 297
Circuit systems, 187
Classification, 36, 91–96, 99, 101, 102, 108, 109, 130, 229, 233, 264–268, 290
malware, 93, 99
with PE headers (ClaMP), 93
Classifier
accuracy, 270
fusion techniques, 232
Cloud
computing, 178, 182, 183, 189, 190, 200, 201
distribution model, 191
network, 182
researchers, 182
storage, 179, 182, 185, 190, 192
Clustering algorithm, 134, 143
Coarse-filtering, 134
Code
breaking, 15
mnemonics, 106
word, 128–134
Coefficient, 32, 39, 58, 75, 77–79, 83, 162
Cognizance base, 273
Collusion attack, 38
Color
image, 9, 10, 215, 229
encryption, 294
information, 7
model, 8
space, 4, 5, 7, 8, 37, 41
tone, 10
Colossal accumulation, 272

Column processing, 32
Combination, 15, 50, 90, 100, 102, 116, 119, 123, 145, 178, 213, 214, 302
  detection, 95
  security techniques, 50
Command
  control (C-C), 89, 99, 108
  line instance, 90
Commitment, 128–132
Communicated, 13, 16, 23, 25, 26, 34, 156, 157, 180, 183, 185, 192, 198, 199, 204, 209–211, 292
  models, 26
  nodes, 195
Company logo, 4
Competency, 271
Compilation time, 92
Compilers, 107
Complex, 28, 38, 87, 88, 96, 127, 157, 296, 301
Composite
  chaotic systems, 15
  media, 9
  signal, 9, 11
Compression, 10, 24, 38, 158, 159, 162, 170, 172, 212, 294, 301, 302
Computational
  burden, 297
  complexity, 12, 36, 49, 291, 294
  cost, 38, 94, 240
  power, 296
  program, 88, 257
  puissance, 256
  science, 57, 87, 115, 155, 177, 203, 229, 255, 257, 289
  system, 16, 178, 182
  time, 296
  tomography, 234
Concatenated codeword, 130
Conceptual data, 275
Condensation, 130
Confidential, 15, 22, 205, 210, 290
  data, 2, 21, 88
  information, 2, 117
Conjunction, 146, 179, 234, 240
Constant
  advancement, 35
  development cycle, 50
Constraints, 12, 268, 290
Contagio Community, 102
Content management, 34
Control
  elements, 107
  flow graph (CFG), 94
Conventional
  authentication systems, 115
  biometric systems, 125
Convolution
  function, 302
  neural network (CNN), 93–95, 101, 233
    long short-term memory (CNN-LSTM), 101
  theorem, 304
Copyright, 4, 13, 35, 45, 46, 48–50, 155, 156, 162, 174
  Act, 35, 48
  laws, 35
Corner pixels, 310
Corpus, 182
Correlation, 60, 61, 75, 79, 92, 223, 296, 321
  analysis, 223
  coefficient graph analysis, 77
  equation, 42
Corruption, 31
Cost saving, 190
Cover
  image, 29–32, 39, 90, 158, 162, 164–166
  media, 26, 28
Cross-matching attack, 135
Cryptanalysis, 15, 17, 289, 297, 298
Cryptographic, 13–16, 18, 22–25, 28, 34, 35, 42, 44, 49, 50, 127, 128, 181, 204, 207–211, 213, 218, 290, 292, 295, 299
  algorithm, 35, 218
  attack, 38
  digital key, 127
  hash, 92, 129
  key, 115, 124, 127, 133, 138, 145, 146
  mechanizm, 290
  puzzle, 88, 178
  tasks, 218
  techniques, 15, 24, 25, 37, 42, 212, 213, 290
Cryptomining, 88
Cryptosystem, 16, 115, 116, 119, 124, 125, 131–133, 136, 145, 203, 204, 210–212, 221, 222, 224, 291, 293, 294, 296, 297, 299–302, 319, 323

Cuckoo sandbox, 100, 106, 110
Custom payloads, 90
Cyber
attackers, 108
security, 1, 3, 155, 187
space, 4, 50
world, 88, 108
Cyclic redundancy check (CRC), 1, 57, 87, 115, 134, 155, 177, 203, 229, 255, 289
Cyclopean educational accumulations, 259

D

Data, 1, 2, 8, 9, 11, 12, 15, 32, 36, 42, 49, 50, 177, 237, 256–259, 295
acquisition, 259
alteration, 49
change, 50
collision, 24
confidentiality, 43
embedding process, 31
encryption, 1, 2, 16, 17, 20, 50, 203, 205, 206
algorithm, 18, 20, 21
functions, 18, 19
model, 18
standard (DES), 17–21, 42, 203, 290, 291
exfiltration, 88
fusion, 233
hiding, 1, 2, 8, 9, 11, 12, 15, 26, 29, 31, 32, 36, 49, 50, 212, 213
system, 12
technique, 9, 11, 12, 15, 30, 49, 50
holding capacity, 9, 30
integrity, 23
leaks, 47
multiple transactions, 191
piracy methods, 3
scrambling process, 42
security, 22, 41, 47, 185
law, 47
set, 96, 100, 102, 103, 110, 261
structures, 92
tracking, 181
Debuggers, 106–108
monitor malicious code execution, 107
Decipher, 13, 15, 291
Decision tree (DT), 94, 100, 109, 110, 268, 277, 278
Decrypted, 14–18, 20–23, 35, 36, 45, 63, 68, 72, 73, 83, 160, 163, 210, 213, 204, 222, 290–292, 294, 296, 297, 308, 309, 319
data, 45
Deemed robust, 120
Deep
belief network (DBN), 102
learning, 87, 94, 99, 101–103, 106, 110
model, 94, 101, 102
techniques, 94
networks, 95
neural network (DNN), 102
Delaunay triangulation template protection approach, 144
Delta
function, 131
value, 30
Denial of service (DoS), 89, 185, 256
Densely infinite-to-one mapping (DITOM), 142
Design consideration, 11, 12, 37, 49
data hiding, 11
capacity, 12
computational complexity, 12
robustness, 11
statistical correctness, 12
transparency, 11
Detected, 4, 11, 37, 40, 49, 87, 88, 93–96, 99–103, 108–110, 160, 194, 200, 203, 229, 235, 272, 273, 303
accuracy, 95
malware, 102
model, 194, 273
rate, 94, 102, 110
statistics, 38
threats, 3
Diagonal, 59, 69, 77–79, 209, 223, 320
Dictionary
attack, 124, 146
unique opcodes, 95
Differential attack analysis, 225
Diffie-Hellman algorithm, 192
Diffusion, 57, 58, 72, 84, 204, 207, 208, 212, 214, 219–222, 290, 292–301, 306, 324
Digital
authentic, 14
forensics, 14, 49

content, 1, 3, 4, 25, 26, 35, 44, 46–49
cryptography technique, 45
image, 4, 37, 39, 156–158, 203, 206, 211, 213–216, 296
media, 2, 27, 28, 34–36, 39, 44, 46–49, 87
protection, 44
security, 2, 49
numbers (DN), 235, 240, 241
products, 35
rights management (DRM), 2, 3, 46–48, 50, 292
scrambling, 2, 14, 40, 41, 49, 50
signature, 22–24, 45, 206
space, 2, 49
stenography, 25, 27
surveillance, 2, 45
technologies, 88
watermark, 2, 4, 13–15, 34–37, 44, 49, 50, 156, 157
robust fragile watermarking, 36
spatial frequency domain, 36
systems, 4
visible-invisible watermarking, 35
watermarking attacks, 38
watermarking classification, 35
watermarking evaluation, 39
watermarking requirements, 37
Direct
stream, 28
visualization, 298
Discrete
cosine transform (DCT), 31, 37, 157, 160–162, 173, 211
fourier transform (DFT), 140, 143, 144, 157, 162
wavelet transform (DWT), 31, 32, 37, 132, 140, 157, 160–164, 173, 235, 242
fractional coefficient, 32
transformation, 32
Distributed denial of service (DDoS), 89, 185
Double-direction XOR operation, 43
Droid detector, 102
Dual
energy X-ray absorptiometry, 234
watermark, 36
Duffing map, 71
Duplication, 44
Dynamic
analysis, 87, 96, 101, 111
behavioral features, 102
class weighting, 102
hand signature, 140
keys, 308
malware, 88, 89, 96, 99–103, 106
analysis, 88, 89, 96, 99, 101, 103, 106, 110
detection, 100, 102
minority class weighting scheme, 102
random growth technique, 294

E

Earth observation (EO), 230, 232
Echo, 25, 34
Ecosystem, 2, 3, 181
Edge malware detection, 95
Effective
performance, 140
template protection technique, 124
Efficient, 8, 12, 17, 20, 21, 27, 29, 32, 34, 37, 42, 157, 179, 182, 200, 213, 223, 226, 232, 289, 298, 299, 324
encryption rate, 20
Ehlers fusion, 237, 242
Electrical
circuits, 188
equipment, 188
Electrocardiogram (ECG), 136
data, 136
Electromagnetic radiation, 231
Electronic voltage regulator, 189
Elevation, 230, 235
Emails, 13, 106
Embedded, 8, 9, 11–13, 25–28, 30–38, 40, 89, 90, 107, 108, 143, 156–163, 165, 167–169, 173, 188, 235
capacity, 31
data, 9, 11, 13, 25, 34
digital
signal, 35
watermarks, 36
information, 11, 12, 36
process, 12, 31, 159, 165
secret information, 27
watermarks, 36, 37
Embryonic information, 260

Empirical
method, 183
proof, 130
Enciphered data, 291
Encryption evaluation metrics, 58
Chi Square value, 63
correlation coefficient, 58
graph, 59
deviation (uniform histogram), 67
global information entropy, 64
gray difference degree (GDEG), 65
histogram
graph, 59
variance, 63
irregular deviation, 66
key
sensitivity, 62
space, 58
local shannon entropy, 65
maximum deviation, 66
mean absolute error (MAE), 64
NPCR-UACI, 67
peak signal to noise ratio (PSNR)-mean square error (MSE) value, 63
structural similarity index, 68
universal image quality (UIQ), 68
Energy, 92, 183
Enforcement, 16, 35, 206
Engender cognizance datasets, 272
Enhanced
performance, 103, 110
smart home development, 177
Enrollment, 117, 127–129, 133, 134, 143
Entropy, 77, 78, 92, 225
analysis, 76, 224, 225
value, 64, 77, 83, 224
Environmental control systems, 179
Equal error rate (EER), 124, 125
Equipollent news, 257
Equivalent, 34, 130, 272, 304, 319
Ergodicity, 299
Error
correction
codeword, 129–133
purpose, 130, 132
tolerance
primitive, 137
purpose, 136
Ethereum, 181
Ethernet, 108
Euclidean
distance, 39, 40
space, 39
Evaluation, 2, 4–6, 39, 57, 58, 72, 73, 83, 84, 95, 168, 225, 229, 234, 244, 248, 249, 275, 276
Evolving hybrid neuro-fuzzy classifier (EHNFC), 103
Executable, 92, 93, 102, 107, 110, 281
commands, 99
files, 93, 101
link format (ELF), 93
programs, 93
Exported functions, 92, 107
Exquisite spectral, 232
External
libraries, 107
program, 107
Extra tree, 268, 277, 279
Extracted, 9, 24–27, 30, 34, 37, 40, 45, 46, 93–96, 100, 102, 107, 117, 130, 131, 134, 135, 140, 144, 157, 158, 160, 162, 163, 165, 166, 168–170, 172, 212, 232, 234, 235, 263–265, 272, 307
algorithm, 27, 172
features, 102
information, 45
process, 9, 25, 26, 30, 37, 45, 46, 166, 272

F

Face
biometric, 136, 138, 140
template protection technique, 134, 140
Fail-safe security, 87
Fair Isaac Corporation (FICO), 269, 270
False
acceptance rate (FAR), 124, 125
cross match rate (FCMR), 122, 123
match rate (FMR), 124
non-match rate (FNMR), 124
positive
detection, 235
ratio (FPR), 100, 197
test, 174
rejection rate (FRR), 124, 125
Fast Hadamard transform (FHT), 160, 162

Feature extraction, 45, 46, 94, 140, 272
Feistel
cipher, 18, 20
structure, 18, 21
File
encryption, 21
fingerprinting, 92
format, 92
header, 94
size, 92
Final
hash, 24
matching score, 123
outcome, 5
permutation, 18–20
Finger
print, 44
biometric database, 130
image, 131, 132, 142, 143
impression, 121, 123, 135, 142
template, 131, 132, 140–145
vein biometric
data, 132
template, 132
First
in-first out order (FIFO), 193
letter algorithm, 25
Fisher discriminant analysis (FDA), 140
Flexible representation of quantum image (FRQI), 213–217
Fourier
spectrum, 131
transform, 241, 242
Fraction, 190, 304
values, 32
Frequency domain, 26, 29, 31, 35–37, 43, 157
algorithm, 29
scrambling, 43
watermarking, 37
Function call graph (FCG), 93, 94
Fuzzy
commitment, 126, 128–132, 145
extractor, 136–138
hashing, 95, 102, 103
import method, 95
logic (FL), 87, 99, 102, 103, 106, 110
oversampling, 102
pattern tree, 95
vault scheme, 126, 133–136, 145

G

Gaussian
data, 267
naive bayes, 267, 280–282
algorithm, 267
Genome project, 102
Genuine, 89, 121, 124, 129, 133–135, 138, 192, 270
acceptance rate (GAR), 124
scores, 121
weight values, 134
Geometric
attack, 38
distortion, 230
invariant features, 140
Geospatial applications, 230
Gerontology, 185
Gingerbreadman map, 71
Global
merchants, 186
standard deviation, 241
Government, 3, 4, 21, 26, 44, 45, 49
e-document, 4
officials, 3, 21, 26
papers, 45
Gradient, 38, 265
attack, 38
Graph
isomorphism algorithms, 93
theory, 32
Gray
difference degree (GDEG), 82
scale, 6, 10, 11, 29, 92, 93, 167, 191, 216
cover image, 29
image, 6, 9–11, 92, 93
Green
cloud computing, 182
computing equipment, 183

H

Hackers, 2, 4, 186
Hadamard, 130, 143, 204, 211, 216, 220
codeword, 130
Reed-Solomon codes, 130
Hash, 16, 23, 72, 91, 127, 129, 135, 137, 140, 142, 160, 193, 301, 302
functions, 16, 23, 24, 142, 301
value, 23, 24, 72, 91, 129, 142, 160

Henon map, 207, 208, 219, 222, 299
Hexagonal grid, 135
Hidden
behavior, 88
communication, 14, 15
data, 27–29, 33, 34, 38
information, 10, 27, 28, 34, 36
malicious instructions, 87
message, 25, 27, 28
secret information, 27
writing, 13
Hieroglyph, 15
High
bit-sized, 30
bright pixels, 10
capacity, 9
data, 9
information hiding scheme, 25
complexity behavior, 43
computation machine, 203, 211, 218
confidential, 2, 3, 15, 34, 49
embedding capacity, 32
fractional values, 32
frequency, 31, 32, 242
band, 32
coefficients, 32
stores, 32
grayscale, 10
independent, 6
information, 30
level
constructs, 106
content, 3
security, 41, 118, 281
malware detection rates, 103
pass filter (HPF), 32, 237, 241, 244, 246, 248
quality printers, 45
randomness, 208, 223
redundancy, 295, 296
resistance, 17
resolution multispectral (HRMX), 231
secrecy, 41
security, 3, 31, 34, 42
mediums, 45
spatial resolution panchromatic, 229
spectral, 230, 242
Hill-climbing attack, 124, 146
Histogram, 59, 62, 74, 76, 223, 224, 237, 238, 322
analysis, 58, 221, 223, 322
plots, 223, 224
variances, 224
Home
automation, 179, 183–185, 188, 189, 199–201
heating, 179
owning proprietors, 186
robotization, 185
Homogeneity, 92
Homomorphic encryption, 116, 119, 120, 144, 145, 147
Horizontal
direction, 7, 321
pixels, 320
resolution, 8
Host, 12, 88, 191
media, 12
name features, 101
Human
behavior protocols, 16
building communication, 183
eye, 7, 10
interpreter, 244
perception, 25, 35
system, 5
verification, 4
society, 256
verification, 11
visual
detection, 10
system, 9, 30, 37, 235
Hybrid, 25, 99, 102, 116, 119, 120, 144, 145, 147, 294
approaches, 116, 119
technique, 102, 145
transforms, 25
Hyperparameter optimization algorithms, 100
Hyperplane, 109, 266
Hyperspherical color
sharpening technique, 243
space (HCS), 237, 243–245, 248, 249

I

Illegal
creation, 48
distribution, 46
Image
authentication, 34
ciphers, 298
classification, 93

dimensional change, 8
encryption, 57, 58, 68, 84, 203, 207, 208, 211–213, 221, 222, 224, 225, 289, 290, 292–302, 304, 320, 324
approach, 294, 299
mechanizm, 292
strategy, 302
enhancement, 249
fusion, 231–234, 240, 249
histograms, 33
pixel histogram, 224
properties, 7
recognition problem, 92
rotation, 8
scrambling, 41–43
algorithms, 41
stenography, 27, 29, 33
Immunity debugger, 107
Imperceptibility, 36, 37, 158, 162, 164, 168–170, 172–174
Import
address tables, 95
hashing, 95
Imposter, 121, 124, 133, 138
scores, 121
Index method, 182
Indian remote sensing dataset, 229
Individual
data hiding techniques, 29
pixel, 310–313
prospects, 49
Indoor regulator, 185
Ineffective payload, 34
Informal phrasing, 187
Infrared face images, 131
Infrastructure, 186
Infringement, 45, 46, 48, 49
Inhabitants, 180, 201
Initial, 18, 19, 72, 219
data, 24
hash value, 24
permutation, 18, 19
seed, 222, 301, 303, 304
Input
biometric data, 137
features, 93
Instantaneous field of view (IFOV), 231
Integers, 32, 207
Intellectual
buildings, 179
property (IP), 18, 34, 88, 100, 156, 157, 199
address, 100
rights, 47
Intelligent
agreement, 181, 192
message, 15
Intended
person, 13, 14, 24, 25, 44
receiver, 27
Intensity, 6, 7, 67, 240, 241, 243, 293, 300, 307, 309–313, 318
Inter block level permutation, 300
Intercept, 13, 29, 210
communications, 88
Interceptor, 26
Inter-class, 130
Interface module, 199
Interference attack, 38
International laws, 47
Internet, 3, 22, 47, 48, 132, 155, 156, 158, 177, 178, 180, 181, 185, 187, 189, 190, 198, 199, 201, 204, 206, 290
internet of things (IoT), 93, 95, 132, 147, 155, 178–181, 185, 186, 189, 191, 194, 198–201, 211, 214, 289, 292
adoption, 185
systems, 132, 186
protocol security (IPsec), 290
technologies, 290
Intra
class vector, 130
subject variance, 135
Intrusion, 2, 3, 49, 190
Invertible
behavior, 139
cipher elements, 19
watermark, 38
Invisible, 35, 36, 38
ink, 25
watermark, 35, 36
Iris, 116, 117, 121, 123, 130, 132, 134, 136, 140, 145, 146
biometric data, 121, 130, 145
codes, 134, 140
Irregular deviation analysis, 79, 84

J

Joe Sandbox document analyzer, 106
JPEG data hiding, 31

K

Kali, 90
  Linux machine, 90
Kalman-Filter fusion, 232
Karhunen Loeve Transform (KLT), 160, 162
Kernel, 99, 109, 265, 302–305, 310, 311
  mode, 99
  trick, 265
Key
  binding scheme, 126
  distribution, 16, 23, 209, 210, 290
  entropy, 136
  exchange, 16, 22, 138
  generation, 23, 72, 136, 137, 204, 210, 211, 213, 219, 226, 294, 304, 307, 309, 313, 322, 324
    scheme, 126, 136, 137
  length, 18
  loggers, 89
  matrix, 219, 220, 306–309, 313, 322, 323
  release process, 125
  sensitivity analysis, 221, 222, 319, 324
  stability, 136
  stream, 17, 219
  transformation process, 19
Kill chain methodology, 100
Kitchen computer, 184
K-nearest
  neighborhood structure (k-NNS), 143
  neighbors (KNN), 93, 94, 109, 110
Kryptos, 13, 14

L

Landcover mapping, 234
Landsat satellite images, 233
Large
  data transmission, 34
  datasets, 95
  information, 24, 41
  prime numbers, 22, 208
Latency, 178, 195, 197
Lattice
  space, 130
  system, 130
Least
  accessible part, 28
  detection, 39
  significant bit (LSB), 25, 29–31, 33, 37, 157, 159, 160
    substitution, 25, 30
Legitimate, 35, 89, 101, 119, 181, 261
  software, 89
Linear
  chaotic map random, 31
  discriminant analysis (LDA), 94, 135, 138
  equations model, 267
  expressions, 297
  feedback shift register (LFSR), 294
  filter, 302
  learning curve, 1, 50
  space, 135
  transformation, 207
Local
  binary pattern (LBP), 131
  block manager processes, 194
  directional pattern (LDP), 131
  markets, 45
  Shannon entropy analysis, 77
  storage, 109, 193
Logarithm, 263
Logical examination, 256
Logistic
  map, 43, 213, 299, 304, 307–309, 319
    seed, 304
  regression, 94, 261, 263, 273, 274, 282
Logo water-marking, 234, 235
Long
  iris code, 130
  key lengths, 17
  short-term memory (LSTM), 94, 101
Low
  bit encoding, 28, 33
  computation
    complexity, 37, 297
    time, 30
  false positive rate., 103
  frequency, 32, 43, 242
    band, 32
    stores, 32
  high sub-bands, 32
  low band, 32
  pass filter, 32
  power consumption, 188

Low resolution multispectral (LRMX), 231, 240
Lucifer design, 17

M

Machine
learning (ML), 87, 89, 99, 233, 255, 256, 258, 259, 261
algorithms, 93, 109, 110, 263, 264
component, 101
model, 100, 258, 269, 273
sanctions, 257
to-machine (M2M), 198
Magnetic resonance imaging, 234
Magnitude, 127, 178, 231, 319
Malicious, 87–91, 95, 96, 100–102, 106, 107, 110, 186
activity, 96
binary file, 100
code, 89
contents, 90
IPs, 100
programs, 89, 107
scripts, 88, 89
software, 87, 88, 91, 96, 106
websites, 106
Malware, 87–96, 99–103, 106–111, 255, 256
analysis, 88, 91, 99, 101–103, 106, 108, 110
tools, 103
categories, 93
classification, 93–96, 102, 109, 110
detection, 88, 91–96, 100–103, 106, 109, 110
evasion techniques, 103
evolution detection, 93
families, 93, 101, 102
images, 109
sample, 93, 94, 96, 100, 101, 106, 108
signatures, 103
threats, 88
variants, 87, 93, 102
Man-in-middle (MitM), 185
Manual detection, 91, 92, 111
Markov chain detection, 100
Mask matrix, 304, 311
Mathematical
concepts, 13
evolution, 209
expression, 5, 207, 214, 219, 222, 224, 225
models, 2, 9, 15, 23, 49
data hiding, 8
mathematical model (color spaces), 9
mathematical model (PSNR), 10
Maximum
bits, 9, 12
capacity, 9
deviation analysis, 78
Mean
absolute error (MAE), 64, 83, 84
image intensity, 92
square error (MSE), 7, 10, 11, 39, 64, 75, 77, 83, 168, 245, 248, 271–277, 282
Media, 1, 4–6, 8–17, 23–29, 34, 36–38, 41, 42, 44–50, 290
alteration, 5, 8
authentication, 44
evaluation criteria, 4, 49
calculative-objective criteria, 5
comparing variance (color spaces), 7
perceptual-subjective criteria, 5
manipulation, 11
security, 1, 2, 15, 34, 44, 46, 50
Medical
diagnostics, 296
images, 33, 211, 235
Message digest algorithm 5 (MD5), 91, 301
Metadata, 36, 47, 92
Metasploit, 89, 90
Metrics, 40, 57, 58, 64, 67, 68, 72, 73, 78, 80, 83, 84, 221, 224–226, 232, 270, 293, 299
Micro-block numbers, 197
Micro-controllers, 189
Micro-phone, 88
Micro-processors, 189
Migration, 183
Mini-servers, 187
Minutia
angles, 135
cylinder-code (MCC), 142
fuzzy vault, 135
locations, 135
point, 134, 135, 141–144
triplets, 131, 140
vicinity decomposition (MVD), 140, 143

Mobile
devices, 188
station, 188
Modality, 25, 132, 234
Modification, 17, 22, 45, 156, 157, 159
intensity hue saturation (MIHS), 237, 244, 246, 248
Most significant bits (MSBs), 160
Multi
assortment information, 259
collection, 259
dimensional, 259
fuzzy vault scheme, 135
directional PVD, 9
faceted issues, 258
layer perceptron (MLP), 93
line code (MLC), 142
media, 1–4, 8, 10, 12–15, 42, 44, 46–50, 57, 87, 108, 115, 155, 156, 177, 203, 211–213, 229, 249, 255, 289, 290, 292, 295
authentication, 44
field, 50
information, 211–213, 290
printed media authentication, 44
scrambling process, 50
security, 1–4, 10, 13, 14, 42, 46, 47, 49, 50, 57, 87, 115, 155, 177, 203, 229, 255, 289
video authentication, 45
model biometric system, 138
resolution data fusion techniques, 235
server-E-healthcare systems, 132
source satellite dataset fusion methods, 233
spectral (MS), 229–232, 235–237, 240–243
colors, 243
data, 241–243
intensity component, 243
Multiple
biometric samples, 137
hidden layers, 92
multimedia security methods, 3

N

Naive bayes (NB), 93, 94, 100, 109, 110, 267
National
Laws for Multimedia Security, 50
security, 2
Native color system, 243
Natural language processing (NLP), 267
Near infrared (NIR), 230, 233, 234, 237, 241
Network, 1, 3, 21, 28, 48, 88, 101, 103, 106, 108, 110, 178–182, 185, 191–195, 197–200, 204–206, 271, 290, 302
analyzers, 108
packet management, 198
protocol, 28
security, 1, 3, 21
stenography, 28
store, 182
traffic, 103, 193
Neural sodality models, 256
Niv Shalev Omri (NSO), 88
Non-invertibility, 123, 140
cancelable biometrics, 139
transformation, 119, 132, 139, 141–144, 146
Non-linear
dynamic system, 43
systems, 291
Non-mated
matching score distributions, 122
score distribution, 122
Non-periodic system, 43
Non-plain image related (n-PIR), 299
Non-redundant, 272
Non-repudiation, 16, 22, 290
Normalized difference
snow index (NDSI), 234
vegetation index (NDVI), 233, 234
water index (NDWI), 234
Novel enhanced quantum representation (NEQR), 213, 215, 216
Number
of iterations, 42
of pixel change rate (NPCR), 58, 67, 80, 81, 83, 213, 225, 289, 293–295, 301, 302, 309–313, 324

O

Obfuscation, 92, 94, 96, 103
Objective
confidentiality, 16
criteria, 4–6, 8
evaluation, 6
method, 5

Observational hazard minimization, 271
Occasion mode, 187
Oddities, 257
Onboard accelerometer sensor, 135
One
dimensional chaotic maps, 294, 299, 300
time
encryption, 324
key (OTK), 298, 299, 302, 303
Online, 47, 89, 135, 137, 145, 190, 255, 281, 290
advertisements, 89
gregarious affiliations, 255
OpCode, 94
sequence, 94, 95
Open
ports, 89, 186, 187
source
dynamic malware analysis system, 106
framework, 89
intelligence tools (OSINTs), 88, 90
tools, 89
Operation code (OpCode), 94
Optimal, 132
accuracy, 93
chaotic map, 298
feature configuration, 100
levels, 295
limit, 265
Oracle attack, 38
Organizations, 106, 110, 181, 182
Orientation values, 134, 144, 145
Original
audio, 43
biometric, 115, 116, 118, 119, 123–126, 128, 133, 138, 139, 141, 146
data, 115, 116, 119, 123, 124, 128, 133, 141, 146
feature, 139
template, 115, 116, 118, 119, 123, 124, 126, 133, 138
data, 4, 13, 37, 40, 41, 43
fingerprint data, 127
gallery template, 117
grayscale image, 10
image, 4, 6, 8, 10, 37, 41–43, 45, 58, 63, 66, 67, 156, 225, 248
media, 5–8, 45
message, 15, 40, 41
plain text tends, 18
signal, 33, 41
software, 3
source, 4
template, 115, 123, 129, 145, 146
text, 18
video, 45, 46
Overlay, 179, 180, 192, 194, 200
knots, 192
Ownership, 4, 13, 34–36, 38, 44, 45, 48, 50, 156, 185, 187
claims, 13
information, 36

P

Packer, 92
detection, 92
Pair
polar
minutiae structures, 135
structure, 135
wise minutiae features, 135
Palmprint
biometric, 135, 145
image, 135
Panchromatic (PAN), 229–232, 234–237, 240–243
images, 232, 240, 241, 243
sensor, 229
Pansharpened, 229–233, 235, 237, 239–243, 248, 249
approaches, 229, 232
image, 229, 232, 240, 242, 244, 249
method, 230
techniques, 230–233, 237, 239
Parallelization technique, 94
Parametric switching chaotic system (PSCS), 295
Parity, 28, 33
bit, 33
check, 18
coding, 28, 33
Passive aggressive, 275
Password, 115, 117, 125, 146
encryption, 23
gravers, 89
key-release, 125
protection solutions, 24

Patched reference tables, 31
Peak signal
  ratio, 4
  to-noise ratio (PSNR), 6, 7, 10, 11, 28, 30, 31, 39–41, 63, 64, 75, 77, 83, 168–170, 172–174
    values, 7, 10, 11, 28, 39, 169
Perceptron, 264, 275, 276
  algorithm, 264
Perceptual, 11, 12, 14, 28
  mask, 40
  transparency, 9, 11, 12, 14, 26–28, 37
  vision, 9
Performance
  biometric system, 124
  measurements matrices (ML algorithms), 269
    accuracy score, 269
    AUC score, 271
    F1 score, 270
    mean square error (MSE), 271
    model training time, 271
    precision score, 269
    recall score, 270
Periodicity concept, 42
Permanent identity loss, 118
Permutation, 18, 123, 208, 300, 301
  logic, 19
  network phase, 218
  phase, 324
Perpetual amendment, 259
Personal
  computers, 179, 189, 199
  data, 49
  privacy, 2
  revenge, 88
Phase
  coding, 28, 33
  dispersion, 34
Phishing, 255, 256
Photon polarization, 209, 211
Photoshop, 35
Physical
  change, 7, 8
  fingerprints, 44
  stenography, 25
  token, 117
  transfer, 1
Physiological, 116, 117, 136, 146
  biometric traits, 117
Pixel value differences (PVD), 9, 30
Plain
  image related (PIR), 298, 299
  text, 15, 17, 18, 21, 23, 290, 291, 294–297, 300–303
Plant canopy liquid water content, 234
Pnscan, 93
Polar logarithmic grid, 131
Polarizations, 209
Polynomial, 133–135, 144, 145
Popularity, 88, 185, 203, 213, 234
Portable executable (PE), 91–96, 106–108
  file, 94, 96
  header, 93, 108
Possession, 88, 117
  recognition systems, 117
Possibility
  detection, 3
  reversibility, 42
Potential, 2, 30, 99, 106, 178, 179, 232, 233
  ease, 3
  intruders, 2
Precision, 6, 58, 73, 270, 271, 273–275, 298, 319
  score, 269, 270, 273–281
Predefined
  laws, 48
  protocols, 16
Principal component (PC), 131, 162, 184, 232, 240, 243, 248, 256, 257, 260
  analysis (PCA), 143, 162, 233, 248
  method, 243, 248
Print
  currency, 45
  media, 44, 45
    authentication, 44
  proof, 46
Private
  data, 3, 281
  information, 88
  key, 22, 23, 191, 291
  space, 2
  watermarking system, 37
Privilege malware, 99
Probe biometric template, 126
Program, 73, 91–93, 103, 106, 107, 260, 261, 267

Proliferation systems, 88
Proof, 16, 45, 178, 209
of work (POW), 178, 179
Proposed watermarking scheme
remotely sensed images, 163
arnold scrambling transform, 163
false positive test, 172
imperceptibility test, 169
performance comparison, 172
preliminaries, 163
quality measures, 168
redundant discrete wavelet transform (RDWT), 163
robustness test, 170
simulation results (watermarking scheme), 167
watermark embedding/extraction process, 165, 166
watermarking embedding-extraction process, 164
Protected
data, 25, 26
IP, 36
rights management, 48
rings, 99
technological measures, 48
Protocol, 26, 28, 38, 108, 145, 210, 290
Pseudo, 25, 31, 43, 72, 121, 140, 162, 256, 291
code, 263–269
imposter scores, 121
key, 25
media, 25
random, 31, 43, 72, 140, 162, 291
bits generator, 17
methods, 31
Public
key, 22, 23, 127, 178, 290, 291
cryptographic system, 22
messages, 25
repository, 88
sector, 181
watermarking system, 37

Q

Quality, 5, 10, 14, 28, 30, 33, 37, 39–41, 43, 57, 63–68, 78, 82, 83, 158, 162, 169, 170, 181, 182, 199, 232, 235, 249, 260, 280
degradation, 30
evaluation, 249
factors, 39
loss, 14, 39, 41
Quantifiable, 256, 271
procedures, 257
strategies, 258
Quantitative
assessment, 248
coronary angiography, 234
evaluation, 249
vascular angiography, 234
Quantum, 203, 204, 209–222, 226
chaos, 204, 218, 219
computation, 214
computers, 214
cryptography, 208–211, 218
key distribution protocol, 210
frame, 217, 218, 226
diffusion, 220, 221, 226
permutation, 219, 220
gates, 204, 211, 212, 216, 220
image representation techniques, 214
information processing, 203
machines, 218
mechanics, 213, 214
representation methodology, 215
secret key matrix, 219
Qubit substitution, 218
Query
biometric
data, 129, 137
features, 133
fingerprint, 127

R

Radiometric, 230, 231, 240
distortions, 230
resolution, 231
Random, 58, 64, 65, 76, 77, 83, 203, 214, 220, 224, 291, 295, 297, 324
array, 127
character sequences, 26
dynamic quantization transformation (RDQT), 130
forest (RF), 94, 96, 100, 109, 110
generated image, 42, 43
key, 131, 137
multispace quantization (RMQ), 140

number generation (RNG), 290
projection, 130, 140
statistical generation, 27
Ransomware, 88, 89, 95, 100, 101, 103, 108
categories, 95
construction kit (RCK), 89
detection, 100
encrypts, 88
program, 88
Raw
dalvik byte, 94
malware, 94, 101
dataset, 101
samples, 94
sequences (API method), 94
Recall score, 273–281
Recognition systems, 117
Reconstruction, 10, 32, 115, 133
quality, 10
Record multiplicity attack (RMA), 123
Recurrent neural network (RNN), 101
Redundant discrete wavelet transform (RDWT), 155, 162–167, 173, 174
Reed-Solomon
code, 130
encoder, 134, 136
Regularized logistic regression classifier, 101
Reinforcement learning, 261
Relaxation, 48
Relevant, 32, 99, 234
data, 32
Reliable, 6, 96, 110, 117, 131, 203, 232
mechanism, 96, 110
Remote
sensing
datasets, 233
systems, 231
systems, 108
Removal attack, 38
Replication variable, 258
Requirements, 29, 37, 119, 120, 124, 156–158, 182, 186, 190, 205, 260, 293
watermarking for protection, 158
imperceptibility, 158
robustness, 158
security, 159
Resilience, 103, 110
Resolution, 7, 8, 229–233, 236, 237, 240–244, 248, 249
merge, 241
technique, 244, 248
Restrictive license agreements, 47
Resultant, 62, 127, 131, 163, 218, 219, 234, 240, 294, 313
binary vector, 131
pattern, 127
Revenue, 89
Reverse
engineer, 48, 87, 106, 111
framework, 87, 106
stenography system, 32
Revocability, 116, 119–121, 132, 134, 136, 140, 142, 146
palmprint template, 140
Revoked
template, 121
user access, 47
Rivest Shamir Adleman (RSA), 22, 23, 290, 291
algorithm, 22
Robotics, 296
Robust, 9, 11, 12, 14, 22, 26, 28, 29, 34–39, 42, 49, 96, 110, 124, 157, 158, 161–164, 168, 170, 173, 174, 211, 213, 234, 299
algorithm, 12
watermarking, 36, 38
system, 38
Rootkit, 89
Round, 19, 22
key transformation, 21
Run-time
behavior, 102
jobs, 183

S

Salting techniques, 139, 141
Sample malware, 88
Sandbox, 100, 101, 103
extracts, 103
Satellite
dataset, 233, 234
image, 229, 230, 232, 233, 235, 240, 249
Saturation, 7, 241
Savvy homebuyers, 185
Scalability, 177, 179, 188, 200

Scale invarient feature transformation (SIFT), 161
Scrambled, 14, 40, 41, 43, 63, 65, 165, 166
algorithm, 41
image decreases, 43
Screen, 34, 47, 89
capture, 47
casting, 34
Secret
codes, 13
data, 27, 28, 30, 34
information, 25, 28, 89
inks, 25
key cryptography, 16
message, 12, 25, 26, 29, 33, 50, 108
communication, 12
Sector encoding, 132
Security
agencies, 14
analysis platform, 99
attack, 201
departments, 45
flaws, 38
management, 183
network systems, 20
process, 42
system, 11, 38, 180, 188, 189, 200
techniques, 50, 119, 290
Self
driving conveyances, 258
similarity attack, 39
sufficient, 260
Semantic data, 14
Sensitive, 10, 39, 63, 83, 212, 221, 222, 225, 231, 235, 297, 299, 302, 319, 320, 324
analysis, 38
information, 49, 89, 213
Sensors, 155, 179, 186, 188, 190, 200, 201, 230, 232, 235, 249
Sequence, 39, 72, 73, 95, 100, 101, 162, 206, 210–212, 216, 219–221, 301, 307, 309
data, 39
Service level agreement (SLA), 182, 192
Shannon entropy, 65, 77, 224
Shortest path graph kernel (SPGK), 94
Sigmoid function, 263
Signal, 3, 8, 9, 14–16, 22, 24, 28, 32–36, 38–41, 43, 136, 162, 168, 292
to-noise ratio (SNR), 33, 34, 39
Signature, 44, 45
detection, 89, 96
Simulation, 195, 235, 313, 319, 320, 324
Singular value decomposition (SVD), 160–162
Skew map, 299
Smart
card, 117, 130, 140
home layer, 179, 185, 188, 191
distribution of blockchain (BC), 191
handling transaction of blockchain (BC), 192
technology, 179, 180
power matrices, 186
water supply, 186
wearable, 135
Snapshot, 272
Social
media, 3, 296
network service (SNS), 108
Software, 3, 4, 35, 87, 89, 106, 108, 158, 187, 190
piracy, 3
program, 87, 106
Solitary quantification, 270
Sophisticated
malware, 88
mobile phones, 183
society, 186
Spam, 255, 256, 272–282
comments, 281, 282
impacts, 256
Spatial, 26, 29, 31, 33, 35–37, 142, 155, 157, 159, 172, 211, 220, 229–232, 235, 237, 240–242, 244, 248, 249, 289, 310
accuracy, 232
domain, 26, 29, 31, 33, 36, 37
algorithm, 29
stenography, 29
watermarking, 37
information, 230, 289, 310
products, 232
Spatiotemporal
data fusion methods, 233
fusion (multisource remote sensing data), 233
pansharpening, 233

Spectral
bands, 31, 233
coefficient, 37
preservation, 248
resolution, 231, 235
system, 33
values, 233
Spyware, 88, 89, 99
Square
blunder, 278, 280
infelicity, 271
error, 248, 271
Standard
deviation, 241, 248
library functions, 107
swapping matrix, 43
Statistical
attack, 212, 223, 298
behavior, 320
correctness, 12, 49
hypothesis, 293
measurements, 245
tests, 100
Steam cipher, 17
Steganalysis, 30, 31
Steganographic, 2, 49, 87, 108, 109
tool, 90
Stego
botnet uploads, 108
image, 27, 30, 31, 88–90, 109
containing bot commands, 108
malware analysis, 89
media, 28
video, 87
Steno, 24
graphic, 10, 13, 14, 24–29, 31, 33, 35, 50
algorithm, 28
approach, 29
field, 10
techniques, 25, 26, 28
Stochastic gradient descent (SGD), 264, 274, 282
Store
data, 33, 191, 193
templates, 117
Straight permutation, 19
Stream cipher, 17
String, 91, 92, 103, 107, 131, 141, 143, 142, 181
matching condition, 103
Structural similarity, 40, 95, 168, 232
index
analysis, 81
metric (SSIM), 40, 68, 81, 83, 168, 170, 172–174, 232, 245, 248
Substitution, 10, 11, 18, 19, 21, 25, 83, 212, 213, 293–295, 301
box, 19
byte transformation, 21
Super
secretive communication, 26
secure, 24
Supervised learning, 260
Support vector machine (SVM), 93, 94, 96, 100, 109, 110, 145, 265, 276, 277
Surveillance, 203–206, 211–214, 220, 292
video, 211, 296
encryption, 211
Swap, 19, 21, 211, 220
gate, 204, 220
matrix, 43
Symmetric, 16, 23, 195
block cipher, 21
key, 17, 20, 23, 24
algorithm, 17, 18, 20
block cipher, 17
cryptographic system, 17
cryptography, 16, 17, 23, 24
Synthetic minority oversampling technique (SMOT), 102
System call execution fingerprints (SEF), 102

T

Tamper, 4, 6, 14, 17, 23, 28, 36, 38, 39, 44, 45, 49, 210
data, 4, 6
detection, 34, 42
evidence, 39
Targeted media, 8
Technique
authentication, 14
decoding, 15
embedding data, 8
encoding, 15
Template protection technique, 120–122, 124, 130, 134, 135, 138, 139, 141–143, 145, 147

essential properties, 120
performance, 124
revocability, 121
security, 123
unlinkability, 121
Temporal resolution, 231
Tent map, 299
Terabytes, 1–3
Text
file, 26, 27
stenography, 25, 26
Thematic mapper (TM), 127, 128, 234, 242
Three-dimensional chaotic maps, 300
Threshold values, 123, 191
Time
slices, 135
stamp, 182, 206
Tinker-bell map, 72
Toasters, 184
Toffoli gates, 220
Topological mixing, 291
Transform, 37, 38, 135, 157, 215
coefficient, 29, 162, 212
domain, 29, 31, 33
stenography, 29, 31
message, 15
Transparency, 9, 49, 181
Transposition, 18
Trapezoidal standard, 271
Trojan, 89, 90, 96, 256
development kit (TDK), 89
ponies, 256
Trojanspy, 96
True negative relationship (TNR), 197
Truncation, 32
Tsunami, 93
Turbo-codes, 131
Twitter, 255
Two
dimensional
chaotic map, 207
space, 133, 299, 304
layer
concatenated codes, 131
error correction scheme, 130
level Arnold transforms, 42
Markov chains, 100

U

U-index scan, 69
Unauthentic
negatives, 270
spurious negatives, 270
Unauthorized
access, 47, 49, 123, 190, 205
alteration, 48
Unbiased assessor, 272
Uncompressed raw video, 27
Unearth designs, 257
Unfaltering quality, 279
Unicode-printable sequences, 107
Unified
average change intensity (UACI), 58, 67, 80, 81, 83, 213, 225, 289, 293–295, 301, 302, 309–313, 324
matrix, 43
Uniform histogram plots, 223
Unintended
intrusion, 50
person, 15, 50
Uninterruptible power supply (UPS), 187
Unit matrix, 43
Unitary
matrices, 215
transform, 215
Universal
Copyright Convention, 48
Image Quality (UIQ), 82
security, 36
serial bus (USB), 189
Unmanned aerial vehicles (UAV), 155
Unsupervised learning, 260
Unwavering quality, 277
User
accessibility, 47
data packages, 193
information, 47
mode, 99
rights, 46, 48
specific
password, 135
questions, 131
Utilitarianism, 183

V

Vacuum cleaner, 184
Vantage score, 270
Vector, 25, 130–132, 134, 135, 137, 143, 144, 214, 263–265, 273
  machines, 25
Verbal expressions, 271
Verification, 117, 125–131, 133, 136–138, 140, 145, 182, 244, 290
Vertex position, 40
Victim, 88–90, 108
Video
  authentication, 44, 45
  camera interface, 191
  encryption, 45
  stenography, 27
  surveillance, 203–206, 211–213, 216, 219, 221, 226
Vigilance, 204–206, 211, 214, 256, 274
  camera, 204–206, 211, 214
Virtool, 96
Virtual
  desktops, 190
  hub, 49
  machines, 190
  private system technology, 187
Virtualization, 111, 182, 183, 190
Virus, 89, 90, 96, 102, 110
  signature database, 96
Visual, 3
  analysis, 249
  inspection, 61, 249
  interceptable, 29
  managements, 257
  media, 37
  similarities, 93
  undetectable, 28
Voltage regulator, 189, 201
Voluminous, 256, 259, 260
Voronoi neighborhood structure (VNS), 135
Vulnerabilities, 89, 99, 296

W

Watermark, 36–40, 156–173, 235
  algorithm, 35, 37, 160
  attacks, 38, 170–172
  detector, 38
  extraction, 235
  image, 37, 39, 40, 158, 160–162, 166, 169, 172
  methods, 39
  scheme, 155–170, 172, 174
  system, 38
  technique, 38, 44, 50, 155, 162
Wavelet
  domain, 39
  method, 232
  resolution merge, 237, 244, 247–249
  technique, 232
Wax tables, 25
Weather monitoring, 156, 158
Web, 1, 13, 186, 187, 190, 199, 259
Windows, 90, 92, 94–96, 100, 102, 103, 106, 107, 221
  API, 100, 103
  binary, 92
  Operating System, 106
  PE files, 95
  specific fields, 94
WIPO
  Copyright Treaty (WCT), 47, 48
  Treaties Implementation Act, 48
Wireless, 1, 21, 108, 188, 204, 206, 292
  security, 21
Wireshark tool, 108
Word
  encryption, 40
  scrambling, 40
Worms, 89, 90, 99, 103, 256
Wrapper method, 108

X

XOR
  operation, 19–21, 33, 42, 43, 159, 160, 213, 301
    scrambling, 42
  whitener, 19
XYZ color space, 8

Y

YARA rules, 102, 103, 106
YIQ color space, 7
YouTube, 255

Z

Z-order scan, 69
Z-mirror scan, 69
Zero, 43, 58, 63, 101, 121, 124, 172, 208, 223, 248, 320
  correlation, 223
  matrix, 43
Zombie, 89